The Gas Mask in Interwar Germany

Exploring the history of the gas mask in Germany from 1915 to the eve of World War II, Peter Thompson traces how chemical weapons and protective technologies such as the gas mask produced new relationships to danger, risk, management, and mastery in the modern age of mass destruction. Recounting the apocalyptic visions of chemical death that circulated in interwar Germany, he argues that while everyday encounters with the gas mask tended to exacerbate fears, the mask also came to symbolize debates about the development of military and chemical technologies in the Weimar Republic and the Third Reich. He underscores how the gas mask was tied into the creation of an exclusionary national community under the Nazis and the altered perception of environmental danger in the second half of the twentieth century. As this innovative new history shows, chemical warfare and protection technologies came to represent poignant visions of the German future.

Peter Thompson is Assistant Professor of History, Philosophy, and Sociology of Science at Lyman Briggs College, Michigan State University.

SCIENCE IN HISTORY

Series Editors

Science in History is a major series of ambitious books on the history of the sciences from the mid-eighteenth century through the mid-twentieth century, highlighting work that interprets the sciences from perspectives drawn from across the discipline of history. The focus on the major epoch of global economic, industrial and social transformations is intended to encourage the use of sophisticated historical models to make sense of the ways in which the sciences have developed and changed. The series encourages the exploration of a wide range of scientific traditions and the interrelations between them. It particularly welcomes work that takes seriously the material practices of the sciences and is broad in geographical scope.

The Gas Mask in Interwar Germany

Visions of Chemical Modernity

Peter Thompson
Michigan State University

Shaftesbury Road, Cambridge CB2 8EA, United Kingdom

One Liberty Plaza, 20th Floor, New York, NY 10006, USA

477 Williamstown Road, Port Melbourne, VIC 3207, Australia

314–321, 3rd Floor, Plot 3, Splendor Forum, Jasola District Centre, New Delhi – 110025, India

103 Penang Road, #05-06/07, Visioncrest Commercial, Singapore 238467

Cambridge University Press is part of Cambridge University Press & Assessment, a department of the University of Cambridge.

We share the University's mission to contribute to society through the pursuit of education, learning and research at the highest international levels of excellence.

www.cambridge.org
Information on this title: www.cambridge.org/9781009314824

DOI: 10.1017/9781009314862

First published 2023

A catalogue record for this publication is available from the British Library.

Library of Congress Cataloging-in-Publication Data
Names: Thompson, Peter, 1988- author.
Title: The gas mask in interwar Germany : visions of chemical modernity / Peter Thompson, Science History Institute, Philadelphia.
Description: New York, NY : Cambridge University Press, 2023. | Series: Science in history | Includes bibliographical references and index.
Identifiers: LCCN 2023000067 (print) | LCCN 2023000068 (ebook) | ISBN 9781009314824 (hardback) | ISBN 9781009314848 (paperback) | ISBN 9781009314862 (epub)
Subjects: LCSH: Gas masks–Germany–History–20th century. | Gases, Asphyxiating and poisonous–War use.
Classification: LCC UG447.6 .T46 2023 (print) | LCC UG447.6 (ebook) | DDC 623.4/592–dc23/eng/20230103
LC record available at https://lccn.loc.gov/2023000067
LC ebook record available at https://lccn.loc.gov/2023000068

ISBN 978-1-009-31482-4 Hardback

For Alison

Contents

Figures

Tables

Acknowledgments

The extent to which other scholars and academic institutions have supported this project is greater than can be properly acknowledged here. Nevertheless, special thanks are due to the members of my University of Illinois dissertation committee, who guided this project from its inception and provided numerous sets of comments that have reached across diverse historical subfields. This exceptional committee was comprised of coadvisors Mark Micale and Peter Fritzsche, as well as David Sepkoski, Harry Liebersohn, and Jimena Canales.

For comments on specific sections of the dissertation project, I would like to thank Craig Koslofsky, Tamara Chaplin, Antoinette Burton, Sabine Hake, Bruno Cabanes, Chris Otter, Roger Chickering, Christoph Weber, Karen Hagemann, and Konrad Jarausch. This message of thanks extends to the members of the 2018–2019 University of Illinois Urbana-Champaign (UIUC) Dissertation Writing Workshops, the UIUC German Reading Group, the 2019 German Historical Institute Transatlantic Doctoral Seminar, and the North Carolina German Studies Seminar. Among the many UIUC history doctoral students who have read and commented on parts of this project, special thanks are due to Ryan Allen for both his incomparable scholarly insights and his friendship.

I would like to thank the UIUC History Department, the Siebel Scholars Program, the Illinois Program for Research in the Humanities, and the Fulbright Program for their generous financial contributions to the dissertation stage of this project. I would also like to thank Ulrich Herbert and the University of Freiburg School of History, which provided me with a welcoming institutional home while I conducted archival research in Germany.

This project required consultation with far too many archives in order to thank them all here, but I would like to express particular appreciation to the staffs of the Federal Militärarchiv, the Evonik Industries archive, the Bayer archive, the BASF archive, the Bavarian Kriegsarchiv, the archive of the Max Planck Society, the German Diaries Archive in Emmendingen, and the Deutsches Literaturarchiv in Marbach.

As the dissertation was transformed into a monograph, it received substantial support through a postdoctoral fellowship at the Beckman Center at the Science History Institute in Philadelphia, PA. I would like to thank Charlotte Abney Salomon, the Associate Director of the Beckman Center; Jesse Smith, the Director of Curatorial Affairs; and my cohort of Beckman Center fellows for providing a welcoming and supportive environment in which to write on the history of modern chemistry.

For shepherding the work of a new author through the publication process, I would like to thank Lucy Rhymer at Cambridge University Press, and I would also like to thank my two anonymous peer reviewers for their invaluable comments. For the right to use several of the book's images, I would like to again thank the Max Planck Society Archive as well as Randall Bytwerk and the German Propaganda Archive at Calvin University.

On a personal note, I would like to thank my parents and sister for endlessly supporting an aspiring scholar. And finally, I would like to thank Alison Masyr, who was there for me every step of the way.

Abbreviations

BASF	*Badische Anilin und Soda Fabrik* (Baden Aniline and Soda Factory)
DChG	*Deutsche Chemische Gesellschaft* (German Chemical Association)
Degea	*Deutsche Gasglühlicht-Aktiengesellschaft* (German Gas Light Corporation)
Degesch	*Deutsche Gesellschaft für Schädlingsbekämpfung* (German Company for Pest Control)
KRA	*Kriegsrohstoffabteilung* (German War Raw Materials Department)
KWG	*Kaiser Wilhelm Gesellschaft* (Kaiser Wilhelm Society)
KWI	*Kaiser Wilhelm Institut für physikalische Chemie und Elektrochemie* (Kaiser Wilhelm Institute for Physical and Electrochemistry)
KWKW	*Kaiser Wilhelm Stiftung für kriegstechnische Wissenschaft* (Kaiser Wilhelm Foundation for the Sciences of Military Technology)
MICC	Military Inter-Allied Control Commission
MPG	*Max Planck Gesellschaft* (Max Planck Society)
OHL	*Oberste Heeresleitung* (German Military High Command)
RLB	*Reichsluftschutzbund* (Reich Air Protection League)
TASCH	*Technischer Ausschuss für Schädlingsbekämpfung* (Technical Committee for Pest Control)
VDCh	*Verein Deutscher Chemiker* (The German Chemists Association)
WILPF	Women's International League for Peace and Freedom

Introduction

On April 22, 1915, the German chemist Fritz Haber and a hand-selected group of technicians coordinated the first large-scale chlorine gas attack of World War I. In the days that followed this assault against Allied troops, the German state celebrated its supposedly successful use of poison gas and Haber was given a military rank with near total control of future chemical weapons development. Historically, Haber's attack signified a massive escalation of chemical warfare, which had previously been either rudimentary or conceptual in nature.[1] Furthermore, the German release of 168 tons of chlorine gas and the involvement of academic scientists in direct warfare efforts proved an unmistakable turning point in World War I's larger historical narrative. According to the Allied nations, Haber and the Germans had broken with the previously accepted rules of warfare and now there would be no turning back. While the other belligerent nations began to bring the full weight of their national chemical industries into the war, the Germans maintained their head start, which was subsequently augmented by Haber's deployment of new chemical weapons such as mustard gas and the German military's adoption of new and improved gas tactics. However, this initial advantage did not quickly or decisively win the war for the Germans and their new weapons tended to bog down ground troops, thereby entering them into a new and uniquely dangerous form of modern warfare. By 1918, these changes fully expressed themselves through the constant gas shell barrages that created a highly toxic world for the average infantryman.

The philosopher Peter Sloterdijk has described Fritz Haber's use of poison gas in World War I as the birth of a new historical epoch, arguing

[1] Poison gas, the term most commonly used during World War I, is one of many chemical weapons. However, not all chemical weapons are gaseous. For instance, mustard gas is a liquid that can become airborne when fired in an explosive shell. Furthermore, more expansive legal definitions of chemical weapons include all of the technologies involved in the dispensing of poisonous chemicals. This study primarily employs the actors' term "poison gas" unless it is intentionally and explicitly referring to the more expansive category of chemical weapons.

that chemical warfare generated a new subjective relationship to the atmospheric environment. By weaponizing the very air that supported human life, poison gas introduced many Europeans to the central precepts of atmospheric terrorism. Sloterdijk writes:

> The attack on humans in gas warfare is about integrating the most fundamental strata of the biological conditions for life into the attack: the breather, by continuing his elementary habitus, i.e. the necessity to breathe, becomes at once a victim and an unwilling accomplice in his own annihilation.[2]

Furthermore, the unique methods of this new weapon forced Europeans to begin considering both the chemical construction of their world and possible methods of protection. For Sloterdijk, such changes represent the birth of the modern mind, since "aesthetic modernity is a procedure of applying force not against people or things, but against unexplained cultural relations (e.g., the previously ignored atmospheric composition)."[3]

For the historian, the birth of a strictly delineated historical "modernity" is difficult to place in 1915. Nevertheless, there is certainly a strong case for dating the beginnings of certain earth-shattering cultural and social changes to the earliest instances of chemical violence in World War I. For this reason, this historical study of German chemical weapons posits the conscious realization of these potential changes among its historical actors as a specifically German "chemical modernity," denoting the distinctive importance of chemical weapons for formulating the contested visions of impending and unseen danger that proliferated in the postwar German world.

In the wake of World War I, these contested visions would help to shape and define Germany's social, cultural, and political worlds. Debates between antigas activists and the scientists and engineers who previously produced chemical weapons brought poison gas production under intense public scrutiny. It was precisely in this moment of heightened anxiety that a loosely affiliated group of scientists and engineers, describing themselves as "gas specialists," took center stage to assure the German public that protection from poison gas was indeed possible.[4] Pulling on their own narration of World War I chemical warfare, these men claimed that

[2] Peter Sloterdijk, *Terror from the Air*, trans. Amy Patton and Steven Corcoran (Los Angeles: Semiotext, 2009), 22–23.

[3] Ibid., 79.

[4] The term "gas specialist" is a translation of a variety of German terms that poison gas experts used to describe themselves in the 1920s, 1930s, and 1940s. This included *Spezialisten*, *Experten*, *Gelehrten*, and *Fachleute*.

Germans could survive in a world permeated by poison gas as long as they maintained the requisite technology, knowledge, and personal discipline.

Through a reliance on distinctly modern technologies, the gas specialists envisioned their own version of a German "chemical modernity" in which the risks of chemical death could be controlled and possibly directed. The inherent dangers of poison gas made it a central concern for interwar scientists and military men who wished to create rational bureaucratic and technological structures that could harness the powers of new military technologies. However, the shifting nature of poison gas and its seeming immateriality ensured that many Germans in the 1920s and 1930s doubted the possibility of such national regulation, let alone protection. Nevertheless, the guidelines and training aimed at producing what interwar experts called "gas discipline" among German citizens fit into high-modernity's broader attempt to mitigate or control the new risks of modern life.[5]

In his well-known sociological studies of risk, Ulrich Beck has asserted that such modern projects of rational protection are only possible through a so-called risk calculus that grants political power to legal and governmental institutions in order to combat unforeseen and complex risks, such as global pollution, newly discovered diseases, urban crime, and industrialized warfare.[6] In this newly insured world, humans are constantly exposed to, or expecting exposure to, such large-scale problems, all of which were created or exacerbated by the process of modernization itself. In fact, according to Beck, modern humans manufacture the very crises that they then build intricate social structures to counteract.[7]

For Beck, this is an ongoing social process that appears to be reaching a critical historical juncture. By the late twentieth century, the risk-mitigating engineering projects of European society had become far too large and dangerous for individuals to sincerely undertake or understand. Thus, large-scale, high-risk projects of science and technology controlled by intricate bureaucratic structures have become increasingly necessary to protect humanity from widespread catastrophe.[8] Simultaneously, the social credibility of these projects is often slowly eroded due to the unanticipated ancillary risks that are invariably part of human-built and

[5] Matthias Beck and Beth Kewell, *Risk: A Study of Its Origins, History and Politics* (Hackensack, NJ: World Scientific Publishing, 2014), 35–37.

[6] Ulrich Beck, *World at Risk*, trans. Ciaran Cronin (Malden, MA: Polity Press, 2009), 26.

[7] Ulrich Beck, *Risk Society: Towards a New Modernity*, trans. Mark Ritter (London: SAGE, 1992), 183.

[8] Charles Perrow, *Normal Accidents: Living with High-Risk Technologies* (New York: Basic Books, 1984), 3.

human-controlled systems.[9] For this reason, the risk-mitigating experts of the early twenty-first century are regularly forced to cede a certain level of institutional and epistemic power as "we increasingly live on a high technological frontier which absolutely no one completely understands and which generates a diversity of possible futures."[10]

With the complexities of contemporary society in mind, this project pushes Beck's insights back to a moment in which certain risk-mitigating technocrats reached the apex of their self-styled institutional power. Since World War I, scientists and engineers under the supervision of Fritz Haber had produced gas protection technologies and guidelines in tandem with new battlefield gases. Combined with a commitment to the martial discipline and training of the trenches, gas protection technologies and plans would serve as the gas specialists' fundamental answer to interwar national poison gas protection. Although national gas protection required massive federal spending as well as constant vigilance and competence from each and every German citizen, the gas specialists' solutions appeared simple and straightforward. Indeed, practically any call for national technological defense provided the German people with a tangible sense of protection while avoiding reliance on what then seemed impotent agreements of international law. At the same time, these prescriptions were not without partisan interest since gas protection technologies and air raid training would provide the gas specialists with significant manufacturing contracts, substantial governmental power, and a new level of professional respect.

The gas specialists' stress on national protection also struck resonant chords with the concurrent desires for rearmament among various militant German nationalists. After their rise to political power in 1933, the Nazis institutionalized the gas specialists' call to action through the creation of the centralized *Reichsluftschutzbund* (RLB), or Reich Air Protection League. Total national protection remained a technical impossibility under the Nazis, but the RLB's effort to universally distribute gas masks after 1937 ostensibly served to provide a material site of individual comfort while simultaneously and collectively militarizing the average German civilian. While some gas specialists expressed concerns over these actions, most were willing to support the measures that made aero-chemical protection a daily concern. As such, the gas specialists easily translated their work to fit under the Nazi regime and briefly thrived in an expanding bureaucratic structure focused on mitigating the risks associated with national aero-chemical defense.

[9] Ibid., 156.
[10] Jane Franklin, ed., *The Politics of Risk Society* (Cambridge: Polity Press, 1998), 25.

However, the ritualized drills of the *Reichsluftschutzbund* were informed not just by the technical knowledge of the gas specialists but also by Nazi conceptions of national belonging and exclusion. Thus, the RLB's activities attempted to realize the dream of a distinctly "Nazi chemical modernity," in which the constant threat of chemical attack was not only overcome through the gas specialists' technological augmentation and steely determination but also mastered and redirected toward those who had failed to protect themselves from the newly poisoned atmosphere.

A wide array of theorists and historians have described the construction of the gas chambers of the Holocaust as the defining moment in the Nazis' search for a technologically assisted method of mass murder.[11] More specifically, the designing of the gas chambers has often provided the ultimate case study for the potential problems of a blind commitment to the practical application of scientific knowledge and the search for technocratic efficiency. For many prominent post-Holocaust thinkers, such an unquestioned search for an efficient means to an immoral end revealed the dark side of "high modernity." For instance, in his 1954 book, *The Technological Society*, the philosopher Jacques Ellul argued that the twentieth century was dominated by what he called "technique," or "the totality of methods rationally arrived at and having absolute efficiency ... in every field of human activity."[12] According to Ellul, humanity's reliance on rational thinking and applied science created social conditions that progressively conformed to the smooth and efficient rhythms of the machine. Consequently, Ellul claimed that we had become slaves to the possibility of technological creation, suppressing human freedom and binding ourselves to an increasingly complex and authoritarian technological system.

Regardless of the ultimate truth of Ellul's larger claims, his original insights should encourage us to both historically and morally interrogate the perceived value of technological creation. Frequently, however, proponents of technocratic efficiency have skirted such questions by claiming that technological objects and systems can be neither inherently good nor evil.[13] According to this view, the ethics of technology are decided not in the moment of creation but in the human use of a given object or system. The early twentieth-century German scientists and engineers who created and promoted poison gas and its various ancillary

[11] Konrad Jarausch and Michael Geyer, *Shattered Past: Reconstructing German Histories* (Princeton: Princeton University Press, 2003), 99.
[12] Jacques Ellul, *The Technological Society* (New York: Vintage Books, 1964), xxv.
[13] Jennifer Karns Alexander, *The Mantra of Efficiency: From Waterwheel to Social Control* (Baltimore: Johns Hopkins University Press, 2008), 2.

technologies frequently employed this line of reasoning to defend their research and development choices. For instance, the interwar gas specialist Rudolf Hanslian wrote, "The High Goddess of Science has a double-edged sword. Depending on the impact of [scientists'] achievements *in practice*, they can serve the good of humanity or lead to their ruin."[14]

In light of the continuing cultural power of such utilitarian claims, techno-skeptical theorists such as Langdon Winner have mounted a response. In his well-known 1977 work, *Autonomous Technology*, Winner wrote:

> Is technology a neutral tool to human ends? No longer can an affirmative answer be given without severe qualifications. The most spectacular of our implements often frustrate our ends and intentions for them. Skepticism greets the promise that our transportation crisis will be solved by a bigger plane or a wider road, mental illness with a pill, poverty with a law, slums with a bulldozer, urban conflict with a gas.[15]

For Winner, the concept of "use" is often too narrowly defined. Proponents of unrestrained scientific inquiry and application have traditionally bracketed the moment of technological use in space and time. To their minds, a human picks up a tool and then decides how it should be used. But as Winner argues, technologies are already interwoven into the cultural fabric of humanity before the moment of praxis. Following on the heels of certain Marxist insights, Winner claims that "human beings do make their world, but they are also made by it."[16] Thus, the mere existence of a given technology presents, if not demands, the very possibilities of its use.[17]

Here, Winner may at times overstate the power of both technological creation and use to shape social structures. In his response to Winner's work, the sociologist Bernward Joerges reminds us to avoid retrofitting dramatic histories of technological creation and application for overly pious parables. To Joerges' mind, the authorial intentions behind technological creation and use are often indeterminate and highly contingent. These moments of technological employment do not necessarily induce particular social relations, but they do tell us something important about the social relations of a given moment and place. Thus, technologies

[14] Rudolf Hanslian, *Vom Gaskrieg zum Atomkrieg: Die Entwicklung der wissenschaftlichen Waffen* (Stuttgart: Verlag Chemiker Zeitung, 1951), 8.

[15] Langdon Winner, *Autonomous Technology: Technics-out-of-Control as a Theme in Political Thought* (Cambridge, MA: MIT Press, 1977), 29.

[16] Ibid., 88.

[17] Anthony Giddens, *The Consequences of Modernity* (Oxford: Oxford University Press, 1990), 139.

should be read as a form of media that expresses social relations through both rights to use and actual uses.[18] They reveal to us the ways in which humans ultimately legitimize particular uses of technologies and the ways in which those uses are then employed as a form of social power.

While the technologies behind poison gas inspired various interwar German interpretations that could mix anything from awe and fear to an insistence on practical control, the continued production and use of these technologies required steadfast supporters such as Fritz Haber and the later gas specialists. In their role as technical experts, these men continually claimed that chemical weapons technologies created a more humane and civilized form of warfare that would inevitably lead to greater scientific progress. Thus, their story reveals the importance of a dogmatic commitment to a vision of scientific progress and technocratic efficiency for the subsequent propagation and uses of such deadly technologies. This ought not, however, suggest that Fritz Haber or the gas specialists should be held historically responsible for the gas chambers of the Holocaust. Rather, it holds Haber and the gas specialists responsible for the interwar proliferation of both chemical weapons technologies and normalized visions of a chemically dangerous world.[19] It is this work that inaugurated the many German understandings of "chemical modernity," including the Nazi vision in which the gas chambers of the Holocaust could be conceived and, to a certain extent, legitimated. For this reason, a simple causative connection between Fritz Haber's chemical weapons program and the gas chambers of the Holocaust cannot, and should not, be made.

Admittedly, one could tell this larger narrative about the gas specialists' scientific and technological commitments solely through a detailed developmental history of militarized poison gas. However, gas was a uniquely ephemeral weapon for several reasons. First, it was difficult to detect and describe when deployed. Second, it was never intentionally used on a European battlefield after 1918, thus largely leaving poison gas to the realm of the interwar imagination. The gas mask, on the other hand, provided a material site for civilian interactions with an interrelated web of both offensive and defensive chemical weapons technologies. As such, a study of both poison gas and the gas mask more effectively encourages

[18] Bernward Joerges, "Do Politics Have Artefacts?" *Social Studies of Science* 29, no. 3 (1999): 424.

[19] This also does not mean that Fritz Haber and the gas specialists were the only historical actors responsible for the propagation of chemical weapons technologies or normalized visions of a chemically dangerous world. Indeed, even our current understanding of a chemically precarious world is perpetually reified by seemingly mundane individual and collective relationships to chemical technologies.

us to expand our definition of "the chemical" and to reevaluate technologies intended for protective use that populate daily life. By examining lived experience with the gas mask through German ego documents such as memoirs, diaries, and letters, this project shows that it was often the mask, rather than the gas itself, that introduced interwar Germans to the pervasive presence of technological possibility, mediation, and failure.

As Germans grappled both practically and philosophically with poison gas and the gas mask, they contended with an insistent and newly unstable technological world. Since the gas mask was claimed as the first line of defense against a weapon that could insidiously permeate the atmosphere at any moment, it further mediated a newly unstable relationship between Germans and their environment. With gassing as an imminent threat in the interwar mind, the simple act of breathing air could no longer be taken for granted. For this reason, the German story of chemical weapons technologies exposes the ways in which the conceptual categories of danger, risk, management, and mastery informed subjective relationships to the environment in the early twentieth century.

Of course, the Germans were not the only ones to develop both chemical weapons and corresponding visions of a chemically impregnated world. In the wake of World War I, the English, French, and Americans also developed industrial-scale chemical weapons programs and similarly grappled with the implications of atmospheric warfare. However, none of these other nations were forced to reckon with their chemical poisons quite like Germany. In the early twentieth century, the German nation relied most heavily on ties to its unmatched chemical industry, while Germans themselves experienced a unique level of chemical fear derived from a sense of weakening geopolitical power after the loss of World War I. This produced a more pressing German interrogation of whether chemical weapons could be controlled and, if so, by whom.

At the same time, it is also true that these concerns maintain importance for a longer history with near-global reach. For this reason, this study reappropriates the term "Chemical Modernity" in capitalized letters to refer to the longer epoch in which industrial chemicals and environmental pollutants have increasingly yet quietly come to impact, and even sometimes define, modern life. Since the onset of the Chemical Revolution (around 1770) and the subsequent industrial scaling of laboratory chemistry, work with and among invisible chemicals has changed the ways in which humans understand their environment.[20]

[20] Sara B. Pritchard and Carl A. Zimring, *Technology and the Environment in History* (Baltimore: Johns Hopkins University Press, 2020), 107.

In the wake of either the material or theoretical recognition of potentially life-threatening and diffusive chemical compounds, contemporary humanity can no longer assume that the Earth's atmosphere will always support life. When fully accepting this state of mental uncertainty, potential death can quietly lurk in every breath. The German author Jörg Friedrich makes this clear in his provocative historical treatment of the bombing campaigns of World War II. He writes:

> Now, through heat, radiation, and toxic gases, the very air was being transformed into something unlivable. The incendiary weapon and the subsequent atomic weapon of World War II introduced the notion of extracting a state of destruction from the workings of general laws of nature. Reality was no longer a place in which to dwell or even do battle. Living space became a death zone.[21]

Here, Friedrich dates the dawn of "atmo-terrorism," or the weaponization of the atmospheric conditions necessary for life, to World War II. This is a fairly common claim that is supported by both the sheer death and destruction caused by World War II aerial bombing and the cultural power of postwar atomic fear. Without denying the real human cost of the aerial bombing campaigns of the 1940s, this project both implicitly and explicitly argues that the imagined possibilities for total destruction that would later characterize atomic fear were already present in the 1920s and 1930s. While poison gas could not, in fact, destroy entire cities and nations during this period, contemporaries certainly fantasized about this prospect and began to reevaluate the meaning of individual, national, and atmospheric security. Thus, from "gas psychosis" to "atomic fear" (and now perhaps "environmental angst"), chemical dangers lurking anywhere from the microscopic to the astronomic have come to express a fundamental feature of what it means to be human in the twentieth and twenty-first centuries.

In the newly perilous world described by Friedrich and others, "Mother Earth" is now envisioned as the ground on which the battles of total war will be waged. In coordination with large-scale environmental changes, advances in weapons of mass destruction would seem to suggest that this partially realized form of warfare will soon become a true struggle for mere existence.[22] We can perhaps see this unfolding in everything from the use of chemical weapons in recent Middle Eastern conflicts to the imagined future wars over clean air and water that the

[21] Jörg Friedrich, *The Fire: The Bombing of Germany, 1940–1945*, trans. Allison Brown (New York: Columbia University Press, 2006), 84.

[22] Roy Scranton, *Learning to Die in the Anthropocene: Reflections on the End of a Civilization* (San Francisco: City Lights Books, 2015), 19.

current environmental crisis portends.[23] In these instances, the conceptual categories of danger, risk, management, and mastery that animated early twentieth-century German debates again take center stage in our minds. Thus, atmospheric total war and the questions surrounding human survival will undoubtedly have massive implications for the way in which we now narrate our current moment, our history, and even our future. Since it would seem that we still live under the imaginary power of imminent atmospheric total war, this project refers to the longue durée "Chemical Modernity" in order to make diachronic connections between the climactic German case study and our more contemporary concerns. Specifically, by reflecting on a historical study of German chemical weapons development, it asks us to consider both the historical roots and ultimate implications of large-scale technological fixes for the magnifying problems of modern systemic risk.[24]

To ultimately reflect on such contemporary concerns, the study will follow a largely chronological history of poison gas and gas mask production as well as the developing cultural visions of chemical threat in early twentieth-century Germany. Chapter 1 will begin by historicizing the German development of poison gases during World War I. The chemist Fritz Haber will serve as the central figure in this story. His commitment to both applied science and the imperial German state informed his advocacy for the development and use of militarized poison gas. Employing files from German military archives and personal papers from various scientists and soldiers, this chapter narrates the way in which the German military's decision to employ chlorine gas on the battlefields at Ypres did not shorten World War I but rather generated an arms race in which all belligerent nations attempted to develop the most lethal gases. This military buildup demanded a massive yet efficient apparatus for furthering both scientific research and industrial production. Haber's skill in building the most expansive and efficient national chemical weapons program thus serves as one of the earliest instantiations of what would later be called the military-industrial-academic complex and Haber's government-funded project can be classified as an early prototype for the nationalized technoscientific projects of the mid-twentieth century.

Chapter 2 provides a developmental history of the modern gas mask, first produced in 1915 at Haber's Kaiser Wilhelm Institute for Physical

[23] Emmanuel Kreike, *Scorched Earth: Environmental Warfare as a Crime against Humanity and Nature* (Princeton: Princeton University Press, 2021), 400.

[24] Katharina Gerstenberger and Tanja Nusser, eds., *Catastrophe and Catharsis: Perspectives on Disaster and Redemption in German Culture and Beyond* (Rochester, NY: Camden House, 2015), 14.

and Electrochemistry (KWI). The German military's manufacture of over 8 million gas masks was both an attempt to protect the average soldier from an increasingly toxic world and an integral factor in the continued realization of such a chemically saturated nightmare. If an airborne weapon of total annihilation was to be created after 1915, then it would now need to be strong enough to break the mask's filter or seal. By intertwining the historical development of gas with that of the gas mask, the moments in which a specifically German "chemical modernity" were conceived and pursued begin to come into focus.

Chapter 3 utilizes soldiers' writings to indicate that, contrary to the developers' view of the gas mask as a life-saving device, combatants were more frequently frightened of both the appearance of the gas mask on others and the physical feeling of the mask against their own skin. While German tacticians hoped to craft chemically resistant soldiers through gas mask training, these newly envisioned "chemical subjects" continued to ruminate on the many ways in which masks could malfunction. Sitting in the trenches, soldiers largely feared both the uncoordinated and creeping nature of gas and the smothering feeling of their affixed gas mask. By examining the sensorial and metaphorical language in a wide array of soldier diaries, trench journals, and troop reports, this chapter seeks to construct the emotive experience of German World War I soldiers as they came to recognize their precarious role in a modern world now seemingly steeped in gas.

Chapter 4 examines how German military doctors attempted to treat the injuries caused by poison gas. World War I medical professionals often downplayed reports of gassing by claiming that the affected men had predispositions to pulmonary illness or constitutional weakness. Furthermore, many physicians and psychologists saw what was then termed "gas-neurosis," or psychological discomfort stemming from poison gas exposure, as an affront to their concept of German national health. Due to both this cultural commitment and a lack of knowledge regarding poison gas exposure, doctors tended to dismiss gas symptoms such as chronic coughing, dizziness, lung inflammation, insomnia, and hallucinations that could develop years after initial contact. Much like the scientists who had developed poison gas at Fritz Haber's Institute, most German physicians continued to view the weapon as something that could be controlled and treated with the proper application of scientific and medical knowledge.

Chapter 5 continues the narrative of German poison gas production into the 1920s. Still tied to Fritz Haber and his protégés, most of this work was either hidden under the guise of pesticide research or conducted in the Soviet Union through clandestine armament deals. Both this illegal poison gas production and national debates surrounding rearmament inspired a significant interwar pacifist response. Focusing on the female scientist Gertrud Woker, this chapter evaluates the rhetorical methods pacifists used to critique poison gas research and production. The antigas movement inspired several international disarmament and peace treaties throughout the 1920s, but the enforcement and popularization of these treaties proved difficult as both fears over national security and visions of future aero-chemical war proliferated. Unfortunately for the pacifists, a 1928 gas leak at a Hamburg chemical plant stoked these fears into powerful calls for greater national security, thus encouraging the swift political rise of the so-called gas specialists. Through their gas protection journals and their public demonstrations, this group of scientists and engineers succeeded in putting pressure on the Weimar government to increase its focus on civilian air and gas protection at the expense of international disarmament agreements.

Chapter 6 then examines German intellectual understandings of chemical warfare technologies. Several of the most influential interwar intellectuals were veterans of World War I, having experienced gas attacks and used gas masks during their wartime service. Revealing the salience of poison gas in the interwar imagination, this chapter explores the numerous literary, artistic, and cinematic works that attempted to grapple with the individual soldier's relationship to chemical weapons. Indeed, the continued contact with relentlessly changing and often dangerous technology such as poison gas and the gas mask exemplified the mental uncertainty and political instability of early twentieth-century Germany. As part of a larger debate surrounding militarized technology, arguments over the controllability of poison gas and the viability of gas discipline most clearly played out in the writings of Ernst Jünger and the joint projects of Walter Benjamin and Dora Sophie Kellner. These three thinkers constructed highly theoretical visions of aero-chemical warfare technologies that neatly represented two of the major political commitments in the continuing debate over Germany's potential rearmament and the use of poison gas.

Chapter 7 examines the writings of gas specialists in their various gas protection journals. Through the publication of articles and books, this group of personally and professionally connected scientists and engineers continued to heighten the German public's concern for gas preparedness, calling for both increased gas drills and civilian familiarization with gas protection technology. This then created a greater public desire for visible steps toward national protection, including civilian gas mask distribution. The Weimar government's inability to provide such a large quantity of gas masks, coupled with feelings of national insecurity stemming from episodes such as the 1928 Hamburg gas leak, contributed to the Nazis' eventual consolidation of power. With the creation of the *Reichsluftschutzbund*, the Nazis intended to centralize national gas protection services and to dramatize the possibility of aero-chemical attack. As part of this theatrical staging, they attempted to provide gas masks for every German civilian. Practically speaking, this endeavor proved impossible, but the distribution of civilian gas masks was also a way to armor the German populace both visually and psychologically. Citizens who received gas masks were required to always keep them nearby, ready to pull them onto their faces at a moment's notice. Thus, by forcing the individual to respond to a seemingly imminent chemical attack, the gas mask could ostensibly mobilize the collective power of a Third Reich comprised of "Nazi chemical subjects."

Chapter 8 returns to poison gas, relating the ways in which the Nazis' efforts in national chemical protection were intimately tied to their search for new offensive chemical weapons. The chapter provides a reading of the Nazi relationship to poison gas and the party's attempt to realize a specifically "Nazi chemical modernity" in which poison gas would serve as a litmus test to determine the masters and victims in a chemically precarious world. Without drawing direct lines of causation between the 1915 German deployment of militarized gas and its reapplication in the Final Solution, this chapter explores the rhetorical and conceptual threads that developed across the interwar technological proposals of the gas specialists, the rhetorical racism of Nazi geopolitics, and the attempted chemical genocide in the gas chambers of the Holocaust. Finally, the short conclusion makes broad connections between this German story and the continued contemporary concern for the chemical construction of the environment.

1 The Structures of Violence: Fritz Haber and the Institutionalization of Gas Warfare

Traditionally, historians have cited the Second Battle of Ypres as the first modern use of chemical weapons. According to the most common telling of the story, on April 22, 1915, outside the small Belgian town of Ypres, German troops released chlorine gas toward French colonial soldiers. Military reports claim that the French lines dissolved in panic as men ran choking from the eerie yellowish-green gas that crept toward them. Two days later, chlorine was again used against Canadian troops, who apparently stood their ground among the deadly swirling clouds. The British military claimed that the attacks killed 15,000 men and injured another 5,000, while London newspapers published illustrations of French soldiers in their bright blue uniforms falling to the ground and grasping at their throats.[1] Numerous related op-eds decried the German use of chemical weapons as the ultimate breach in military conduct. In response, the Entente powers ordered the manufacture and distribution of thousands of cloth pads saturated with gas-neutralizing chemicals, hoping that this stopgap measure would help their boys withstand the perfidious German weapon.

While this story certainly creates an evocative image, the realities of the Second Battle of Ypres do not conform to such a stark morality tale. The German chlorine gas was not nearly so effective as the Allied press claimed, and far fewer men were injured or killed.[2] Certainly, the attack at Ypres had massive implications for the methods of chemical warfare in World War I, the technological development of the gas mask, and the propaganda war that proved exceedingly important during and after the war, but the strategic military value of the chlorine attack was dubious at best. Furthermore, contrary to the reports of many politicians, military commanders, and common soldiers, this was not, in fact, the first use of chemical gases in World War I, and it was certainly not the first European

[1] Rudolf Hanslian, *Der chemische Krieg* (E. S. Mittler & Sohn: Berlin, 1937), 89.
[2] The Germans insisted that their medical services only treated 200 gas casualties on April 22 and that only twelve men died.

meditation on chemical weapons. Rather, Ypres was a partly realized fragment in a long process of scientific, industrial, and military collaboration that developed through fits and starts in the nineteenth and twentieth centuries.

For precisely this reason, Ypres serves better as a study of the process by which the German military collaborated with and absorbed both research scientists and industrial firms into its own institutional structures. Certainly, scientists had helped in providing military technologies well before 1915, and national armies had required supplies and armaments from industrial factories throughout the nineteenth century. But Ypres provides us with a moment in which a formerly academic scientist was given a military rank in the hope that his scientific knowledge could win a war outright. In fact, German military divisions were created, and tactics were drafted to conform to the ideas of a man who had no previous training in military strategy. Through this position, this same scientist was able to strengthen his already intimate connections to the leading chemical firms in Germany, thus creating a formal German military bureaucracy that encompassed members of the scientific community and leaders in industrial chemical production. Other industrialized nations copied this same process to a certain extent, but it was the sheer scale, fluidity, and swift development of the German institutional hierarchy that allowed the Germans to dominate in chemical warfare until 1917. Thus, Ypres and the subsequent development of the German chemical weapons program was an important initial step in the realization of what would be later termed the "military-industrial complex," and the nationalized large-scale scientific projects that helped to define the political, military, and scientific worlds of the mid-twentieth century.

Prophecies of Poison: International Law before the War

In the nineteenth century, most European armies routinely used smoke to both conceal troop movements and to drive sappers out of their mines. Several militaries further considered the use of cyanide on bayonet blades or the deployment of sulfur fumes against city defenses.[3] However, none of these preindustrial chemical warfare plans required laboratory or factory-produced poisons. The development of industrial-scale chemical weapons

[3] Guy R. Hasegawa, *Villainous Compounds: Chemical Weapons and the American Civil War* (Carbondale: Southern Illinois University Press, 2015), 108–109; Charles Stephenson, *The Admiral's Secret Weapon: Lord Dundonald and the Origins of Chemical Warfare* (Rochester, NY: Boydell Press, 2006), 80.

required substantial advancement in the field of organic chemistry and the expansion of national chemical industries. This process began with scientists' investigation of the molecular structures of various chemical compounds, toxic or otherwise. In the nineteenth century, industrial and academic projects in the fields of pharmaceutical drug research and dye manufacturing tended to cross-fertilize with one another, often leading to surprise discoveries of new compounds.[4] While both British and French scientists made substantial contributions to the field, German chemical concerns dominated international markets in the late nineteenth century through expansive and effective cooperation between industrial firms, academic institutions, and government agencies.[5]

German companies such as Bayer AG, BASF (Badische Anilin und Soda Fabrik), and Hoechst AG grew from small dye works to massive chemical firms.[6] This was due largely to a favorable national patent system, supportive universities, and an advantageous geographic position on the Rhine River, downstream of the large steel and coal sites of the Ruhr Valley.[7] By 1913, German chemical production was valued at approximately 160 million dollars, out-producing its closest rival, the United States, by more than 60 million. Germany further exported about 28 percent of the world's chemicals, outpacing its nearest rival, Great Britain, by 10 percent.[8] The massive growth of the three largest companies and the German chemical industry more broadly also led to a stockpiling of chemical resources that could be isolated or synthesized. For instance, dye manufacturers such as Bayer, BASF, and Hoechst amassed large stores of chlorine as a byproduct of their dye production. The industrial scale of chemical stores encouraged greater consideration of their potential role in any future industrial use. While nineteenth-century Europeans certainly could not foresee the extent of World War I's

[4] Eric Croddy, *Chemical and Biological Warfare: A Comprehensive Survey for the Concerned Citizen* (New York: Springer, 2002), 134.

[5] L. F. Haber, *The Chemical Industry, 1900–1930: International Growth and Technological Change* (Oxford: Clarendon Press, 1971), 128–136.

[6] Smaller firms like AGFA, Cassella, Chemische Fabrik Kalle, Chemische Fabrik Griesheim-Elektron, and Chemische Fabrik vormals Weiler-ter Meer produced more specialized dyes, and together with the big three firms produced about 90 percent of the world's dyestuffs. Fred Aftalion et al., *A History of the International Chemical Industry: From the Early Days to 2000* (Philadelphia: Chemical Heritage Press, 2001), 475.

[7] Peter Murmann, "Knowledge and Competitive Advantage in the Synthetic Dye Industry, 1850–1914: The Coevolution of Firms, Technology, and National Institutions in Great Britain, Germany, and the United States," *Enterprise and Society* 1, no. 4 (2000): 702.

[8] Haber, *The Chemical Industry, 1900–1930*, 108.

industrialized combat, they were keenly aware of the national advantages that German chemical production offered.[9]

Initial European handwringing over the potential development and use of chemical weapons turned to international action at the First Hague Peace Conference of 1899. The conference, initiated by Czar Nicholas II of Russia, was aimed at both setting an international standard for the rules of warfare and creating an overarching international body for adjudicating disputes between member states. Twenty-six nations, including all major European powers and the United States, signed three main treaties and three additional declarations on July 29, 1899. Convention II with Respect to the Law and Customs of War on Land included a prohibition on the employment of "poison or poisoned arms."[10] The second declaration further prohibited the "Use of Projectiles the Object of Which Is the Diffusion of Asphyxiating or Deleterious Gases."[11] Still more regulations prohibited "treacherously" killing or wounding enemies and employing "arms, projectiles, or material calculated to cause unnecessary suffering."[12] All present nations except the United States, which saw these statutes as attempts to weaken the military ingenuity of newly industrial powers, subsequently ratified the conventions.[13] They entered into force on September 4, 1900.

Four years later, US President Theodore Roosevelt proposed a Second Hague Conference that intended to expand upon preexisting agreements and to create similar settlements for naval warfare. Continued international unrest, especially related to the Russo-Japanese War and the First Moroccan Crisis, delayed the conference until 1907 while simultaneously reaffirming the importance of international legislation and mediation. The prohibition of the use of "poison or poisoned arms," now moved to Convention IV, remained in the new treaty's language. Attending states signed the new set of agreements and declarations on October 18, 1907, and enacted them on January 26, 1910. Simultaneously, the British pushed to create national limitations on armaments and to set up a court of arbitration for international disputes. But it was the Germans who now mustered enough support to defeat

[9] John J. Beer, "Coal Tar Dye Manufacture and the Origins of the Modern Industrial Research Laboratory," *Isis* 49, no. 2 (1958): 131.

[10] Convention (II) with Respect to the Laws and Customs of War on Land (Hague, II) (July 29, 1899).

[11] The Avalon Project, Laws of War: Declaration on the Use of Projectiles the Object of Which Is the Diffusion of Asphyxiating or Deleterious Gases (July 29, 1899).

[12] Dieter Martinetz, *Der Gaskrieg 1914–1918: Entwicklung, Herstellung, und Einsatz chemischer Kampfstoffe* (Bonn: Bernard & Graefe Verlag, 1996), 9.

[13] Hugh R. Slotten, "Humane Chemistry or Scientific Barbarism? American Responses to World War I Poison Gas, 1915–1930," *Journal of American History* 77, no. 2 (1990): 478.

these further proposals, claiming that they ultimately aimed to enthrone the British atop the international order.

In many ways, these diplomatic discussions revealed the growing political struggle between Britain and the newly emerging industrial powers of Germany and the United States. But regardless of their omissions and limitations, the Hague Conventions appeared as a successful means by which international diplomacy could limit the scale and ferocity of future conflicts. In the case of poison gas, the wording of the declaration appeared unequivocal at the time. While Europeans had never actually experienced industrialized chemical warfare, they made it clear that they preferred to proactively censure and prevent any future considerations.

Leviathan Wakes: Chemical Applications in the Early Stages of World War I

At the outbreak of World War I in 1914, the major belligerents quickly mobilized their national chemical manufacturers for military production. But early intentions were primarily restricted to uniform dye manufacturing and explosives production. Moreover, in the initial stages of fighting, the aforementioned large German chemical firms had little desire for war. Industry leaders such as Carl Duisberg, the CEO of Bayer, knew that war-related restrictions on exports would significantly hurt their bottom line. It was precisely this strong economic desire to avoid war that left firms such as Bayer relatively unprepared in 1914.[14] Nevertheless, the combined industrial force of the large German chemical firms was able to supply the German War Ministry with almost all its initially required materials. The same could not be said for the Allies who realized the unsatisfactory state of their own national chemical industries after losing access to German products.[15] Both Britain and France scrambled to import dyes

[14] Carl Duisberg, *Meine Lebenserinnerungen*, edited by Jesco v. Puttkamer (Leipzig: Philipp Reclam, 1933), 98.

[15] The success of the German chemical industry in its collective war effort has been largely attributed to the standing cooperation between various firms before the war and the centralization of war production by the German government. Agfa, BASF, and Bayer had already formed a loose cooperative group in 1904 referred to as the "little IG." In the same year, Hoechst, Casella, and Kalle formed a Frankfurt am Main regional alliance. In 1914, the Germans created the KRA, or War Material Department, which served to centralize and streamline the chemical industry's war effort until the institution of the Hindenburg Program in the summer of 1916. This centralization was furthered in late 1915 and early 1916 when the IG group expanded to include more firms and to share knowledge and methods related to war production. Jeffrey Allan Johnson, "Technological Mobilization and Munitions Production: Comparative Perspectives on

from Switzerland while concurrently setting up government supply boards and kick-starting newly nationalized dyeworks.[16]

The supply of nitrogen, an essential component for explosives, was a different story for the Germans. Prior to the war, German firms imported about one-half of their required nitrogen from places such as Chile. The war's trade embargos and the British naval blockade now cut German industry off from their foreign nitrogen sources. This left the nation with slightly less than 140,000 tons of nitrogen at the start of the war, essentially setting a limit on its ability to produce artillery shells and bullets.[17] In September 1914, the academic chemists Emil Fischer and Fritz Haber made the German War Ministry painfully aware of this shortage and pointed to the BASF plant at Oppau and its newly implemented Haber-Bosch Process as a solution to German deficiencies.[18]

In 1909, Fritz Haber and his laboratory assistant Robert Le Rossignol had first succeeded in producing ammonia from nitrogen in the air using high-pressure devices and various catalysts. In 1910, the chemist Carl Bosch scaled Haber's process for industrial production at BASF.[19] By 1913, the BASF plant had a production capacity of 36 tons of ammonia per day and a store of 7,200 tons of synthetic nitrogen. The company increasingly raised its production capacity over the course of the war and the growth in demand from the German military led to the construction of a second production site at Leuna in 1917. By the end of the war, the Haber-Bosch process accounted for almost 50 percent of Germany's nitrogen production, certainly saving the Germans from early surrender.[20] Furthermore, the success of the Haber-Bosch process set the tone for future scientific aid to the German state. Hailed as a major national contribution that could be easily converted to peacetime applications, Haber's synthesis of nitrogen suggested to many in the scientific community that wartime research and production was indeed a major boon to the longer process of incremental scientific discovery and, by extension, the progress of humanity.

Germany and Austria," in *Frontline and Factory: Comparative Perspectives on the Chemical Industry at War, 1914–1924* (Dordrecht: Springer, 2006), 3–10.

[16] Haber, *The Chemical Industry, 1900–1930*, 188, 194.

[17] The physicist Max Planck later claimed that without Haber's nitrogen fixation process, "Germany would have collapsed economically and militarily in the first three months of the World War." Rüdiger Hachtmann, *Wissenschaftsmanagement im Dritten Reich* (Göttingen: Wallstein, 2007), 392.

[18] Haber, *The Chemical Industry, 1900–1930*, 200.

[19] Bretislav Friedrich, "Fritz Haber Und Der 'Krieg Der Chemiker'," *Physik in Unserer Zeit* 46, no. 3 (2015): 118–125.

[20] Haber, *The Chemical Industry, 1900–1930*, 203.

Unlike the Germans, the French government displayed little concern for state cooperation or intervention with their chemical industry early in the war. Nevertheless, the French did find the first direct battlefield application for the products of industrial chemistry.[21] Beginning in 1913, French scientists in André Kling's *Laboratoire Municipal* first produced what were then called *cartouches suffocantes* (an early form of tear gas projectiles). These *cartouches suffocantes* were intended to be used as a method of civilian riot control that could skirt the restrictions of the Hague Conventions.[22] The French military claimed that this tear gas was not strictly poisonous, and its stated aim was pacification through irritation of the eyes, skin, and lungs, rather than outright destruction.[23] Furthermore, the ethyl bromoacetate inserted into the projectiles was not a gas but rather a liquid that was aerosolized on impact.[24] As early as August 1914, the French military filled rifle projectiles with this newly developed tear gas and fired it at German lines. However, these first projectiles reportedly created such a weak gas cloud that German soldiers rarely noticed it. Nonetheless, the French continued to intensify the effect of their tear gases and to develop their gas grenades through 1915.[25]

As tear gas developed, official German troop reports began to differ on its effectiveness. Reflecting on fighting in the Vosges mountains in December 1914, the German infantryman Waldus Nestler wrote, "[The French] had thrown hand grenades or shot rifle grenades shortly before the attack, spreading a fog that brought tears to the eyes ... When we heard this, we were penetrated with a terrible feeling. What a nasty weapon that was now being used! What a terrible form of warfare!"[26] The suffering of soldiers like Nestler would certainly suggest that this tear gas was effective and the French military continued its research and production. By January 15, 1915, German Captain Hermann Geyer was reporting that the French used 12 tons of tear gas grenades in a particular engagement. By April of that same year, the French army provided its soldiers with the improved Bertrand No. 1 tear gas grenade, which was said to be more effective in trench warfare conditions.[27] And by the end of the war, Captain Geyer was

[21] Sophie Chauveau, "Mobilization and Industrial Policy: Chemicals and Pharmaceuticals in the French War Effort," in *Frontline and Factory*, 21.
[22] Martinetz, *Der Gaskrieg 1914–1918*, 9.
[23] These early tear gases can be toxic to humans if inhaled in sufficient concentrations.
[24] Throughout World War I, the French and all other belligerents primarily used xylyl bromide as their main tear gas agent.
[25] Olivier Lepick, *La Grande Guerre Chimique, 1914–1918* (Paris: Presses Universitaires de France, 1998), 55.
[26] Waldus Nestler, *Giftgas Über Deutschland* (Berlin: Nebelhorn-Verlag, 1932), 4.
[27] Ulrich Trumpener, "The Road to Ypres: The Beginnings of Gas Warfare in World War I," *The Journal of Modern History* 47, no. 3 (1975): 463.

now recounting the fact that the French Ministry of War claimed that their tear gas could in fact asphyxiate a soldier if he was exposed to a considerable density over a prolonged period of time.[28]

In 1913, the British also began seriously discussing the use of battlefield gases. Unlike the French, however, most British military officials initially returned to nineteenth-century visions of smoke screens and sulfur dioxide attacks. While these ideas were not immediately realized, the British did begin irregular testing of chloroacetone, benzyl chloride, and other tear gases.[29] However, the British War Office chose not to create dedicated chemical weapons laboratories until after the Germans had already unleashed their own chemical weapons. Indeed, the British appear to have been the most unwilling to develop and use chemical weapons, perhaps because they had the most invested in the international legislation that previously banned them. In fact, under the command of Lord Herbert Kitchener, the British military even hesitated to retaliate against the Germans in 1915, citing the already disregarded restrictions in the Hague Conventions.[30] It was only after the completion of a scientific investigation and a government report on the German chlorine attack that the British began to form a gas warfare service. This so-called Special Brigade was placed under the leadership of Major Charles Howard Foulkes of the Royal Engineers. By the summer of 1915, both the British and the French had organized their gas warfare divisions and began readying their own offensive gas attacks.[31]

Like the Allies, the Germans also studied the effects of chemical agents before the outbreak of World War I. However, according to the German artillery specialist, Colonel Max Bauer, these tests demonstrated that chemical weapons had little military value.[32] This initial futility impeded

[28] German Captain Hermann Geyer was an important staff officer for Colonel Max Bauer. He drew up several of the attack plans for German gas pioneer troops and contributed to the tactics used in the Kaiser's Offensive of spring 1918. After the war, he wrote several tracts on chemical warfare, was promoted to General, and served under the Nazis in World War II. Geyer-Denkschrift über den Gaskampf BA-MA N/221.

[29] Martinetz, *Der Gaskrieg 1914–1918*, 10.

[30] Ulf Schmidt, *Secret Science: A Century of Poison Warfare and Human Experiments* (Oxford: Oxford University Press, 2015), 26.

[31] James Steinhauser et al., *One Hundred Years at the Intersection of Chemistry and Physics: The Fritz Haber Institute of the Max Planck Society 1911–2011* (Boston: De Gruyter, 2011), 31.

[32] Max Bauer (1869–1929) first served as the head of Section II of the German High Command, responsible for artillery, mortars, and fortifications. Bauer received an honorary doctorate from the University of Berlin in 1915 and the Pour le Mérite in 1916 after he helped develop seventeen-inch howitzers to destroy outdated fortifications in Belgium and northern France. This success led to a surprisingly quick rise through the military ranks, eventually landing Bauer a position as an advisor on economic and political questions to General Erich Ludendorff. Throughout the war and into the

preliminary research, development, and employment.[33] Furthermore, historian Eric Dorn Brose has pointed out that sociocultural conflicts, waged in the Prusso-German army from 1871 to 1914, also retarded the employment of new military technologies such as chemical weapons.[34] With the development of modern technologies such as artillery guns, machine guns, and rifles, the future tactics of the German military became a site of contestation between representatives of different military branches. According to Dorn Brose, artillery and army engineering officers largely favored an embrace of these new warfare technologies, which they saw as both the key to future battlefield successes and an avenue to new personal prestige and institutional standing in the military. On the other hand, the cavalry, which had previously enjoyed the greatest honor and esteem in the Imperial German Army, feared its obsolescence in the face of technologies that could destroy a cavalry charge in mere minutes. Infantry officers similarly distrusted many new battlefield technologies, in part for their potential to shake up the preexisting military hierarchy but even more for their demonstrably violent effect on the common soldier's body.

As such, representatives from both the cavalry and the infantry used three major arguments to criticize and prevent the increased use of new technologies on the battlefield. First, they claimed that many technologies (like early tear gas) often failed to work as desired; these tools could not be trusted to achieve a desired tactical outcome. Second, certain technologies were not part of the inherited rules of war, either because they were prohibited by international treaties such as the Hague Conventions or because they simply broke with military tradition. Third, they insisted that battles were won and lost through sheer willpower. According to many cavalry and infantry officers, it was both the discipline and the élan of men that decided victory.[35]

Given the entrenched status and privilege of the German cavalry and infantry, these arguments apparently succeeded in holding off serious military engagement with many previously untapped areas of industry and science until the fall of 1914. This status quo finally changed in

1920s, Bauer served as a vital link between the German military, heavy industry, and nationalist political groups, allowing him to also play a central role in the creation and production of poison gas. Martin Kitchen, "Militarism and the Development of Fascist Ideology: The Political Ideas of Colonel Max Bauer, 1916–18," *Central European History* 8, no. 3 (1975): 199–200.

[33] Trumpener, "The Road to Ypres," 463–464.

[34] Eric Dorn Brose, *The Kaiser's Army: The Politics of Military Technology in Germany during the Machine Age, 1870–1918* (Oxford: Oxford University Press, 2001), 3.

[35] Ibid., 66–69.

September of that year, when the German military was defeated by Allied forces at the First Battle of the Marne, thus forcing the Germans to formally abandon their initial overarching strategy known as the Schlieffen Plan. The loss at the Marne prevented the Germans from continuing their march toward Paris and induced them to dig into their positions on the Western Front, setting the stage for the years of trench warfare that would follow. Disgruntled factions of the officer corps began searching for someone or something to blame for this unexpected military failure. High-ranking advocates of new military technologies such as Colonel Max Bauer attributed the setback to the army's willful avoidance of technological change, stating that the days of cavalry charges were now over.[36] The subsequent removal of General Helmuth von Moltke as chief of the general staff, the appointment of Lieutenant General Erich von Falkenhayn, and the reevaluation of German tactics in late 1914 now opened the door for a new breed of officer who either promoted or quietly accepted the arrival of new military technologies on the battlefield.

Besides the German army's broader acceptance of new military technologies, Erich von Falkenhayn decided to seriously revisit the possibility of employing chemical weapons for three distinct reasons. First, as the war on the Western Front became less mobile, the Germans sought any technological advantage that might break through the Allied lines. Reports from the front stated that traditional shells and explosives were not completely effective in destroying entrenched enemy positions. Second, facing a shortage of steel in 1915, the Germans were forced to manufacture their artillery shells using cast iron. Because cast iron was weaker than steel, German shell walls were so thick that they did not allow much space for interior explosives. German High Command hoped to find an alternative shell filling that could expand outward from such a small container.[37] Third, in early fall of 1914, the Haber-Bosch process was not yet fully integrated into industrial war production. The Germans were still facing a rapidly diminishing supply of nitrogen and,

[36] Many of the technological progressives like Bauer were officers in the artillery or army engineer corps. In contrast to the often-aristocratic cavalry and infantry officers, these men were largely of middle-class origin. This difference is important for explaining the sociocultural changes occurring in the German military during this period. Middle-class men could hope to achieve greater military and societal standing by entering a branch of the army that required a level of practical scientific education and allowed for advancement through the application of specialized knowledge. This prerequisite education then helped to foster connections between these branches of the military, academia, and industrial firms. Gerald Feldman, *Army, Industry, and Labor in Germany 1914–1918* (Princeton: Princeton University Press, 1966), 36–37.

[37] Ian Hogg, "Bolimov and the First Gas Attack," in *Tanks and Weapons of World War I*, ed. Bernard Fitzsimons (New York: Beekman House, 1973), 17.

thus, a rapidly diminishing supply of explosives and bullets. Chemical aerosols and gases appeared as a possible replacement that could be abundantly produced by a preexisting German industry.[38]

As early as September 1914, Falkenhayn had already given Colonel Max Bauer orders to reinvestigate the applicability of chemical weapons on the Western Front. A month later, Bauer gathered a group of scientists and army officials at the Wahn artillery range outside of Cologne to discuss the creation of a new chemical shell that could dislodge entrenched enemy positions.[39] Proposed by the internationally renowned academic chemist, Walther Nernst, the first experimental weapon produced a nonlethal nasal irritant that could be fired in a standard shell casing.[40] The German military first deployed 3,000 of these so-called Ni-shells on October 27 at Neuve-Chapelle.[41] However, they apparently had so little effect that French soldiers later reported that they never detected the chemical irritant. Indeed, according to a popular story, the gas was so weak that General Falkenhayn's son won a case of champagne after betting that he could stand in a cloud of Ni-shell irritant for five minutes.[42] The German army received 20,000 of these Ni-shells before the project was deemed a total failure.[43]

By November of 1914, Bauer's chemical research committee had moved on to a suggestion from the chemist Hans Tappen.[44] Tappen proposed filling artillery shells with the liquid irritant xylyl bromide to create an effect similar to that of the French tear gas grenades. Bauer was initially skeptical of shells filled with liquid, but Tappen eventually proved the feasibility of such a weapon by testing them on himself.[45] These so-called "T-Shells" were first evaluated at the Kummersdorf artillery range outside Berlin and then again at Wahn in January 1915.[46]

[38] Werner Plumpe, *Carl Duisberg 1861–1935: Anatomie eines Industriellen* (Munich: C. H. Beck, 2016), 443–444.

[39] This expert committee included the director of the Bayer dyeworks factory at Leverkusen, Carl Duisberg, and the influential chemist Walther Nernst, best known for his pioneering work on thermodynamics and electrochemistry.

[40] Max Bauer, *Der grosse Krieg in Feld Und Heimat* (Tübingen: Osiander'sche Buchhandlung, 1922), 67–68.

[41] The "Ni" in Ni-shells stood for *Niespulver*, or sneezing powder.

[42] Hogg, "Bolimov and the First Gas Attack," 17.

[43] "Herstellung Und Lieferung von Geschoss-Fullung" (BA) 201-005-001.

[44] Hans Tappen was a scientist at the Kaiser Wilhelm Institute for Physical and Electrochemistry. He was also the brother of Gerhard Tappen, a close advisor to General Falkenhayn.

[45] Tappen's personal war diary, November 6–9, 1914 (BA-MA) Nachlass Gerhard Tappen.

[46] The shells were named "T-Shells" in honor of Tappen, "Herstellung Und Lieferung von Geschoss-Fullungen III" (BA) 201-005-003.

Falkenhayn himself was present at the Wahn tests and later approved the production and use of Tappen's new chemical weapon. The chemical firms of Bayer, Hoechst, and the Kahlbaum factory in Berlin were contracted to produce the T-Shells and only a few weeks later, they were issued to Colonel-General August von Mackensen's Ninth Army on the Eastern Front.[47] Mackensen was ordered to produce a diversion with the gas shells, while Field Marshal Paul von Hindenburg arranged for a major offensive against Russian troops. In preparation for the diversionary chemical attack, German soldiers were outfitted with cotton pads that could be tied in front of their mouths and noses. The pads were moistened with a chemical retardant that had to be reapplied regularly.

On January 3, 1915, the Germans attacked Russian troops at Bolimow, between Lodz and Warsaw. The T-Shells were fired into the Russian lines, but the cold weather suppressed the xylyl bromide and the shells had little to no effect.[48] Returning to the drawing board, Tappen added bromoacetone (B-*Stoff*) to the shells to prevent their failure in cold weather.[49] These new T/B-Shells were then sent to Duke Albrecht of Württemberg and his Fourth Army in Flanders. In the early spring of 1915, the new T/B-Shells also achieved little effect at Nieuport and in the region surrounding Ypres. This encouraged Bauer's advisory committee to turn its attention toward yet another potential chemical weapon that had been in development since December 1914 under the direction of the chemist Fritz Haber.

A Devoted Man of Science: The Early Life of Fritz Haber

Fritz Haber was born in 1868 in Breslau, Prussia, to a wealthy Jewish family whose collective story reflected that of many upwardly mobile Jewish families in the German-speaking lands of the late nineteenth

[47] By the autumn of 1915, the Bayer and Hoechst dyeworks were transitioning much of their industrial production to war purposes. The companies had realized that the war was not going to end quickly and that this was their best chance at wartime expansion. While they received wartime government contracts and subsidies, they continued to complain that these were too irregular and insufficient. BASF avoided the production of weapons likes chlorine gas, but they did provide intermediary materials like ethylene chlorohydrin and thiodiglycol that were then used for poison gas production. Werner Abelshauser et al., *German Industry and Global Enterprise. BASF: The History of a Company* (Cambridge: Cambridge University Press, 2004), 165.

[48] Robert E. Cook, "The Mist That Rolled into the Trenches: Chemical Escalation in World War I," *Bulletin of the Atomic Scientists* 27, no. 1 (1971): 36.

[49] Bromoacetone was yet another irritant that proved more stable at lower temperatures. Hanslian, *Der chemische Krieg*, 57.

century. The extended Haber family first made their money as wool merchants in Kempen, but by the time Fritz was born, they had obtained respected positions in politics, law, and other lines of mercantile business. Fritz's father, Siegfried Haber, owned a successful firm that sold pharmaceuticals, dyes, resins, oils, fats, and paints.[50] Siegfried reportedly had a loving relationship with his wife, Paula, who died giving birth to Fritz. According to biographies of Haber, Paula's early death created a lasting rift in the relationship between Siegfried and Fritz and the young boy was quickly given over to the care of his aunts. As a child, Fritz also developed a close and affectionate relationship with his stepmother, who eventually became his primary caregiver.

While the Haber family remained tightly tied to the broader Jewish community through family connections in Breslau, they were not noted as particularly religious or observant Jews. Consequently, young Fritz was sent to the Johanneum primary school and the St. Elizabeth Gymnasium, both of which were open to students of different religious backgrounds who desired a humanistic education in the German tradition. While Fritz was not noted as a particularly gifted student while attending either of these schools, he nevertheless passed his gymnasium exams at St. Elizabeth's in 1886. Upon graduation, Fritz's father preferred that his son remain in the family dye trade, but the notably gregarious boy saw university as a path to the wider world. With grudging assent from his father, Fritz attended Friedrich Wilhelm University in Berlin to study chemistry, a field that could prove valuable to the family dye trade.

In Berlin, Haber was initially disappointed with the work of his advisor, the chemist A. W. Hofmann.[51] This encouraged him to first transfer to Heidelberg University in order to work under Robert Bunsen for the duration of the 1887 summer semester. He then moved back to Berlin to attend the Technical College of Charlottenburg for the remainder of his degree.[52] In the summer of 1889, Haber's studies were put on hold so that he could serve a one-year term of Prussian military service in the Sixth Field Artillery Regiment back in Breslau.[53] Haber was immediately attracted to the military lifestyle, and he continued throughout his life to

[50] Siegfried Haber's company was one of Germany's largest importers of natural indigo.

[51] August Wilhelm von Hofmann (1818–1892) was a respected German organic chemist at the Royal College of Chemistry in London and the University of Berlin. He was a cofounder of the German Chemical Society in 1867.

[52] Most known for his development of the Bunsen burner, Robert Bunsen (1811–1899) was a pioneer in photochemistry.

[53] Volunteer service was normally three years, but Haber only served one because his father was willing to pay for his horse and military equipment. This was a common practice for wealthier families in Prussia.

affect the clipped but informative speech, upright posture, and brisk marching gait that he picked up in service. In fact, he was so pleased by this new lifestyle that he competed for an officer's commission with the intention of finding a permanent career in the military. While Haber apparently did prove himself worthy enough to become a serious candidate for the commission, he was ultimately passed over. Greatly affected by this disappointment, he believed that his Jewish background had hurt his candidacy, as no Jew had ever been selected as a reserve officer in a combat service.

After this difficult personal failure, Haber returned to the Technical College of Charlottenburg to begin studying organic chemistry under Carl Liebermann.[54] In 1891, he finished his thesis on the derivatives of piperonal and received his doctorate from Friedrich Wilhelm University.[55] With a degree in hand, Haber returned to Breslau to honor his agreement and work in his father's chemical company. Once home, Fritz continuously fought with his father, eventually choosing to spend most of his time in practical apprenticeships at distant chemical factories. Feeling that his poor relationship with his father would always lead to conflict, Haber finally left the family business in 1892 and took an academic position as an unpaid assistant under the chemist Ludwig Knorr at the University of Jena. Over the next two years, Knorr assigned Haber to repetitive laboratory tasks that he quickly came to resent.[56]

While Haber's scientific work in Jena was relatively uneventful, he did undergo one major personal change. In November of 1892, Haber converted to Protestantism. It remains unclear exactly why Haber chose to convert, but this move certainly helped him to secure a place in the academic world. Finding an *Assistent* position would have been made undoubtedly more difficult by the obstacles put in place to bar Jews from the academy. Whatever the combination of motivations, most theories about the nature of Haber's conversion argue for a practical cause rather than a real change of faith.[57] In fact, Haber's personal doctor and close

[54] Carl Liebermann (1842–1914) was a student of Adolf von Baeyer, taking over his chair in organic chemistry at the University of Berlin once Baeyer left for the University of Strasbourg in 1870.

[55] The Technical College of Charlottenburg was not yet able to grant doctoral degrees so Haber received his from the Friedrich Wilhelm University.

[56] Ludwig Knorr noticed the young Haber's resentment and thought little of his work. Haber, in turn, wanted to switch to physical chemistry and work with Wilhelm Ostwald. He was granted a meeting with Ostwald in 1893, but this did not lead to a place in his Leipzig laboratory. Morris Goran, *The Story of Fritz Haber* (Norman: University of Oklahoma Press, 1967), 16.

[57] Only Haber's second wife, Charlotte, claimed that Fritz converted based on religious convictions, even though Haber rarely attended church. Charlotte also believed that the conversion may have been Fritz's way of symbolically breaking from his father. Daniel

friend, Rudolf Stern, saw the conversion as a continuation of Haber's true faith, a belief in the insolubility of the German nation. Stern claimed that Haber sincerely desired to become more "authentically German" and to profess his loyalty to the Imperial Reich through his religious profession.[58] Certainly, through his early education and his time in the military, Haber had become a staunch German patriot with a particularly strong desire to align himself with the imperial regime. As such, Haber's godson, Fritz Stern, once described him as "German in every fiber of his being, in his restless, thorough striving, in his devotion to friends and students, in his very soul and spirit."[59]

In the spring of 1894, the recently converted Haber was finally able to leave Jena after obtaining a paying *Assistent* position at the Technical University of Karlsruhe. Karlsruhe was not a major center of academic chemistry at the time, but it did have strong funding avenues through ties to the BASF chemical company in nearby Ludwigshafen. Once situated in Karlsruhe, Haber's two main advisors, Carl Engler and Hans Bunte, gave Haber a great level of research freedom.[60] This allowed him to move into the field of electrochemistry, where he gained attention for his reportedly tireless work ethic. Haber slowly built a name for himself through research in both organic and electrochemistry in the late 1890s and early 1900s, and in 1906 he received a full professorship in physical chemistry.

Upon taking up his university chair in Karlsruhe, Haber and his *Assistent* Gerhard Just began an exceptionally varied research agenda that brought significant recognition and widespread approval to his laboratory. By 1908, Haber was concentrating on nitrogen fixation and signing a collaborative contract with BASF. This work would ultimately garner Haber his first taste of national renown and the 1918 Nobel Prize in Chemistry. On the eve of World War I, Haber thus stood in a fortuitous position between the academy and industry, between the worlds of "pure science" and technological application.

Charles, *Master Mind: The Rise and Fall of Fritz Haber, the Nobel Laureate Who Launched the Age of Chemical Warfare* (New York: HarperCollins, 2005), 28–29.

[58] Ibid., 29–32.

[59] Fritz Stern, *Dreams and Delusions: The Drama of German History* (London: Weidenfeld & Nicolson, 1988), 54.

[60] Carl Engler (1842–1925) directed work on dye production and fiber technology in his laboratory at Karlsruhe. Hans Bunte (1848–1925) was Engler's appointed successor, working both on dye technologies and gas illumination.

The Foundation of the Kaiser Wilhelm Society for Physical and Electrochemistry

Of course, Fritz Haber was not the only German chemist who straddled the presumed intellectual and institutional divides between pure and applied research. As early as 1905, the celebrated German chemists Emil Fischer, Walther Nernst, and Wilhelm Ostwald had sought to create a national chemical research institute that would be able to foster academic collaboration on large-scale research projects.[61] They set up an association to solicit funds for the new "Imperial Chemical Institute," and while they received donations from individual scientists and industry leaders, they failed to obtain backing from the Prussian Ministry of Finance. The nascent plan was then renewed in 1908 with the intention of including various other areas of natural science in a larger research organization that would be called the Kaiser Wilhelm Gesellschaft (The Kaiser Wilhelm Society or KWG). The Society, originally envisioned as the German version of Oxford University, would now operate independently of the German academy. However, it was still intended to foster significant research connections between academics, especially at Humboldt University in Berlin. Fischer reallocated the funds originally collected for the standalone chemical institute to the Kaiser Wilhelm Society and requested that one-third of the annual budget go to physical chemistry. The Prussian Ministry, however, remained hesitant to proffer state funds, and it was unclear whether the KWG could strictly operate on private donations. As late as September 1910, the Prussian government was unsure if it would provide financial support, even though Kaiser Wilhelm II was set to announce the opening of the scientific society in October of that year.

At a September 17th organizational meeting, a wealthy Saxon banker named Leopold Koppel assisted with the KWG funding problem by agreeing to independently fund the Kaiser Wilhelm Institute for Physical and Electrochemistry (KWI), a subsidiary physical chemistry institute within the overarching Kaiser Wilhelm Society. Koppel had made his money through the ownership of several hotels, an independent banking firm, and a gas light company that he started in 1892 with the Austrian chemist Carl Auer von Welsbach. Located in Berlin, the gas light company

[61] At the time, Emil Fischer (1852–1919) was the Director of the Institute for Chemistry at the University of Berlin. Walther Nernst (1864–1941) was the director of the Institute for Physics at the University of Berlin and the major scientific rival of Fritz Haber in the field of organic chemistry. Wilhelm Ostwald (1853–1932) was the director of the Institute for Physical Chemistry in Leipzig.

was named the Deutsche Gasglühlicht-Aktiengesellschaft (Degea), later changed to the Auer Gasglühlicht Gesellschaft, or Auergesellschaft.[62] The Auer company primarily worked with gas mantles, luminescence, rare earths, and radioactive materials. It was this work that brought Fritz Haber to the Auer offices in Berlin as a research consultant. From the moment of his arrival, Haber greatly impressed Leopold Koppel and the two men quickly became friends.[63] They shared not only professional interests in the field of physical chemistry but similar personal backgrounds. Koppel, born Jewish, converted to Christianity both to marry his wife and to garner future business advantages.

Between 1908 and 1910, Koppel attempted to woo Haber to a permanent position at the Auer company. Haber repeatedly declined these offers, stating that he was pleased with the scientific freedom that he enjoyed in Karlsruhe. Koppel was not easily deterred, however, and he saw several business opportunities in providing the initial 1 million Reichsmark endowment for the Kaiser Wilhelm Institute for Physical Chemistry and Electrochemistry.[64] First, Koppel made sure that the KWI could apply for scientific patents, which could then be inexpensively licensed to his Auergesellschaft. Second, the KWI was another avenue for bringing Haber closer to the Auer factories in Berlin. By asking Haber for his employment demands in 1910, Koppel made it clear that he intended to install Haber as the head of the newly founded institute. Expecting that his requests would be too great and that he would remain at Karlsruhe, Haber stipulated that the KWI had to be made a first-rate research institution. Thus, Haber required a sufficiently large budget, total control over hiring, a professorial position at the University of Berlin, membership in the Royal Prussian Academy, a tenured position as a state official, the transfer of his Karlsruhe research assistants, and an annual salary of 15,000 Reichsmark. To Haber's surprise, Koppel agreed to these hefty demands and Haber was installed at the head of the KWI for Physical and Electrochemistry in the summer of 1910. Two months later, Kaiser Wilhelm II gave Haber the title of *Geheimer Regierungsrat* (privy councilor).

Haber, now a leading chemist in German academia, busied himself with plans for building his institute in the Berlin suburb of Dahlem. With input from BASF, Haber created laboratories that were more modern than

[62] Degea Aktiengesellschaft (Auergesellschaft) Berichte, Korrespondenz (EIA) Degussa IW 24.5./2–3.
[63] Charlotte Haber, *Mein Leben mit Fritz Haber* (Düsseldorf: Econ Verlag, 1970), 116–117.
[64] Charles, *Master Mind*, 121.

those of most universities.[65] He further made sure to include space for guest researchers, revealing his desire to attract great scientific talent to his institute. In his role as KWI director, Haber would prove himself a great organizer of science and a uniquely personable diplomat. In 1917, he even helped to lure Albert Einstein to serve as the director of the KWI for Physics.

Haber and Einstein first met in 1911 at the annual meeting of German natural scientists in Karlsruhe. According to later reflections, Einstein was impressed by Haber's "uncanny agility" and "fertility of ideas."[66] Haber, in return, enthusiastically supported Einstein's theory of relativity, which helped to convince Einstein to join the Prussian Academy of Sciences in Berlin in 1914. This mutual admiration turned into yet another strong and lasting friendship for Haber. For many German scientists, including Einstein and the chemist Richard Willstätter, Haber was perhaps best defined by the loyalty and respect that nourished his numerous friendships. Writing on his travels with Haber, Willstätter once reflected, "The most beautiful trips were the ones I took with Fritz Haber; they were hours of friendship in which I came to know and understand his individuality, his noble mind, goodness of heart, wealth of idea, and his boundless, extravagant drive."[67]

Throughout 1911, Haber continued solidifying both his personal relationships in the scientific world and the physical buildings that would serve as his institutional home. In the span of eleven months, construction crews erected two large academic buildings, which officially opened their doors on October 23, 1912. Kaiser Wilhelm II attended this opening ceremony and made a short speech that was influenced by a recent mine explosion. The Kaiser encouraged the scientists of his new institute to synthesize chemicals that could alert miners to the presence of dangerous gases and ultimately save more Germans from the hazards of airborne toxins. Such a call to national duty appealed to Haber, and the Institute's staff of five scientists, ten technicians, and thirteen assistants began working on a methane gas detector. At the end of 1912, Haber received great acclaim from both the Kaiser and the larger Kaiser Wilhelm Society when he presented his prototype for the marsh gas whistle, an instrument that changed tone based on the presence of toxic gas. Rights to the production of the marsh gas whistle were signed over to Koppel's Auergesellschaft, but further potentially life-saving research was abandoned with the outbreak of the war in 1914.

[65] Steinhauser et al., *One Hundred Years at the Intersection of Chemistry and Physics*, 15–16.
[66] Fritz Stern, *Einstein's German World* (Princeton: Princeton University Press, 1999), 103.
[67] Stoltzenberg, *Fritz Haber*, 205.

Taming the Air: Haber's Early Chlorine Gas Research

On the eve of World War I, Fritz Haber was in the middle of a six-week vacation at Karlsbad to recover from "gallstones and moodiness."[68] When the war was announced in the summer of 1914, Haber immediately ended his sojourn and returned to Berlin to volunteer for active duty as a noncommissioned officer. This was quite clearly a quixotic renewal of the exceptional patriotic spirit that Haber had displayed in his short stint in the Prussian military. But at the age of forty-four, Haber was rejected for active military service. Overcoming his initial disappointment, Haber soon found other ways to contribute to the war effort. First, he signed the Manifesto of the Ninety-Three, which unequivocally supported Germany's war aims and its military actions.[69] Even among fellow signees, Haber was particularly supportive of the imperial German state, displaying a distinct desire to serve in the hierarchical structure under the Prussian Kaiser and his military generals. Indeed, the physicist J. E. Coates once claimed that Haber's greatest wish was "to be a great soldier, to obey and be obeyed."[70]

Like many of the prominent signees of the Manifesto of the Ninety-Three, Haber equated a potential German victory with the triumph of *Kultur*, or a more spiritually edifying form of learning that was said to be the wellspring of the German mind. But for Haber, this concept of Germanic *Kultur* extended beyond the writings of Goethe and Schiller. Like many other contemporary German scientists, Haber viewed German natural science as equally entangled in the supposed cultural struggle in which the German nation was now engaged. As such, science could now be viewed as a purely national endeavor rather than the international and collective enterprise of the nineteenth century. Boiling this logic down to a singular phrase, Haber later asserted that a true German scientist must "serve mankind in peace and the fatherland in war."[71]

[68] Ibid., 127.

[69] Klaus Hoffmann, *Otto Hahn: Achievement and Responsibility* (New York: Springer New York, 2001), 83. From the field of chemistry, Adolf von Baeyer, Emil Fischer, Walter Nernst, Richard Willstätter, and Wilhelm Ostwald all joined Haber in signing the Manifesto. On the whole, only strict pacifists like Albert Einstein expressed disagreement with the nationalist bellicosity that now pervaded German science.

[70] Margit Szöllösi-Janze, "The Scientist as Expert: Fritz Haber and German Chemical Warfare during the First World War and Beyond," in *One Hundred Years of Chemical Warfare: Research, Deployment, Consequences* (New York: New York, 2017), 21.

[71] Henry Harris, "To Serve Mankind in Peace and the Fatherland in War: The Case of Fritz Haber," *German History* 10, no. 1 (1992): 31–34.

With such a patriotic disposition and personal connections throughout the world of German chemistry, Haber was quickly installed as the head of the chemistry department for Walther Rathenau's Board of Wartime Raw Materials. Often referred to as the "Haber Office," this position allowed Haber to exert his influence over both the scientific academy and industrial firms. His specific task was to adapt current chemical research projects for military purposes. The Kaiser Wilhelm Institute for Physical and Electrochemistry was an integral piece of this transition, and Haber directed his departments to begin researching new explosives and shell fillings, including tear gases. But by the end of 1914, Haber and Rathenau's personal relationship had deteriorated, and Haber left his position with the Board of Wartime Raw Materials to focus solely on a new role as head of the Chemistry Section in the Ministry of War.

Given his chemical expertise, Haber had served as a consultant on Hans Tappen's T-Shell project since the fall of 1914. During the testing of these projectiles, Haber realized that tear gases would never be effective in small quantities. To achieve the requisite density of gas, a magnitude of shells would need to be fired in tight clusters.[72] Shell casings were still scarce for the German military in 1914 so there was little that could be done to strengthen T-Shell attacks. To avoid similar logistical problems, Haber now suggested to Colonel Max Bauer and General Erich von Falkenhayn that perhaps chlorine gas, a widely available byproduct of many industrial chemical processes, could be released from cylinders. According to this plan, Haber would insert siphon tubes into steel drums that each held twenty kilograms of liquid chlorine. He would then attach a lead pipe to each siphon tube and each pipe would be placed on top of the German trench. When wind conditions were favorable, a valve could be opened, thus mixing the liquid chlorine with air and turning it into a gas that could waft toward the enemy trench. In theory, the chlorine would form a yellow-green cloud with a greater density than air. Thus, the cloud would roll along the ground, corroding enemy weapons and forcing a troop retreat.

Part of the initial attractiveness of Haber's plan was its legal and moral ambiguity. Because Haber claimed that his chlorine gas was intended to irritate rather than kill the enemy, the German military could argue that they were adhering to the rules of war set down in the Hague Conventions. Falkenhayn and Bauer also claimed that any clause in the First Hague Conventions' articles regarding "poison or poisoned weapons" referred to the intentional poisoning of drinking water or foodstuffs rather than

[72] Harold Hartley, "Report on German Chemical Warfare, Organisation and Policy, 1914–1918" (MPG) Va 5, 528.

independent poison gas weapons. Further still, the wording of the Second Convention's ban on poisoned gas specifically mentioned projectiles, but Haber's cylinders would not employ bullets or shells. Finally, the wording of the agreements on poisoned weapons repeatedly mentioned the inducement of an enemy's "unnecessary suffering." Haber and Max Bauer continually claimed that chlorine gas was no worse than the explosives that permanently maimed and disfigured so many soldiers in the first year of the war.[73] If none of the aforementioned explanations appeased the Allies, the German commanders at least felt confident that the French had already tested these legal boundaries through their use of potentially lethal tear gases.[74] The German Ministry of War thus gave Haber's plan a green light for development and both Haber and his KWI were subsequently brought closer to the Prussian military bureaucracy.

In the following months, Haber proved extremely skillful in both transforming his KWI laboratories into a military research institute and integrating his endeavors into the preexisting military hierarchy.[75] Toward the end of 1914, he began to assemble what would later be called *Pionierkommando* 35 and 36 (Pioneer Regiments 35 and 36). These were two army engineer regiments responsible for the deployment of poison gas and all other future battlefield tasks related to gas warfare. While Colonel Otto Peterson served as the regimental commander of the gas troops, Haber and Friedrich Kirschbaum of the KWI were made military advisors.[76] In this role, Haber had notable power in staffing, and he began to hand-select 500 promising young chemists and physicists to join the ranks. As one member of the regiments later reported: "Any officer selected by Haber, regardless of where he was at the time, was soon transferred by Lieutenant Colonel Bauer to Peterson's organization ... Haber sought out all physicists and chemists of whom he had heard."[77] Haber's select group included a number of later-famous scientists including Otto Hahn, Gustav Hertz, James Franck, and Hans Geiger.[78]

[73] General Falkenhayn specifically requested a gas that could produce "lethal poisoning." Isabel V. Hull, *A Scrap of Paper: Breaking and Making International Law during the Great War* (Ithaca: Cornell University Press, 2014), 233.

[74] Nick Lloyd, *The Western Front: A History of the Great War, 1914–1918* (New York: W. W. Norton, 2012), 102.

[75] Jeffrey A. Johnson, *The Kaiser's Chemists: Science and Modernization in Imperial Germany* (Chapel Hill: University of North Carolina Press, 1990), 189.

[76] Otto Peterson was a regular army officer in his mid-50s who led the engineers of Königsberg before taking command of the Gas Pioneers.

[77] Franz Richardt, Report (MPG) Dept Va, Rep 5, 1494.

[78] Otto Hahn (1879–1968) was a chemist who would later win the Nobel Prize for splitting the atomic nucleus. Before the war, he was already working on radiation at the KWI for

As the pioneer regiments grew throughout the war, additional young scientists and engineers such as Rudolf Hanslian and Hugo Stoltzenberg joined as a way to advance their careers and gain access to Haber. By April 27, 1915, the pioneer troops had grown from 500 to 1,600 men. While some of these scientists would later feel guilty for participating in the early development of chemical warfare, Haber was reportedly able to convince many of them to join through the assurance of his respected position in chemistry and his sheer force of personality. Furthermore, Haber held personal meetings with several hesitant men in which he talked of his unshaken belief in the German ability to win the war swiftly with a massive gas strike that would save countless lives. Otto Hahn remembered how Haber gave him a lecture about:

> how the war had now become frozen in place and that the fronts were immobile. Because of this situation, the war now had to be fought by other means in order to be brought to a favorable conclusion ... I interjected that the use of poisonous substances was certainly universally condemned, whereupon he replied that the French had already tried something similar with shells in autumn of 1914. We would thus not be the first to use this kind of weapon. Anyway, in war, methods have to be used that lead to its rapid conclusion.[79]

By late December 1914, Haber and his men were already conducting small-scale tests of chlorine cylinders referred to as "F batteries" at the Wahn artillery range. For gas protection, the pioneer troops were outfitted with the *Dräger Selbstretter* breathing apparatus, a cumbersome canister of oxygen that was attached via rubber tubing to an adjustable mouthpiece.[80] Hampered by the *Selbstretter*, the gas troops would slowly embed a few gas cylinders at a time and practice the synchronized release of the gas. While technical problems still delayed the development of a coordinated piping system for large-scale chlorine attacks, the German

Chemistry. Gustav Hertz (1887–1975) was an experimental physicist who would later win a Nobel Prize for his work on inelastic electron collision in gases. He was the nephew of Heinrich Rudolf Hertz. James Franck (1882–1964) was a physicist who would win a Nobel Prize for discovering the laws governing the impact of electrons on the atom. He would later direct the Chemistry Section at the Metallurgical Laboratory of the University of Chicago, a significant research sector of the Manhattan Project. Hans Geiger (1882–1945) was a physicist who helped in constructing the Geiger-Müller counter and in describing the atomic nucleus. Several future leaders of the IG Farben chemical concern, like Ludwig Hermann, were also gas pioneers. Auszug aus dem Kriegstagebuch von Otto Gerhard Lt. im Res Reg 201 und ab Marz 15 bei den Gastruppen, 1956 (IZS) N 60.14/1–16.

[79] Johannes Jaenicke, Memorandum on a conversation with Otto Hahn in January 1955 (MPG) Dept Va, Rep. 5, 1453.

[80] Otto Hahn, *Mein Leben: Die Erringerung des grossen Atomforschers und Humanisten* (Munich: Piper, 1986), 118.

High Command now chose to clear the weapon for use on the Western Front. Interestingly, the German generals never witnessed the weapon's effectiveness because they feared that a large-scale test could spread chlorine for miles around Wahn.[81]

Both the German High Command and Haber were insistent that surprise was a key component in the potential military success of chlorine gas.[82] The gas would need to create a considerable level of chaos and confusion in the Allied lines if the Germans were to achieve a complete breakthrough. However, German commanders were also skeptical about the feasibility of secretly installing large cylinders near the enemy's trench. Furthermore, the speed required for this attack was made doubly difficult by its reliance on the weather. The plan needed a slow and steady wind moving in a westward direction in order to carry the gas toward the Allied troops.[83] Beyond these technical problems, several military commanders harbored lingering distrust of the sociocultural changes that gas might bring to warfare. A close advisor to Falkenhayn, General Gerhard Tappen, described chlorine gas as "initially repugnant" and of an "unchivalrous nature," while Major General Berthold von Deimling, who would later superintend the attack at Ypres, stated, "I must admit that the task of poisoning the enemy as if they were rats went against the grain with me as it would with any decent sentient soldier."[84]

Regardless of these objections, the German High Command knew that breaking through the enemy's trench lines was now essential to a victory in the West. The early stages of trench warfare had made the achievement of such a breakthrough seem exceedingly difficult, and technological solutions appeared increasingly attractive. In the process of deciding where to employ this potential wonder weapon, High Command asked each of the German Army commanders if they were interested in testing gas on their front. All but Duke Albrecht of Württemberg refused, and thus Albrecht's 4th Army at Ypres was chosen to bear witness to the first chlorine attack. Unfortunately, Ypres was not the ideal location for the attack since the ground was uneven and the wind direction tended to shift throughout the year. Nevertheless, the Germans convinced themselves that westward winds would prevail and that the uneven ground would not greatly hinder the gas.

[81] January 9, 1915, Kriegstagebuch (BA-MA) Nachlass Gerhard Tappen.
[82] Mit der Länge des Kriegs wachsen, 1918 (MPG) Va 5, 522.
[83] Prior to World War I, meteorology was not a particularly standardized science, but the Germans quickly began to expand their meteorological knowledge for the deployment of poison gas. By 1917, the *Frontwetterdienst* (front weather service) was sending weather reports to the German artillery every two hours.
[84] Tappen to Reichsarchiv, July 16, 1930 (MA-BA).

Over the course of January and February 1915, the German army ordered 6,000 cylinders each holding forty kilograms of chlorine gas and another 24,000 cylinders with twenty kilograms. By early April, the Bayer chemical concern had laboriously delivered about 700 tons of chlorine gas to the pioneer troops in Flanders.[85] Due to the absence of filling stations, cylinders were filled at railheads and then freighted to the Ypres front. Loading and unloading the heavy cylinders was exceedingly difficult and extra soldiers were requisitioned for the duty. None of this process was particularly stealthy and the Allies shelled the Germans after hearing the noise produced by nightly cylinder placement. In early March, Allied shells finally hit entrenched chlorine cylinders, injuring several German soldiers. In yet another successful Allied shelling, three men were killed and fifty were injured by the escaping gas. Even Colonel Max Bauer and Fritz Haber nearly died when they accidentally rode into a gas cloud during a final test of the cylinders at Beverloo on April 2.[86]

The Malevolent Cloud: The Attack at Ypres

German commanders initially believed that their soldiers did not need any special protection against the effects of the chlorine gas so long as the wind blew toward the Allies. Haber's near-death experience at Beverloo convinced him otherwise, and he procured 3,000 *Selbstretter* for the pioneer troops and ordered another 3,000 more for machine gunners and other troops deemed essential.[87] At the last minute, in mid-April, regular German infantrymen were issued cotton mouth pads seeped in sodium thiosulfate.[88] This method of protection was not ideal since soldiers would have to hold these pads to their noses and mouths while simultaneously engaging in combat.[89] Furthermore, the German General Staff did not believe that chlorine gas could cause ocular damage and the pads did nothing to protect soldiers' eyes.[90] Beyond flawed medical knowledge, this lack of protection was also due to the impatience of the German commanders who wanted to conduct the attack before

[85] By 1915, Carl Duisberg had cultivated a close relationship with Max Bauer and was in regular contact with Haber. Due to the loss of international markets at the onset of the war, Duisberg eventually agreed to have Bayer supply the German military with the raw materials of chemical warfare.

[86] Bauer, *Der grosse Krieg in Feld und Heimat*, 69.

[87] Letter from Haber to Dräger in: Rudolf Hanslian, *Der deutsche Gasangriff bei Ypren am 22. April 1915* (Berlin: Verlag Gasschutz und Luftschutz, 1934).

[88] Chef des Generalstabes des Feldheeres, 1915 (BA-MA) PH3/1012.

[89] Gasschutzwesen, Allg.u.Besonderes 1915–1918 (BHK).

[90] Generalkommando I. Armee-Korps (WK) 1276.

they were redeployed to the Carpathian mountains in order to repel increasing Russian pressure. The German commanders first alerted the pioneer regiments for attack at 10:30 pm on April 14 but canceled the attack later that night due to imperfect weather conditions. After yet another aborted offensive four days later, both the pioneer regiments and the assault troops reportedly appeared on edge, worrying that the Allies had learned of the plan and taken measures to protect themselves. And indeed, the French had heard of the attack from a captured German reserve infantryman named August Jager. Several other German prisoners corroborated Jager's report, which was also confirmed by the visibly gaseous explosion produced by the Allied shelling of German cylinders in March.[91] Nonetheless, the French chose to disregard the information, assuming that Jager was planted by the Germans to spread false information. The British and Canadians apparently took these reports more seriously but still did little to prepare themselves, assuming that the new weapon would fail to produce serious results.

The third German gas alert came on April 21 but was again postponed until the next morning. Finally, on April 22, the men waited until just after 5:00 pm when their commanders finally gave the order for attack and 6,000 cylinders were simultaneously opened along 7,000 meters of the northern Ypres front.[92] An awe-inspiring 150 ton white cloud of chlorine rose to 30 meters in the air and slowly wafted at about one mile per hour toward the French lines, gradually changing to a yellowish-green color. Following the prevailing winds, the cloud eventually reached the French trenches, predominantly manned by Franco-Algerian troops. The French colonial soldiers who did not immediately suffocate on the ground retreated from their positions in terror. A French colonel later reported that "people were fleeing everywhere ... all distressed and without their caps, with open shirts, running like mad, going in all directions, crying out for water, spitting up blood, some even rolling on the ground in desperate attempts to breathe."[93] In turn, the Germans warily advanced behind their cloud, trying to avoid any lingering gas. Reportedly, they shot or stabbed many of the French soldiers that they found writhing in agony. One Private Fischer wrote of the attack:

> Woe be unto the enemy when they encounter our perfume. It's pure mass murder and with almost devilish joy we put an end to them quickly. It is easy to explain

[91] William Moore, *Gas Attack! Chemical Warfare 1915–19 and Afterwards* (London: Leo Cooper, 1987), 23.

[92] Rudolf Hanslian, *Der deutsche Angriff bei Ypern am 22. April 1915* (Berlin: Verlag Gasschutz und Luftschutz, 1934), 11.

[93] Lloyd, *The Western Front*, 108.

why we didn't take many prisoners, we beat to death everyone left alive in the trenches ... With poison we fight for Kaiser and fatherland and against the whole world.[94]

In about an hour, the Germans had taken the northern outlying areas of Ypres. By dusk, they had captured around 2,000 Allied soldiers and fifty-one artillery pieces. But with the arrival of night, the German soldiers stopped at a local canal rather than pursuing the retreating French army. This allowed the remaining French troops to regroup with Canadian soldiers and counterattack the new German positions during the night.

After holding their ground for two days, the Germans then released 15 more tons of chlorine toward the 2nd Canadian Brigade. The Canadians held firm for a short period of time but eventually retreated toward the town of Ypres. Again, the Germans stopped their advance after they reached their initial objective and dug in.[95] This slow and steady form of advance counteracted Haber's insistence on speed and surprise, revealing both the soldiers' inexperience with the advantages of their new weapon and their distinct fears about its indiscriminate capacity for killing.

Over the course of the next month, the Germans would employ poison gas four more times. On each occasion, they used it either to weaken Allied morale directly before an attack or to delay Allied reinforcements from town. But soon after the first gas attack against the Canadian Brigade, the British began outfitting their troops with their own gas-retarding chemicals and cotton pads that could be tied to the face. On May 24, the Germans conducted their final chlorine gas assault on Ypres, creating a very limited push to the south of the first attacks. In this final offensive, the British experienced few casualties and maintained their positions, knowing by now how to resist the choking clouds. By the end of May 1915, the Germans had realized that their chlorine gas would not be the kind of wonder weapon that could swiftly win the war.

Seeing such a realization as a threat to his fledgling military career, Fritz Haber claimed that the German High Command had not allocated sufficient resources for the attack, nor had they acted quickly enough on April 24 when the element of surprise was at its greatest.[96] It was not the fault of his weapon, Haber argued, but rather the folly and timidity of both commanders and soldiers that had prevented the attack from

[94] Gerhard Hirschfeld and Gerd Krumeich, *Deutschland im Ersten Weltkrieg* (Frankfurt am Main: S. Fischer Verlag, 2013), 24–25.
[95] Trumpener, "The Road to Ypres," 474–475, 477.
[96] Otto Lummitsch, "Meine Erinnerungen an Geheimrat Prof. Dr. Haber" (MPG) Dept Va, Rep 5, 1480.

succeeding. On the other hand, German commanders learned somewhat different lessons from Ypres. For them, gas was another new technology that required systematic preparation and technical know-how. Modern officers would now need to assess the right terrain, weather, and air density for future gas clouds.[97] Additionally, regular German infantrymen needed gas training and a form of protection against both their own gas clouds and any retaliatory attacks. Crown Prince Rupprecht, commander of the Sixth Army, had pointed out that the prevailing winds in France headed eastward toward the German lines. If and when the Allies adopted similar gas weapons, they could deploy them "ten times more often than [the Germans] could."[98]

Regardless of the relative failures of the Ypres attack, the chlorine gas did claim its victims. Casualty reports varied widely due to both the inability to assess the damage of a new weapon and the desire to over- or under-inflate casualty numbers for propaganda purposes. Comparative reports estimate that there were about 350 gas deaths and thousands more injured over the course of the Second Battle of Ypres.[99] Furthermore, while the attack did little to break the Allied front, it had a great effect on the psyches of the Allied soldiers. The poison cloud was an imposing visual spectacle, and the injuries that it produced appeared particularly gruesome to the uninitiated. Reflecting on the uniquely grisly scene, the German soldier Rudolf Binding wrote that "The effects of the successful gas attack were horrible ... All the dead lie on their backs, with clenched fists; the whole field is yellow."[100]

A Supposedly Unprecedented Crime against Humanity: Allied Perceptions of Ypres

Subsequent narrations of the death and destruction at Ypres infuriated the British both at the front and back home. In no uncertain terms, *The London Times* reported on April 26, 1915, that:

> the Germans were employing stifling gases, which were released by special devices in the German trench and driven into the French trench by the northerly wind. [This] was forbidden by the International Declaration in the Hague on July 29, 1899, and Germany agreed to this declaration ... It has not been proven whether missiles have been used at the attack to the north of Ypres, but it has been shown that suffocation was the sole purpose, and that Germany

[97] Anlagen zum Kriegstagebuch Juli 1915–April 191 (LBWK) 456 F 55, 392.
[98] Eugen von Frauenholz, ed., *Kronzprinz Rupprecht von Bayern: Mein Kriegstagebuch* (Munich: Deutscher National Verlag, 1929), 304–305.
[99] Trumpener, "The Road to Ypres," 460.
[100] Rudolf Binding, *A Fatalist at War* (London: George Allen & Unwin, 1929), 64.

has violated the Hague provision, and thus placed a new crime on the great list of its perfidy.[101]

Twelve days later, *The Times* further described the chlorine attack as "devilish tricks by which soldiers are poisoned and condemned to slow, agonizing death."[102] The same piece continued on to chastise the secrecy of the plan, stating that:

Nothing is more remarkable than the complete silence of the German press, evidently in the pursuit of strict secrecy about the use of poisonous gases in battle ... The whole world must now understand that the Germans and their base means cannot destroy the work of civilization that has been built up over centuries of human life.[103]

Contrary to the claims of *The Times*, German newspapers did in fact report on the deployment of gas at Ypres.[104] On April 26, the *Frankfurter Zeitung* described the attack but also claimed that poison gas was far preferable to the destruction wreaked by conventional shells.[105] Parroting the same justifications that had been previously crafted by Haber and Falkenhayn's staff, the *Kölnische Zeitung* argued that:

The basic idea of the Hague agreements was to prevent unnecessary cruelty and unnecessary killing when milder methods of putting the enemy out of action are possible. From this standpoint, the letting loose of smoke clouds, which, in a gentle wind move quite slowly towards the enemy, is not only permissible by international laws, but is an extraordinarily mild method of war.[106]

Still more German articles pointed out that the earlier French use of *cartouches suffocantes* proved that the German gas was only a matter of retaliation. For instance, an article in the *Vossische Zeitung* claimed that "our opponents have used this means of combat for months."[107] Since they largely viewed standing international law as a tool of British power, the German press and the German public had a relatively easy time adhering to these justifications. Conventions that restricted modern

[101] 4. Armee, 1915 (BA-MA) PH3/569 L.
[102] Marion Girard, *A Strange and Formidable Weapon: British Responses to World War I Poison Gas* (Lincoln: University of Nebraska Press, 2008), 140.
[103] 4. Armee, 1915 (BA-MA) PH3/569; English Reports, 1915 (BA-MA) PH 3/577.
[104] Wolfgang Wietzker, *Giftgas im Ersten Weltkrieg: Was konnte die deutsche Öffentlichkeit wissen?* (Saarbrücken: Verlag Dr. Müller, 2007), 69–73.
[105] "Through German Eyes: Poisonous Gases: A Quick and Painless Death," *The London Times*, April 29, 1915, p. 6.
[106] *Kölnische Zeitung*, June 26, 1915.
[107] Tim Grady, *A Deadly Legacy: German Jews and the Great War* (New Haven: Yale University Press, 2017), 68.

weapons like poison gas thus appeared outdated, only serving to uphold British military advantages.[108]

It was precisely this battle of words, played out between national presses, that made the Second Battle of Ypres a historic event. According to the Allied press, April 22 was the day on which modern terror weapons were first employed by any nation. Such a reception forced many British, French, and even German citizens and soldiers to reevaluate their assumptions about honor and fairness on the battlefield.[109] However, at this same moment, the military men and scientists who had significant knowledge about the history and nature of gas warfare were substantially less scandalized.[110] For these specialists, April 22 was certainly not the birth of chemical warfare or terror weapons. Humans had contaminated drinking water, poisoned arrows, and deployed chemical explosives for centuries. Furthermore, Ypres was not even the "modern" inception of chemical warfare because both the French and the Germans had used tear gases for months prior to Haber's attack.[111]

Instead, for experts, April 22 was remarkable in the history of chemical warfare simply for the industrial scale of the attack. The 150 tons of chlorine gas used at Ypres could only have been produced by an industrialized nation that had engaged large chemical companies in military production. Several tacticians and technicians, including Fritz Haber, further realized that the scale of the attack was important for its devastating psychological effects. Haber would go on to note that gas "troubles the mind with fresh anxieties of unknown effects and further strains the soldier's power of endurance at the very moment when his entire mental energy is required for battle."[112]

The Warrior in the White Coat: Fritz Haber, the KWI, and Militarized Chemistry

The German testing of chemical weapons continued unabated during and after the Second Battle of Ypres. Even in the midst of one of the first gas attacks against the British, the gas pioneer James Franck was found

[108] See, for instance, *Berliner Tageblatt*, April 26, 1915.
[109] Gerhard Kaiser, "Wie die Kultur einbrach Giftgas und Wissenschaftsethos im Ersten Weltkrieg," *Merkur* 56 (2002): 220.
[110] Lepick, *La Grande Guerre Chemique, 1914–1918*, 65.
[111] Olivier Lepick, "Une Guerre Dans La Geurre: Les Armes Chimiques 1914–1918," *Revue Historique Des Armees* no. 203 (1996): 73–86.
[112] Matthew Stanley, *Einstein's War: How Relativity Triumphed amid the Vicious Nationalism of World War I* (New York: Dutton, 2020), 125.

fearlessly taking air samples in a shell crater.[113] The enthusiasm of Haber's scientists was later matched by gas researchers in Britain and France. These men cited four major reasons for their eagerness to engage in such militarized research. First, in 1915, the long-standing link between science and progress remained relatively unquestioned, especially among scientists themselves. This encouraged the view reiterated by Haber that poison gas must intrinsically be a more humane weapon.[114] Indeed, scientific idealists even envisioned future knock-out gas that could pacify entire armies without a shot fired. Second, many scientists defended their work with poison gas as a patriotic necessity, relying on national propaganda to demonize the enemy and suspend the international nature of the scientific community. Third, many believed that gas research would lead to advances in tangential scientific and technical fields that would eventually benefit humanity.[115] Reflecting this belief, Curt Wachtel, a pharmacologist under Haber and the future founder and director of the Institute of Industrial Hygiene and Professional Diseases, would write:

> A science guilty of creating the means of immeasurable destruction and human suffering may be considered more than fully exculpated through the great contributions to public hygiene and welfare which resulted from the same research on war gases. Certainly, many more lives have been saved through these and other accomplishments resulting from the wartime study of gas attack than were sacrificed by the latter during the World War.[116]

Lastly, many gas researchers saw the war as a way to advance themselves professionally or to enhance the institutional power of scientific organizations, university laboratories, or chemical firms. All of these arguments and self-justifications revealed the extent to which the scientific development of gas warfare reflected what the writer Diana Preston has called the "conflicts and ambiguities between expediency and morality in [the First World War]."[117]

When faced with such "conflicts and ambiguities," it is clear that the political stakes were quite high for these men of militarized science. Shortly after reports of Haber's April 22 gas attack returned to Germany, the Kaiser celebrated what he perceived as a great victory and declared "Ypres Day" as

[113] Jost Lemmerich and Ann Hentschel, *Science and Conscience: The Life of James Franck* (Stanford: Stanford University Press, 2011), 55.

[114] Slotten, "Humane Chemistry or Scientific Barbarism?," 477.

[115] World War I gas research did in fact advance countless other scientific and technical fields, including meteorology, pest control, firefighting, mining, underwater exploration, medicine, and pharmaceuticals.

[116] Curt Wachtel, *Chemical Warfare* (London: Chapman & Hall, 1941), 12.

[117] Diana Preston, *Before the Fallout: From Marie Curie to Hiroshima* (New York: Walker & Company, 2005), 52.

a time of commemoration. In a rather extraordinary move, Wilhelm II ignored the standard procedures of military advancement and promoted Haber from a sergeant in the militia reserve to a regular captain in the German army. In doing so, the Kaiser fulfilled Haber's early dream of entering the Prussian careerist ranks. Now even more beholden to the imperial regime, Haber began to regularly wear an officer's uniform and to demand the formal signs of respect that normally came with the position.[118] At the same time, Haber's new military role came with a great deal of additional work, which Haber apparently tackled with zeal, often at the expense of his personal relationships and his physical health.[119] In the summer and fall of 1915, Haber further reorganized the Kaiser Wilhelm Institute for Physical and Electrochemistry to meet the increasing demands for war-related research. While over half of the KWI's expenditures were related to military projects in 1915, by February of the following year, practically its entire budget was dedicated to military research.[120] The KWI's advisory committee, which included Emil Fischer, Walther Nernst, and Richard Willstätter, now brought on Bayer's CEO Carl Duisberg to facilitate further cooperation between laboratory research and industrial production (Figure 1.1).

On July 4, 1916, Leopold Koppel proposed the creation of a new fund to the Ministry of War that would finance the continued research of warfare technologies. Fritz Haber was most likely the real voice behind the suggested creation of the Kaiser Wilhelm Foundation for the Sciences of Military Technology (KWKW). Two management boards would direct this new institute, comprised of both scientists and military men. At its core, Haber saw this new institute as a way to place the two groups on equal footing and as a method to garner military honors for war researchers. The KWKW began operation in the spring of 1917 but achieved little research success and was eventually transformed in 1920 into the Kaiser Wilhelm Foundation for Technical Science.[121] Haber also envisioned the creation of an Institute for Applied Chemistry and Biochemistry as a way

[118] Harris, "To Serve Mankind in Peace and the Fatherland in War," 35.

[119] When World War I began, Haber's first wife, the chemist Clara Immerwahr, openly objected to Fritz's nationalistic military contributions, finding them contrary to the international ideals of science. On May 2, 1915, after seeing Fritz return from Ypres, Clara shot and killed herself with his service pistol. Many have speculated on the meaning of this act, but it should be noted that Clara had been dealing with serious depression for a long time prior. Bretislav Friedrich and Dieter Hoffmann, "Clara Immerwahr: A Life in the Shadow of Fritz Haber," in *One Hundred Years of Chemical Warfare*, 55–60.

[120] Bretislav Friedrich and Jeremiah James, "From Berlin-Dahlem to the Fronts of World War I: The Role of Fritz Haber and His Kaiser Wihlem Institute in German Chemical Warfare," in *One Hundred Years of Chemical Warfare*, 32.

[121] Manfred Rasch, "Science and the Military: The Kaiser Wilhelm Foundation for Military-Technical Science," in *Frontline and Factory*, 187–192.

Figure 1.1 Fritz Haber in his German military uniform (1916).
Courtesy of the Max Planck Gesellschaft, Berlin. Bild Nr I/5, Sammlung Jaenicke

to continue his research on chemical weapons after the war.[122] Through such proposals, Haber was already readying himself for the war's end by trying to secure peacetime funding and cementing his power both in the scientific world as well as in the German military hierarchy. In 1917, he wrote, "In peacetime, the development of weapons should be given over to the military, but research on the chemical makeup of these weapons should be carried out in a new scientific institute, while the Kaiser Wilhelm Institute for Physical Chemistry and Electrochemistry must return to its original goals."[123] Ultimately, the board of directors for the larger Kaiser Wilhelm Society refused to create Haber's proposed Institute for Applied Chemistry and Biochemistry, realizing that it was

[122] Kaiser, "Wie die Kultur einbrach," 217.
[123] Fritz Haber, Memorandum, September 18, 1917 (MPG) Dept Va, Rep 5, 1616.

permanently intended for military purposes. Nevertheless, before World War I was over, the German War Ministry had already agreed to provide 6 million marks for future applied weapons research. Even the Kaiser Wilhelm Society eventually compromised and granted Haber a special department within his original KWI for Physical and Electrochemistry for continued weapons research.

By 1917, overarching control of Haber's KWI had shifted from the Central Chemistry Office to Department W8 of the Ministry of War.[124] Organized under what was termed Section A10, Haber continued to direct both theoretical and applied chemical warfare research.[125] At his KWI, Haber would eventually come to control nine departments working on various aspects of chemical warfare throughout Berlin. In order to staff these departments, Haber continued to recruit younger scientists through his many long-standing academic and industrial connections.[126] The various departments held weekly meetings where all research information would be efficiently funneled up to Haber. At the ground level, scientists were given a surprising level of research freedom, but the need for national secrecy and Haber's desire for a strict institutional hierarchy often reduced communication between the departments and stifled creativity across the larger institute (Table 1.1).

By September 1917, Haber's institute employed 2,000 people, 150 of whom had been made temporary state officials. The KWI's budget increased by 3 million marks and the institute spread from its original buildings in the Berlin suburb of Dahlem to occupy various other Kaiser Wilhelm Society buildings.[127] Physically representing the distinct military purpose of the original KWI structures, impromptu guard shacks sprang up around the large laboratory buildings and a fence was built to surround the compound. The year 1917 also saw a major shift in KWI research priorities when German gas cylinder operations were halted in favor of gas shells. As part of this shift, Haber's scientists worked tirelessly on developing arsenic compounds such as Clark I and II as well as mustard gas that could be placed within the shells. Haber saw these new chemical weapons as ways to thwart the effectiveness of new Allied gas

[124] Ministry of War, Dept W8, October 1916 (MPG) KWG, Dept I, Rep 1a, 36, General Administration, 1153.

[125] A10 was originally placed under the artillery, but it eventually became an independent department of the German General Staff.

[126] For more detailed descriptions of the KWI departments, see Jeremiah James et al., *One Hundred Years at the Intersection of Chemistry and Physics: The Fritz Haber Institute of the Max Planck Society 1911–2011* (Boston: De Gruyter, 2011), 32–33.

[127] Jeffrey A. Johnson, "The Scientist behind Poison Gas: The Tragedy of the Habers," *Humanities* 17, no. 5 (1996): 25–29.

Table 1.1 *The different departments in Fritz Haber's KWI during World War I*

Department	Director	Primary purpose(s)
Dept. A	Professor Reginald Oliver Herzog	Gas mask design and testing
Dept. B	Dr. Friedrich Kerschbaum	Poison gas testing
Dept. C	Dr. Ludwig Hans Pick	Gas protection technologies
Dept. D	Professor Heinrich Wieland	New poison gas development
Dept. E	Professor Ferdinand Flury	Pharmacology and toxicology
Dept. F	Professor Herbert Freundlich	Gas mask filters
Dept. G	Professor Wilhelm Steinkopf	Gas shells and explosives
Dept. H	Professor Otto von Poppenberg	Explosives/trench mortars
Dept. J	Professor Paul Friedländer	Industry cooperation
Dept. K	Professor Erich Regener	Cloud particulates

masks. This, he believed, was another turning point in the chemical war in which the strength of the combined German chemical industry and research institutes could hand the military a decisive victory.

In the end, arsenic compounds never had a major military impact, but mustard gas proved relatively effective when the Germans first used it in July 1917. Mustard agent could severely irritate the eyes and lungs, leading to temporary blindness or infection of the respiratory system. More importantly, it attacked bare skin, creating serious chemical burns over relatively brief periods of time. While German sanitary troops were quickly outfitted with KWI-manufactured rubber suits to protect them from the gas, the Allied soldiers remained exposed, suffering greatly when mustard gas was first employed at the Battle of Passchendaele in the summer and fall of 1917. By then, Haber had realized that the key to this new weapon, much like chlorine gas, was exploiting its capabilities before the Allied scientists could mimic its production.

To meet the massive demand for arsenic and mustard gas shells, Haber directed the building of a new shell filling plant at Breloh. Named the Klopper Works, the plant's operation was placed under the supervision of a wounded gas pioneer named Hugo Stoltzenberg.[128] Prior to this assignment, Stoltzenberg had been a research assistant at the Chemical Institute of the University of Breslau, a chemical weapons researcher at Haber's KWI, and a director at the Adlershof shell filling plant. To both Haber and Stoltzenberg's chagrin, the nascent Klopper Works faced numerous building setbacks, failing to become fully

[128] Gasschutz Lummitzsch, Otto: Meine Erinngerung an Geheimrat Haber, 1955 (MPG) Va 5 534.

operational until early 1918. This delay allowed the Allies enough time to synthesize their own version of mustard gas, often referred to as Yperite. By July 1918, the French had produced a substantial number of mustard gas shells and distributed them to both their artillery divisions and those of the British and Americans.[129] By the summer of 1918, the Germans had lost yet another technological advantage, this one now coming at a moment in which raw materials shortages were beginning to take their toll. While German gas shells were rarely in short supply, essential materials for gas protection such as rubber and chlorinated lime became quite scarce. Furthermore, the arrival of mustard gas on the battlefields of 1917/18 made life at the front exceedingly difficult for German infantrymen. Troop morale plummeted as soldiers ignored correct gas protocol and struggled to avoid gas poisoning during their sporadic retreats. Thus, to a certain extent, Haber's chemical war ultimately contributed to the final German surrender on November 11, 1918.

The Impossibility of Defeat: The Legacy of Fritz Haber's KWI at War's End

Ever the believer in the infallibility of the imperial German state, Haber seemingly refused to imagine the possibility that the Germans would lose the war. Even in September of 1918, Haber was preparing for continued chemical warfare in the upcoming year. But when the armistice did come in November, Haber quickly set his plans in motion to return the KWI to peacetime activities. Departments were separated and laboratories were returned to nonmilitary research projects. Haber did this with the funds that the Ministry of War had granted him in 1916, which were never rescinded even in the face of severe material shortages. It appeared as if the KWI would endure the war in some form, but Haber himself was forced to flee to Switzerland when he was placed on the Allied war criminals list. During this exile, Haber experienced a nervous breakdown brought on by both sheer exhaustion and anxiety over the Allied reception of his wartime work.[130] Yet after only a few very difficult months in Switzerland, Haber was able to return to Germany after the Allies stopped requesting his extradition.

[129] By 1918, chemical shells made up about 50 percent of all fired German shells. Friedrich and James, "From Berlin-Dahlem to the Fronts of World War I," 42.

[130] Unlike Carl Duisberg and Walther Nernst, Haber never admitted to feeling guilt over his role in the death and destruction of World War I. In fact, in the post-war years, Haber kept a framed picture of the gas attack Ypres in his study to remind him of his achievements. Plumpe, *Carl Duisberg 1861–1935*, 467.

Once back in Berlin, Haber began to find new applications for his chemical weapons research, employing his expansive knowledge of poison gases for pest control research at the KWI.[131] During this immediate postwar period, Haber's scientific reputation suffered substantially as scientists with pacifist and universalist commitments continued to question both his nationalism and his militarism. For instance, the German chemist Hermann Staudinger publicly criticized Haber during the war, later writing that "we chemists in particular have an obligation in future to draw attention to the dangers of modern technology."[132] The physicist Max Born also drew harsh lessons from Haber's actions, later noting that Haber served as a "lightning rod" for the ethical concerns surrounding chemical weapons. Even Albert Einstein, who never fully severed his friendship with Haber, continued to personally distance himself over the course of the war and its immediate aftermath.[133]

Regardless of these criticisms, Haber won the 1919 Nobel Prize in chemistry for his contribution to the Haber-Bosch nitrogen fixation process. While the awarding of the prize remained controversial, it went a long way in rehabilitating Haber's previously lofty reputation in the German scientific world. Haber now became the driving force behind the Emergency Committee for German Science, an organization that raised funds for research projects through contacts with the federal and state governments as well as private industry.[134] He also became a regular face at the Prussian Academy of Sciences, striking the pose of a wise elder statesman in the world of German science.[135]

Four years later, in 1923, a German parliamentary inquiry examined Haber to determine if his actions during the war adhered to the rules of international warfare. Revealing that many of his initial positions on chemical warfare had not changed, Haber returned to his arguments about the supposed humane aspects of chemical weapons and the retaliatory nature of the attack at Ypres. This logic was replicated in some of his other major lectures from the 1920s. For instance, Haber wrote:

Gas weapons have experienced the bitterest denigration. The allegations of a violation of international law and of sheer cruelty have been raised against this alleged German invention in all languages. There is no remedy for the hostility of a press that does not take the facts seriously even after the patient restating of the

[131] Margit Szöllösi-Janze, "Pesticides and War: The Case of Fritz Haber," *European Review* 9, no. 1 (2001): 97–108.

[132] Kaiser, "Wie die Kultur einbrach," 215–217.

[133] Stern, *Einstein's German World*, 116.

[134] Helmut Maier, *Rüstungsforschung in der Kaiser-Wilhelm-Gesellschaft und das Kaiser-Wilhelm-Institut für Metallforschung, 1900–1945/48* (Göttingen: Wallstein, 2007), 224.

[135] Stern, *Einstein's German World*, 130–131.

truth. We did not invent gas weapons. Their origins go back to the days when men learned to ferret the enemy out of their fixed positions, and chemical weapons have been mentioned since Plataea in the Peloponnesian War. We did not even invent the more effective gases of modern chemistry; for it is undisputed that France provided its soldiers with gas-rifle grenades when the war broke out.[136]

Haber further insisted that future generations had nothing to fear from the arrival of modern chemical weapons. The committee accepted this defense and deemed that Haber had done nothing to break international law during the war, especially since no other nation chose to indict him.

The political and societal power of scientists, while often difficult to measure, is often best revealed through their personal and professional networks.[137] Early in the war, Fritz Haber's level of political power did not appear substantial outside the world of academic chemistry. While he was tapped to serve as the head of a KWI institute in 1914, he only directed five employees in two academic buildings. However, historian Margit Szöllösi-Janze reminds us that scientists such as Haber can often be "flexible enough to take on tasks that cut across fields," allowing them to quietly control "astonishingly extensive domains."[138] In Haber's specific case, he maintained fortunate prewar connections to the major German chemical firms such as BASF and Bayer, thus allowing him to serve as a mediator between his own academic world, the chemical industry, and the German military. It bears admitting, however, that cooperation between academic chemistry, military structures, and industrial firms was not an entirely new phenomenon in World War I Germany. Personal and professional connections between the academy and industry, such as those that Haber maintained prior to 1914, provided the necessary channels for relatively seamless cooperation during the war.

While industrialists such as Carl Duisberg did not relish the idea of war in 1914, the national need for chemicals and explosives helped to earn large chemical firms such as Bayer, BASF, and Hoechst a leading position in Germany's war economy.[139] This, of course, came with a level of financial uncertainty that spurred the subsequent development of the expanded IG Farben advocacy group in 1916, but it also fostered technological innovation and cooperation between the various firms.

[136] Fritz Haber, *Aus Leben und Beruf: Aufsätze, Reden, Vorträge* (Berlin: Julius Springer Verlag, 1927), 15.

[137] Margit Szöllösi-Janze, "Losing the War, but Gaining Ground: The German Chemical Industry during World War I," in *The German Chemical Industry in the Twentieth Century* (Boston: Kluwer Academic), 94.

[138] Ibid., 94. [139] Ibid., 121.

Cooperation, in turn, allowed the companies to focus on specific chemical products, thus streamlining production and delivery throughout most of the war.[140] Between 1915 and 1918, German chemical companies produced an estimated 99,500 tons of chemical weapons. Trumpeting this production ability, Carl Duisberg retrospectively pronounced that "in chemistry, the superiority of German science and technology over our adversaries has been demonstrated. Hopefully, this will always be the case, not only during the war, but in the future as well." (See Table 1.2.)[141]

Historian Jeffrey Allan Johnson has maintained that the German level of structured and efficient cooperation between academic scientists, the chemical industry, and the military was quite unique. He argues that the German chemical industry's rapid growth in the late nineteenth century, coupled with its fierce internal competition and inability to garner national funding, encouraged Emil Fischer to seek private funds for the creation of a national chemical institute. This privatization created a necessary link to industrialists such as Leopold Koppel and an enduring requirement that academic chemistry justify its value through a research agenda that could be applied to national or corporate endeavors.[142] Given his extensive experience with numerous dyeworks firms, it is no surprise that Fritz Haber was hand-selected by Koppel to serve as the lynchpin between German science and industry. Nevertheless, it is important to point out that Haber's success in merging industrial manufacturing and scientific research into the military apparatus was not inevitable.

The German military commanders' misgivings about using poison gas at Ypres were part of a larger military resistance to external influence, especially the influence of a man they viewed as a Jewish scientist. But changes in military command over the course of World War I and the rise of more technologically minded officers such as Max Bauer allowed for closer cooperation with Haber. The German military continued to maintain the upper hand in their incorporation of science and industry and tensions continued to cut along institutional lines, but the structures were now in place to pacify these personal and disciplinary differences in the service of the German nation. Thus, in Haber's KWI, with its large research departments, big budgets, and connections to outside influence, we see the initial development of both Big Science and a "military-

[140] As part of the Hindenburg Program, the Board of Wartime Raw Materials was reorganized under the WUMBA (*Waffen- und Munitionsbeschaffungsamtes*) and later the *Kriegsamt*. This led to a shortage in chemicals as increased demand overtaxed the production capacities of the major chemical firms.

[141] Abwehr von Stinkgeschossen etc, 1916– (BA) 201-008-001.

[142] Johnson, *The Kaiser's Chemists*, 133.

Table 1.2 *Total amounts of chemicals (in tons) and gas shells produced by Great Britain, France, and Germany between 1915 and 1918*

	UK	France	Germany
Chlorine	21,156	12,500	87,000
Phosgene	1,384	15,700	23,655
Mustard gas	522	1,968	7,659
Tear gas	1,819	481	3,710
Blue cross gases	98	NA	8,027
Gas shells	8.3 million	17.1 million	38.6 million

For further statistics, see: L. F. Haber, *The Poisonous Cloud: Chemical Warfare in the First World War* (Oxford: Clarendon Press, 1986), 261.

industrial-academic complex" decades before Americans became concerned by a similar phenomenon at the end of World War II.[143]

By bridging the military, academic, and industrial worlds, Haber's KWI should be viewed as one of the major steps in the scaling up of increasingly bureaucratized technoscientific projects.[144] Not only did the KWI provide one set of blueprints necessary for future large-scale research projects, but it also encouraged many researchers trained in so-called pure science to begin openly working on technological application.[145] In this way, Haber and his fellow researchers also took a major step in developing what would later be referred to as "technoscience," or the persistent blending of theoretical and applied research, especially within specific institutions and with specific practical aims.[146] Reflecting on both this institutional and disciplinary blending in 1916, the first president of the Kaiser Wilhelm Society, Adolf von Harnack, prophetically wrote:

Our enemies ... brought German science and military strength as close together as possible. Of course we knew all along that these two pillars ..., deep down,

[143] Andrew Ede, "Science Born of Poison, Fire and Smoke: Chemical Warfare and the Origins of Big Science," in *The Romance of Science: Essays in Honour of Trevor H. Levere*, eds. Jed Buchwald and Larry Stewart (Cham, Switzerland: Springer International, 2017), 200.

[144] Szöllösi-Janze, "The Scientist as Expert," 16–17.

[145] Between 1914 and 1918, an estimated 2,000 German scientists worked on chemical warfare. Thus, Fritz Stern referred to Haber's KWI as "a kind of a forerunner for the Manhattan Project." Fritz Stern, "Freunde im Widerspruch: Haber und Einstein," in *Forschung im Spannungsfeld von Politik und Gesellschaft, Geschichte, und Struktur der Kaiser* (Munich: Deutsches Verlag-Anstalt, 1990), 529.

[146] David F. Channell, *A History of Technoscience: Erasing the Boundaries between Science and Technology* (London: Routledge, 2017), 71–74.

have a hidden connection; but we did not know that this connection is so immediate that military strength can be directly promoted by science and constantly open new connections with it … Create, organize, discipline: in this triad of German spirit and German labor, military strength and science come together.[147]

[147] Adolf Harnack Papers (SBB) KWG, 3.-5. Jahresbericht, pp. 3–5, Section IV, Box 23.

2 The Man in the Rubber Mask: World War I and the Development of the Modern Gas Mask

While Fritz Haber's Kaiser Wilhelm Institute for Physical and Electrochemistry (KWI) provided a model for cooperation between science, industry, and the military, the ultimate success of its chemical weapons research program remains debatable. Haber and his scientists continually sought to develop a poison gas that would singlehandedly win the war for the Germans, but their work led to the discovery of relatively few new chemical compounds with any military value. In fact, all of the major poison gases used in the war had been synthesized prior to 1915. Rather, most of the KWI's successes were found in the recombination and reapplication of previously known compounds. For precisely this reason, it can be persuasively argued that the scientific history of poison gas research in World War I is better told as a technological history of ancillary gas protection devices and methods of delivery.

It would be these technologies, especially the gas mask, that determined how people would live in the "prolonged climate of fear" and the "permanent state of readiness" that chemical warfare generated.[1] The careful design of the mask could allow humans to live, and perhaps even thrive, in this new environment, where healthy air could not be taken for granted. Thus, the development of the gas mask was both a resistance to the fulfillment of Fritz Haber's vision of a chemically saturated modern battlefield and an integral part in the gradual realization of such a reality. If an airborne weapon of annihilation was to be created after the introduction of the modern gas mask, it would now have to be able to break the mask's presumably airtight seal. By intertwining the historical development of poison gas with that of respirators and gas masks, the moments in which both the full import of modern chemical warfare was appreciated and visions of its power partially realized come into focus. In these moments of conception and reckoning, questions of corporeal protection and discipline, on both the individual and

[1] Sloterdijk, *Terror from the Air*, 28.

collective levels, penetrated German discussions of chemical warfare and persisted long after World War I.

The Development of Respirators: A Long History of Incremental Development

While World War I certainly appears as a dramatic moment in the history of chemical hazards, humans have long sought to provide breathable air in toxic or otherwise harmful environments. The age-old fear of poisons motivated many premodern attempts to technologically mediate or pacify noxious air. For instance, the ancient Greeks reportedly used sea sponges to protect against natural airborne toxins while, centuries later, Leonardo da Vinci proposed the use of a finely woven textile dampened with water. However, this technological interest in artificial respiration did accelerate in the modern period as numerous state-supported inventors attempted to create and patent a respirator that could be used for eighteenth-century mining, firefighting, and underwater diving operations. Technological design remained rather open-ended during this period, as inventors tinkered with repurposed items such as knight's helmets and hearth bellows.[2] Yet it was through the increasing demands of industrial mining concerns in the mid-nineteenth century that a cohort of competing Anglo-American inventors began to patent respirators that employed charcoal filters, thus eliminating the need for external sources of oxygenated air.[3] Subsequent designs of standalone respirators incorporated greater protection for the face and eyes, as well as more effective filters. By the turn of the twentieth century, respirators were largely successful at combatting the kinds of toxic fumes that scientists, engineers, miners, and firefighters were likely to encounter.

Throughout the nineteenth century, major German respirator manufacturers were established in northern cities such as Hamburg, Lübeck, and Bremen, chiefly due to their proximity to the sea. Companies such as Ludwig von Bremen, Hamburg, founded in 1870 as the first German manufacturer of diving equipment, provided most of the respirators necessary for various underwater industries. Then, in 1899, Neufeldt & Kuhnke, later renamed the Hanseatische Apparatebau-Gesellschaft, supplanted Ludwig von Bremen as the dominant manufacturer. As such, the

[2] John Bevan, *The Infernal Diver: The Lives of John and Charles Deane, Their Invention of the Diving Helmet, and Its First Application to Salvage, Treasure Hunting, Civil Engineering and Military Uses* (London: Submex, 1996), 14–18.

[3] Invented in 1847, Lewis Haslett's "Inhaler or Lung Protector" is often regarded as the progenitor of the modern gas mask, although there were several other American and British designs that maintain similar claims.

Hanseatische Apparatebau-Gesellschaft, or Hagenuk, produced the great majority of turn-of-the-century German diving equipment on production sites in Bremen, Hamburg, and Kiel.

In 1889, Johann Heinrich Dräger, a Lübeck-based mechanic, founded the Dräger and Gerling company to produce beer tap technology and high-pressure valves. The company began working on oxygen supply devices in the late 1890s under the direction of Bernhard Dräger, Johann's son and the company's chief engineer. By 1902, the firm had produced an oxygen supply device for aviation, an anesthetic delivery device, and a portable respirator.[4] In that same year, Bernhard Dräger entered the management of the company and the name was changed to Drägerwerk: Heinrich und Bernhard Dräger. As part of his mission to diversify and grow the company, Bernhard Dräger designed the new Dräger Model 1904/09 breathing apparatus for mining work.[5] The Model 1904/09 received positive reviews when rescue workers used it extensively in the Courrières Mine Disaster of 1906.[6] Following this success, Bernhard Dräger turned his attention to diving rescue equipment, creating a portable diving system that served as one of the predecessors to SCUBA gear. These effective devices and their successful marketing allowed Dräger to open a subsidiary company in the United States named the Draeger Oxygen Apparatus Company in 1907.

Through their use in nineteenth-century industrial accidents, respirator masks from companies such as Dräger already stimulated mental associations with danger and tragedy. When a respirator was required, the wearer presumably knew that they were in a potentially lethal situation. In fact, as technological developments and capital investment allowed the mining, diving, and chemical industries to flourish, the respirator was used in progressively more deadly scenarios. Industrial workers and chemists increasingly donned respirators to work in dark mines, cold waters, or enclosed spaces with deadly gases, such as arsenic and sulfur. Thus, these devices required a substantial level of technological trust from the wearer, who could not afford any malfunctions while at work. Nevertheless, at the dawn of the twentieth century, the steady development of the respirator appeared to most contemporaries as part and parcel of greater technological progress. For most wearers and observers, the respirator was merely a safety device that would both save human life and allow humans to conquer new and dangerous worlds.

[4] Heinrich Dräger, *Heinrich Dräger Lebenserinnerungen* (Hamburg: Alfred Janssen, 1917).

[5] Lisa Dräger, ed., *Von Biermaschine zum Rettungswesen: Die Aufbaujahre des Drägerwerks* (Lübeck: Dräger Druck, 2007).

[6] Wilhelm Haase-Lampe, *Sauerstoffrettungswesen und Gasschutz Gerätebau und Organisation in ihrer internationalen Entwicklung: Gerätebau* (Lübeck: H. G. Rahtgens, 1924), 13.

Developing the Modern Gas Mask: A New Technological Necessity

Shortly after the proclaimed success of the first chlorine cylinder operations at Ypres, the Germans began to test new applications for their poison gas. In early May 1915, Fritz Haber and his gas regiments headed to the Eastern Front, where they hoped that the wind conditions would prove more consistently favorable. Pioneer Regiment 36 was assigned to General August von Mackensen's Ninth Army, now west of Warsaw. Throughout the summer of 1915, the Germans released a mixture of chlorine and phosgene gas in multiple offensives against Russian troops. A new addition to the German chemical arsenal, phosgene had been industrially produced since 1883 in the manufacture of crystal violet dyes. Haber had hoped to use the gas since December 1914, requesting a phosgene-chlorine mix from Carl Duisberg and the Bayer company in March of the following year.[7] Haber wanted phosgene because it was deadlier than chlorine and invisible to the naked eye, but it was also much denser and difficult to control. After being poisoned during the gas filling process, Carl Duisberg reflected on the difficulties associated with phosgene:

> Privy Councilor Haber, at my suggestion, also wants his chlorine bombs at least 25% phosgene – he cannot use any more. Phosgene's nastiness is best seen in the fact that I spent almost eight days in bed after I only inhaled this gruesome stuff a few times. The same thing happened to my assistants, who had to recover for four weeks at Wiesbaden because the lung damage they sustained was worse than mine. If you treated the enemy for hours with this most poisonous of all gaseous products, and they did not (which would probably be case) immediately die, they would later become ill and get feverish bronchitis.[8]

In an attempt to better control the dangerous phosgene gas, Haber and his employees at the KWI mixed it with chlorine, which would provide greater density and help to gradually carry it through the air.

The shallower trenches of the East did little to protect the Russian soldiers from the new low-hanging gas, and they suffered immensely.[9] The gas pioneer Otto Hahn described the German offensive:

[7] Over the course of the war, Bayer and Hoechst produced the most phosgene gas. Starting in September of 1916, Bayer averaged 300 tons and Hoechst 139 tons per month. Michael Freemantle, *Gas! Gas! Quick, Boys! How Chemistry Changed the First World War* (Brimscombe Port, Stroud: The History Press, 2012), 129.

[8] Hahn, *My Life*, 121.

[9] Johannes Martini, Errinerungen aus dem Weltkrieg, 1914–1915 (DTA) Sig 1847 Reg 1562.

During the attack, we encountered a significant number of gas-poisoned Russians who could no longer escape from the cloud. They had been surprised by the gas without a protective mask and were now lying or squatting in a lamentable state. We tried to rescue one or another with our breathing equipment, but we were unable to prevent their deaths.

While the chlorine-phosgene attacks on the Eastern Front were successful in producing Russian casualties, they again did little to advance the German forward line. As such, German commanders began to believe that gas could not be effectively used in coordination with an infantry attack. This was primarily due to the fact that military meteorological science was not sophisticated enough to predict the seemingly erratic winds, and gas would inevitably blow back on the German troops. This did indeed occur regularly, such as on June 12, 1915, when the wind changed direction during a cylinder gas attack, leading to total panic in the German lines. Officers later counted 1,100 combined Russian and German dead and another 350 injured Germans.[10] After several more of these self-inflicted blunders, the German military finally concluded that gas served better as a psychological weapon that could harass the enemy and weaken resistance before a traditional infantry advance. As an added effect, poison gas forced soldiers to don gas masks, which were difficult to wear while fighting and irritating when worn for hours at a time.

Not surprisingly, the psychologically harrying nature of gas soured infantrymen and their officers to the new weapon. In an attempt to embolden troops, the proponents of chemical weapons now turned their attention to gas protection and training. In May 1915, the German army expanded its issue of protective cotton pads.[11] Experts such as Haber knew that these pads provided little defense, but they had no other options until they could produce a fully functioning and cost-effective respirator. In order to familiarize troops with available protective devices, the German medical services began to conduct training sessions and military pharmacists were tasked with the distribution of cotton pads and neutralizing solution.[12] In their earliest incarnation, gas protection protocols stressed the importance of confidence in the rudimentary mask. Thus, medical personnel were instructed to tell the soldiers that if they remained calm and put the cotton pads over their nose and mouth

[10] Hahn, *My Life*, 120.

[11] In some instances, German soldiers were also provided with goggles. Gastruppe S. 18-51 Manuscript für Buch "Otto Hahn–Mein Leben," 1915 (MPG).

[12] Fritz Haber and his KWI took over the distribution of gas protection in 1916.

then they would undoubtedly survive a gas attack.[13] Not surprisingly, however, the pads were difficult to keep fastened to the face and the chemical solution often dried up, leading to German casualties. The lack of effective protection remained a problem on the Western Front until the German gas regiments moved east and gas operations were suspended. During this lull in gas warfare on the Western Front, the Auergesellschaft gas mask factory and the Griesheim, BASF, and Bayer chemical factories began to construct formalized courses in gas protection. The ultimate product of these courses was a cohort of trained gas protection officers who were added to each German regiment in order to drill soldiers in what was termed *Gasdisziplin* (gas discipline), or the approved protocols for surviving a gas attack.[14]

In the search for a more reliable protective device in the summer of 1915, Fritz Haber assigned Departments C and D of the KWI to design a new gas mask. Concurrently, he turned to the Dräger company to reduce the weight and increase the amount of available oxygen for the gas pioneer's *Selbstretter* (self-rescuer). The *Selbstretter* now served as a blueprint with which scientists and engineers could conceive of a new gas mask. While the large oxygen tank apparatus appeared wildly impractical for the common soldier, its reliability in a gas attack was unparalleled. Thus, developers needed to increase portability and significantly reduce the respirator's weight while retaining continued oxygen supply. The charcoal filters used in mining masks looked like an ideal solution. Further gas mask test projects were pursued at Bayer, Schering pharmaceutical, Kahlbaum pharmaceutical, and Leopold Koppel's Auergesellschaft.[15] As an electrical engineering firm, the Auergesellschaft was in a uniquely favorable position to create gas masks because they owned proprietary machines that could manufacture the sockets for electric lamps. This threaded socket design was repurposed for the connection between the gas mask mouthpiece and the filter.

The first air filter prototype was a small cartridge filled with a layer of diatomite, or wood burned with diatomaceous earth. The small cartridge, about the size of a modern diving regulator, could be raised to the mouth at the moment of attack. A subsequent prototype employed a

[13] Generalstabes des Heeres on Gas, 1915 (BA-MA) PH3/252.

[14] Gasschutzwesen, Allg.u.Besonderes, 1915–1918 (BHK).

[15] At the Auergesellschaft, the gas mask project was led by Professor Karl Quasebart (1882–1949). Quasebart was the son of the director of the Chemical Factory Rhenania in Aachen and a professor at the Technical University in Aachen. Margit Szöllösi-Janze, *Fritz Haber, 1868–1934: eine Biographie* (Munich: C. H. Beck Verlag, 1998), 342.

layer of pumice that was saturated with potash to absorb chlorine as well as a layer of charcoal powder to protect against tear gases. The organic chemist Richard Willstätter and his assistants Adolf Pfannenstiel and Friedrich Joseph Weil devised a third filter at the Kaiser Wilhelm Institute for Chemistry.[16] As a committed pacifist, Willstätter refused to work on any KWI project that would contribute to the production of weapons, but he ultimately agreed to obliquely engage in the war effort. Feeling obligated to help since his good friend Fritz Haber was instrumental in securing his position at the KWI for Chemistry, Willstätter ultimately chose to work on gas protection devices. In a matter of five weeks in the summer of 1915, Willstätter's team came up with the "three-layer cartridge," which incorporated layers of diatomite, granulated active charcoal, and either pumice or diatomaceous earth saturated with hexamethylenetetramine or piperazine.[17]

In August 1915, KWI scientists tested Willstätter's cartridge at the Auergesellschaft gas tower and found it reasonably effective in counteracting both tear gases and chlorine.[18] While Haber and Willstätter had dealt with raw materials shortages and high costs during the research process, they had managed to coordinate and direct the creation of an effective filter in just three months.[19] Now, to protect the eyes and nose of the wearer, the cartridge had to be affixed to a mask. Accordingly, Hans Pick of the KWI's Department C, Reginald Oliver Herzog of Department D, and engineers at the Auergesellschaft fashioned the first *Linienmaske* (line mask), named after the airtight seal that the rubberized mask attempted to create with the lines of the soldier's face.[20]

The *Linienmaske* covered the entire face and was strapped around the back of the head. A longer additional strap allowed the mask to hang around a soldier's neck in moments of impending attack. Willstätter's filter could be screwed on or off the mouthpiece, allowing for the replacement of oversaturated filters. German High Command initially ordered 2,750,000 masks and by September 1915, they announced the *Linienmaske* as

[16] The Kaiser Wilhelm Institute for Chemistry was yet another division of the Kaiser Wilhelm Society, separate from Haber's Kaiser Wilhelm Institute for Physical and Electrochemistry.

[17] Willstätter received the Iron Cross II in May of 1917 for his creation of the three-layer filter. Over 30 million filters were used during the war.

[18] Wilhelm Willstätter, *Aus Meinem Leben* (Winheim: Verlag Chemie, 1949), 238. Such filter tests continued throughout the war. Scientists such as Otto Hahn and James Franck recalled sitting in a tear gas filled chamber in 1916 and breathing until their gas mask filters failed. Hahn, *My Life*, 121.

[19] Fritz Haber to Carl Duisberg, August, September, and October 1915 (MPG) Dept Va, Rep 5, 856, 961, 962.

[20] PH/2/ (BA-MA).

standard issue for frontline soldiers in the West.[21] However, this announcement did not take production delays into consideration, and by November of that same year, only 1,200,000 masks had been delivered.[22]

The Auergesellschaft, Drägerwerk, and Hagenuk served as the major producers of gas masks throughout the war. But to meet the high level of demand, smaller firms such as Metzler and R. Roeckl also received military contracts.[23] While these firms assembled the masks, myriad other companies were required for the raw materials and smaller assembled pieces. For example, twenty different firms were involved in the creation of Auergesellschaft masks. Haber's KWI served as the central mask distribution point. In a large factory in Berlin, a workforce predominantly made up of German women tested and packed the masks ready for delivery to the front.[24]

Through Haber, gas mask companies such as the Auergesellschaft and Drägerwerk became increasingly valuable to the German Ministry of War.[25] In return, these companies earned lucrative government contracts in a moment when they had lost substantial business abroad.[26] For instance, Dräger first received an order for 50,000 *Selbstretter*, and by the end of 1915, they were producing 500 *Selbstretter* per day. In July of that same year, they received their first contract for 650,000 *Linienmasken* and by the end of the war, Dräger had delivered 4,573,310 masks. Over the course of the war, the Dräger factory significantly expanded, and employment rose from 300 to 1,500 workers. Profits concurrently rose, and while Dräger made about 2.4 million Reichsmark per year in 1914, sales had risen to 9.3 million Reichsmark by 1918.[27] Interestingly, Bernhard Dräger openly criticized the military use of poison gas while remaining a staunch nationalist and a supporter of Germany's war aims. Often referring to his own son's near-death battlefield experience with poison gas, Dräger

[21] Armeeoberkommando 6, 1917–1918 (BA-MA) 2427.

[22] Fritz Haber Briefe, 1915 (MPG) Va 5, 856.

[23] Auer produced approximately 60 percent of all masks during the war. Dräger and Hagenuk each manufactured about 20 percent. According to a licensing agreement for the threaded filter attachment, Dräger had to pay Auer 22 Pfenning for each mask produced.

[24] By the end of the war, about 10,000 people were employed in the manufacture of gas masks. Bestimmungen Ausstattung des Heeres mit Gasschutzgerat, 1920–1924 (BA-MA) RH/12/4.

[25] Dietrich Stoltzenberg, *Fritz Haber: Chemiker, Nobelpreisträger Deutscher, Jude* (Weinheim: VCH Verlagsgesellschaft, 1994), 149.

[26] About 40 percent of Dräger's pre-World War I sales came from the United States.

[27] Bernhard Lorentz, *Industrieelite und Wirtschaftspolitik 1928–1950: Heinrich Dräger und das Drägerwerk* (Paderborn: Ferdinand Schöningh, 2001), 37.

insisted that by manufacturing devices that would save soldiers' lives, the Dräger company was engaged in one of the few truly moral tasks in the armament industry.[28]

Visions of Widespread Destruction: From Military to Civilian Gassing

By July 1915, the British were beginning preparations for their own gas attacks. After much debate, British commanders had finally decided that the German use of gas at Ypres temporarily invalidated the Hague Conventions. Now that international restrictions were apparently nullified, the British felt that they could not risk falling behind in developing a technology that might produce a major battlefield advantage. The Germans were well aware that the British were now moving gas cylinders to the Western Front, and they began to issue gas masks and gas mask instructions to their frontline soldiers. On September 25, 1915, Major Charles Howard Foulkes and his three British gas companies opened 168 cylinders of chlorine gas at the Battle of Loos. The Germans retreated from the gas and the British took the town of Loos, but they failed to eliminate all resistance in the surrounding area. On September 27, the British unleashed their remaining gas cylinders near the French village of Neuve-Chapelle, but after just two days of fighting, the Germans succeeded in pushing the British back to their original lines.

At the Battle of Loos, the Germans reported about 26,000 men killed or wounded. Of this total, 106 men were reported as sick from gas. The British, on the other hand, counted about 59,000 British casualties, making their attack a comparative failure.[29] However, just like the Germans at Ypres, this was not how the British perceived the battle. Foulkes and the other British commanders were ultimately pleased with their first foray into gas warfare. Conversely, the Germans now experienced the utter chaos and fear of a directed gas attack. Their gas masks continually failed, their rifles and artillery jammed, and officers' orders were muffled in the masked confusion.[30] Nevertheless, laudatory reports of German bravery and cool-headedness in the face of gas swept through the military.[31] These recurrent stories of courage suggested to German

[28] Michael Kamp, *Bernhard Dräger: Erfinder, Unternehmer, Bürger 1870 bis 1928* (Kiel/Hamburg: Wachholtz Verlag, 2017), 385.

[29] Haber, *The Poisonous Cloud*, 57.

[30] Edgar Jones, "Terror Weapons: The British Experience of Gas and Its Treatment in the First World War," *War in History* 21, no. 3 (2014): 355–375.

[31] Erfahrungs aus den Gasangriffen der Englander am 25, September 1915 (BA-MA) PH 6-I/36.

commanders that while gas was surely a uniquely vexing weapon, it could, in fact, be resisted by soldiers who were correctly outfitted, well-trained, and mentally prepared.

Through the winter of 1915–1916, chemical warfare was exclusively waged with chlorine gas cylinders and tear gas grenades. By the summer of 1916, properly worn gas masks largely protected soldiers from available poison gases, thus making these early means of chemical warfare far less fearsome than in the previous year. Furthermore, variable weather conditions restricted the regular deployment of chlorine gas clouds. Experts claimed that chlorine was not effective on completely calm days or in overly powerful winds, heavy rain, strong sunshine, or cold weather.[32] In fact, chlorine gas cylinders were said to be only truly effective on overcast days with a gentle breeze. With these many atmospheric limitations, armies could not effectively utilize gas during any of the winter months. Most soldiers welcomed the winter respite from chemical warfare, but German scientists and engineers back home only accelerated their gas research, hoping to provide a technology that could circumvent these limitations. Wild fantasies of future gas weapons such as machine gun bullets filled with gas, defoliants for use against enemy crops, and aerial gas bombs abounded.[33]

This last vision, while never fully realized in World War I, proved to be the most enduring. Given the fact that poison gas had often been deemed a wonder weapon by the German military and the German press, it only appeared natural that its effects could be further amplified.[34] Still more, the immaterial and creeping nature of gas suggested the possibility of an unlimited chemical spread. Revealing the imaginative power of such assumptions, Friedrich Kerschbaum's KWI Department B tried twice to produce effective aero-chemical bombs, but German scientists continually failed to hit test targets and to produce a dense enough gas cloud.[35] This dream also permeated the collective consciousness of German civilians, especially those living near the western border with France, who particularly feared the possibility of mass aerial gassing. When a press report from October 23, 1916, stated that civilians in the

[32] Schutzengraben-Markblatt Beachte fur den Gaskampf, 1915 (BA-MA) PH 5-II/61.

[33] Gefechtsakten des GasdienstOffiziers, January 1918–September 1918 (LBWK) 456 F 16, 423.

[34] Isabel V. Hull, *Absolute Destruction: Military Culture and the Practices of War in Imperial Germany* (Ithaca: Cornell University Press, 2005), 258.

[35] Rolf-Dieter Müller, "Total War as a Result of New Weapons? The Use of Chemical Agents in World War I," in *Great War, Total War: Combat and Mobilization on the Western Front, 1914–1918*, eds. Roger Chickering and Stig Förster (Cambridge: Cambridge University Press, 2000), 109.

city of Metz had died from gas exposure, Germans around the country fretted over whether the Allies had begun an aerial gassing campaign.[36] On December 11 of the same year, the *Stuttgarter Neues Tagblatt* published a long-form piece on the lack of preparation for a civilian attack. The writers pointed out that civilians would not know if they should find cover or flee and that, even in the few places that did have fallout shelters, civilians feared being trapped in underground tombs.[37]

In this growing atmosphere of fear, some German citizens claimed that they too should be outfitted with gas masks. But even if the state wished to provide masks, it was not in the position to fulfill this request due to raw material shortages and inadequate manufacturing capacity.[38] Thus, authorities tried to pacify the population with vague gas emergency plans.[39] Ironically, calls for the mass creation and distribution of gas masks allowed the German War Ministry to justify the allocation of still more raw materials to the war effort instead of the home front. Practically all of Germany's supply of rubber (and later leather) was earmarked for gas mask production, leaving civilians without various necessities.[40]

While aerial gas bombs remained a recurrent fear on the home front, European scientists and engineers gave most of their attention to developing mid-range gas shells. In the winter of 1915–1916, all of the major belligerents began to fabricate both chlorine and phosgene shells in order to make gas a more accurate and controllable weapon. To make these shells effective, militaries needed ever-thinner casings that could house greater quantities of gas and release it on impact. These new shells

[36] Historian L. F. Haber estimates 5,200 civilian gas casualties, including 111 deaths, throughout the war. These numbers include workers at shell filling plants. Haber, *The Poisonous Cloud*, 249.

[37] Schutzmassnahmen gegen feindliche Fliegerangriffe, Juni 1916–Mai 1917 (LBWH) M 77/1 Bü 629.

[38] Similar fears, compounded by the German use of zeppelins, proliferated among French and British city-dwellers. The British War Cabinet considered distributing gas masks to civilians, but they felt that civilians could never properly don the masks in moments of danger. Susan R. Grayzel, "The Baby in the Gas Mask: Motherhood, Wartime Technology, and the Gendered Division between the Fronts during and after the First World War," in *Gender and the First World War*, eds. Christa Hämmerle, Oswald Überegger, and Birgitta Bader Zaar, 127–144 (New York: Palgrave Macmillan, 2014), 130–132.

[39] On April 4, 1917, the War Ministry ordered 50,000 gas masks for civilians who lived or worked within 8 kilometers of the French and Belgian borders and emergency sanitation groups were set up in major cities. Anweisungen und Verfugungen uber die Gasschutzlager, die Ausbildung, den Bedarf von Gasschutzmitteln und den Gaskampf, 1915–1917 (LBWK) 465 F 3, 620.

[40] Belinda Davis, *Home Fires Burning: Food, Politics, and Everyday Life in World War I Berlin* (Chapel Hill: University of North Carolina Press, 2000), 15–16.

also required a lead lining to prevent the gas from eating away at the iron or steel casing.[41] Due to regular exposure to both the lead linings and the chemicals themselves, shell filling proved to be incredibly dangerous work.[42] For instance, the gas pioneer Otto Hahn wrote of phosgene splashing into his eye while trying to fill a shell, an accident that would ultimately require surgery.[43] Cracked shell linings were also common, leading to extended exposure to both chlorine and phosgene. Medical reports recorded regular instances of fatigue, headaches, indigestion, spasms of the eyelids, and shortness of breath, alongside frequent cases of bronchitis, asthma, lung infections, heart arrhythmias, and depression.[44] In an attempt to combat these problems, factory workers were outfitted with the German *Linienmaske*, and soon the Germans were churning out 24,000 gas shells per month.

Technological Developments and Tactical Changes: Gas Shells and the Escalation of the Chemical War

In the spring of 1916, the French were the first to employ phosgene gas shells at Verdun. This attack took the Germans by surprise, and many soldiers still did not yet have a gas mask due to long-standing production delays.[45] Colonel Max Bauer and Fritz Haber felt that the Allies were now catching up in the race for decisive gas weapons, and they claimed that ideological resistance within the German High Command had wasted their opportunity for chemical superiority.[46] By the Battle of the Somme in the summer of 1916, the Allies were now firing thousands of gas shells before nearly every attack. However, the importance of this change in weaponry was not fully integrated into tactics until 1917, and both the Germans and British continued to supplement gas shells with the older cylinder attacks.

The value of cylinders was first called into question in April 1916 when the Germans lost nearly 1,500 men to their own gas near the French village of Hulluch. During this series of attacks, the pioneer troops had released phosgene and chlorine gas toward the British, but the wind direction

[41] Lead shell linings were later replaced by ceramics and then glass.

[42] The Hoechst and Bayer chemical companies employed large numbers of women to perform most of the gas shell filling for the German military.

[43] Hahn, *Mein Leben*, 122–123.

[44] Gas exposure was apparently so frequent that a contemporary observer claimed that, at any given time during the war, as many as one-third of all Hoechst factory workers were absent due to illness. Gerard J. Fitzgerald, "Chemical Warfare and Medical Response during World War I," *American Journal of Public Health* 98, no. 4 (April 2008): 119.

[45] Wilhelm Wittekindt, Meine Kriegserlebnisse 1916–1918 (DTA) 1229.

[46] Fritz Haber Briefe 1916 (MPG) Va 5, 857.

quickly changed and blew the gas back onto the Second Bavarian Corps. This tactical mistake was followed by a massive failure of the German gas mask, largely due to improper fitting and filter oversaturation. Haber subsequently ordered further development of the *Linienmaske*, and the responsible KWI departments soon delivered the improved *Rahmenmaske* (frame mask), named for its tighter connection to the wearer's face.[47] This improved fit was achieved with a fabric lining inside the mask that could conform to the contours of the face. The mask was also made in three sizes (large, medium, and small), and each soldier was required to know their face size during distribution. From September 1915 to September 1916, 12 million *Rahmenmasken* were delivered to German troops.[48] At the height of this production, the largest manufacturing firms such as the Auergesellschaft delivered approximately 2.3 million masks per month.[49]

The *Rahmenmaske* was issued with Richard Willstätter's three-layer filter already screwed on to the mouthpiece and the entire mask carefully folded within a metal carrying case. Thus, the mask could be quickly removed and donned without fumbling to attach a filter.[50] Such technological developments were coupled with an increased attention to gas discipline. Each regiment and independent battalion was now required to have two gas officers. Usually, this would be a junior officer such as a captain, and an enlisted man, often a sergeant or corporal. These chosen officers were required to participate in a gas school course at least once per year. Formalized army gas schools were first established in October of 1916 in Berlin and Leverkusen, but independent gas courses were also taught closer to the front lines and in chemical factories.[51]

Upon arrival at the gas schools, officers would learn how to best drill their regiment in gas preparedness.[52] A syllabus from a 1917 gas course included instruction in: throwing gas mines, shooting artillery gas shells, the nature of blistering gases, the enemy gas arsenal, ground rules for gas protection, the resources of the gas protection service, the sizing and

[47] Ibid.

[48] Soldiers referred to both the *Rahmenmaske* and the *Linienmaske* as *Gummimasken* (rubber masks) because their face pieces were made of rubberized cotton.

[49] Max Schwarte, *Die militärischen Lehren des Grossen Krieges* (Berlin: Ernst Siegfried Mittler und Sohn, 1920), 458.

[50] Wilhelm Hoffmann, *Die deutschen Ärzte im Weltkriege: Ihre Leistungen und Erfahrungen* (Berlin: E. S. Mittler & Sohn, 1920), 389.

[51] The Berlin gas school, associated with the KWI, could teach 400 officers every ten days. Ryan Mark Johnson, "A Suffocating Nature: Environment, Culture, and German Chemical Warfare on the Western Front" (Dissertation, Temple University, 2013), 112.

[52] Infanterie-Divisionen (WK) 1245, 1915–1918 (BA-MA).

fitting of gas masks, the nature of the front weather service, gas exposure in a gas chamber, gas protection practice, and first aid for gas sickness.[53] Gas officers were further required to keep track of the condition of all regimental gas protection equipment and to conduct gas mask exercises once a month. These exercises usually consisted of ten minutes of singing and simple aerobic movements while wearing a gas mask.[54] Finally, the gas officers would report any use of gas in the field to both the meteorological services and the War Ministry. A third enlisted officer would write and file these monthly reports, which were then funneled upward to Haber's Kaiser Wilhelm Institute when necessary.[55]

By the early fall of 1917, the Germans had largely abandoned their chlorine cylinder attacks and embraced gas shells as the future of chemical warfare. Besides addressing the problems associated with friendly fire, this became necessary because the British Small Box Respirator gas mask proved particularly adept at defending against chlorine clouds. Furthermore, in April 1917, the British unveiled the Livens projector, a small mortar-like weapon that could quickly and accurately fire gas shells at a range of 500 to 1,800 meters.[56] The Livens projector circumvented the extensive delivery apparatus of the gas cylinders, and its shells offered much greater resistance to variable weather conditions. Most importantly, the Livens projector substantially increased the attacker's level of surprise. Mortars could deliver gas at any time and almost any place on a given battlefield. These advantages being apparent, the Germans quickly developed their own rifled projector in the summer of 1917 and from then on, soldiers on the Western Front lived in constant preparation for gas.[57]

Gas projectors were not the only significant technological developments that would alter the face of chemical warfare in 1917. By July of that year, the Germans were firing shells filled with compounds such as Clark I, Clark II, and Adamsite (the so-called Blue Cross gases).[58] These gases irritated the upper respiratory tract, causing both sneezing and vomiting. This effect, in combination with the German belief that Clark particles

[53] Preussische Heeresgasschule Offiziers Ausbildungskurs 1. February 1917 (BA-MA) PH/21/.
[54] RH/12/9 Sig 32 (BA-MA).
[55] Dienstvorschrift fur den Gaskampf und Gasschutz 1917 (BA-MA) PH/2/.
[56] Lichtbild-Erkundung englischer Gaswerfer, 1917 (BA-MA) PH 5-II/267.
[57] Vortrag uber Gasschiessen in Verteidigung und Angriff, 1917 (BA-MA) PH 5-II/379.
[58] Hoechst was the main supplier of Clark. Clark I, or diphenylchloroarsine, was originally synthesized in 1878 by the German chemists August Michaelis and Wilhelm La Coste while Clark II, or diphenylchloroarsine, was developed for the German military in 1918 by the chemists G. Sturniolo and G. Bellinzoni. Adamsite, or diphenylchloroarsine, was first synthesized by Heinrich Otto Wieland in 1915.

were fine enough to permeate gas mask filters, was the reason that Clark received the colloquial name *Maskenbrecher* (mask breaker). German artillery units would fire Clark shells followed by phosgene, hoping that Allied soldiers would remove their masks to sneeze or vomit and then breathe the deadly phosgene. New tactics such as these required precise cooperation between field gunners, who became increasingly central to the objectives of gas warfare.[59]

During this same period, Haber's KWI laboratories also recognized the value of filling shells with mustard gas, which had been produced back in the nineteenth century but previously ignored as an effective chemical weapon because it was not viewed as sufficiently lethal. First fired by the Germans at the Third Battle of Ypres in July 1917, mustard gas fit perfectly into the new belief that chemical weapons should be used to psychologically break the opponent.[60] When soldiers came in contact with small droplets of mustard gas, which was in fact a liquid, it could severely irritate the eyes and induce vomiting. In large quantities, mustard gas could cause internal and/or external bleeding, often destroying the lungs' mucous membrane over the course of four to five weeks. Furthermore, the so-called King of the Battle Gases was perfect for circumventing the gas mask since it could permeate the environment and create a blistering effect on any exposed skin.

The development of new technologies such as mustard gas led to new battlefield tactics. Achieving a surprise attack from a distance was no longer difficult and artillery batteries began firing great quantities of gas shells at small quadrants of enemy territory, hoping to achieve dense gas clouds in seemingly random patterns. These *Gasüberfälle* (gas assaults) were often conducted at night to further confuse and demoralize enemy troops.[61] Additional tactics included firing mustard gas across a front in order to create an impenetrable smoke screen, firing gas shells into the rear of an enemy and fragmentation shells into their front line, and firing high

[59] Feldartillerie-Regiment Nr 18 Gasschiessen aus dem Jahre 1918 (BA-MA) PH/21.

[60] German soldiers often referred to mustard gas as Lost – a combination of the last names of the two men who proposed using it as a warfare gas: Wilhelm Lommel (of Bayer) and Wilhelm Steinkopf (of Department G of the KWI). Mustard agent was synthesized by European scientists as early as 1822, but its irritating properties were not noted until it was again produced in 1860 by the British chemist Frederick Guthrie. The German Meyer-Clarke method of mustard gas production combined the 1886 method of the German chemist Viktor Meyer and the 1913 work of British chemist Hans Thatcher Clarke. At the time, Clarke was working in a Berlin laboratory alongside Emil Fischer. Fischer reported Clarke's discovery to the German Chemical Society, which then passed on this information to the KWI.

[61] Formationen des ingeneurskorps und Pionierkorps der Preussischen Armee, 1917 (BA-MA) PH/14/.

concentrations of mustard gas into recently ceded territory to prevent enemy recursion. Clearly referring to the unpleasantness of this final tactic, the German artillery called these areas of the battlefield *Gassümpfe* (gas swamps).[62]

The tactics of 1917 were primarily defensive in nature, and they required large numbers of gas shells to be effective. Due to the long-standing stalemate on the Western Front, firing a weapon that could harass enemy troops without committing German soldiers to a dangerous advance became increasingly attractive to commanders. Furthermore, these novel tactics were particularly effective against soldiers who were new to the front and did not yet know the difficulty of remaining both calm and alert during a gas attack. This was especially difficult for the unexperienced American troops who first started arriving in Europe in 1918 and suffered particularly high gas casualties.[63] But American and all other Allied gas casualty numbers decreased significantly by the end of 1918, suggesting that previous exposure to gas and knowledge of gas protocol was indeed the best form of protection.

Unfortunately for the Germans, new artillery tactics put significant strains on their supply of raw materials. Thousands of new shells needed to be filled, only to then be quickly distributed and fired at the enemy. As scientists and engineers developed new protective technologies, old gas masks and filters also had to be returned and refashioned with state-of-the-art materials. Due to a national shortage of rubber in August 1917, the Germans were eventually forced to manufacture leather gas masks. The so-called *Ledermaske* (leather mask) reduced the dead space between the face and the mask and featured double-layered celluloid eyepieces. Reflecting the perceived need to fit the mask to a soldier's daily life amid poison gas, spectacles could now be inserted. Such small technical changes then allowed the German War Ministry to claim that the *Ledermaske* was a purposeful technological improvement rather than a desperate search for available materials.[64] To the consternation of German soldiers, the leather mask actually struggled to retain its tight fit and the angled eyepieces made it difficult to shoot.

In May 1918, the KWI further introduced the *Sonntags-Einsatz* (S-E filter), which featured a greater amount of filter charcoal to protect against Blue Cross gases. Filters remained well supplied by companies such as Bayer, Schering, and Riedel de Haen through the end of the war, but gas masks were more difficult to produce and maintain. Gas officers

[62] Gas/ 21 Juni 1918–30 September 1918 (LBWH) M 33/2 Bü 407.
[63] Gasangriff an der amerikanischen Front in Frankreich 1918 (BA-MA) RH/8/.
[64] PH 2 Preussisches Kriegsministerium 139–197, 1917 (BA-MA) PH/2/.

insisted on constant mask cleaning and correct storage to try to extend the life of each gas mask.[65] Furthermore, soldiers were instructed to use their mask until it was utterly useless, resulting in an army that wore a hodgepodge of worn-out rubber and leather masks.[66] By the end of the war, functioning gas masks were reportedly available to only one in four German soldiers.[67]

Adding to this material shortage, by late 1917, the intermittent firing of gas shells heightened the apprehension that soldiers already felt about chemical weapons. The continual appearance of new gases, gas delivery devices, and defensive technologies inspired wild hypothesizing about the future of chemical weaponry. Among the Germans, Blue Cross gases inspired particularly great exaggeration, with soldiers falsely claiming that no gas mask filter could stop their particulate clouds.[68] Unsubstantiated stories of effective hydrogen cyanide gas and flammable gas that could be dropped into trenches and then set ablaze were common in the trenches. To quell such fears, gas officers continued to distribute gas protocol pamphlets that stressed staying calm and trusting the gas mask during an attack. Point seven of one protocol instructed soldiers to: "Breathe calmly and slowly and do not let the heat inside the gas mask annoy you. It will not hurt you. Avoid running and shifting the mask by bumping it: your own restlessness is the enemy's best friend."[69]

If 1917 was the year of significant technological development in chemical warfare, then the summer of 1918 finally saw these technologies fully employed on the battlefield. By then, the Germans had mostly stopped implementing technological improvements for the gas mask, instead focusing on gas tactics and gas drills. Major Georg Bruchmüller devised tactics in which large quantities of various gases were fired at a small target area. This strategy was called *Buntschiessen* (colorful shooting), named for the color code names for various chemical weapons.[70] This proved particularly effective both in demoralizing enemy troops and

[65] 1914=1918 Dokumente M Wolfen (IZS).
[66] Infanterie-Divisionen (WK) 1245, 1915–1918 (BHK).
[67] Yvonne Sherratt, *Hitler's Philosophers* (New Haven: Yale University Press, 2013), 23.
[68] Stephanie Fredewess-Wenstrup, *Mutters Kriegstagebuch: Die Aufzeichnungen der Antonia Helming 1914–1922* (Münster: Waxmann, 2005), 206.
[69] Gasschutz, April 1916–January 1917 (LBWK) 456 F 86, 219.
[70] As part of the rhetorical work involved in obscuring the potential deadliness of chemical warfare, the Germans used colored crosses as codenames for their various types of chemical weapons because colors did not easily convey their lethality. *Weisskreuz* (white cross) referred to all tear gases. Pulmonary agents such as chlorine and phosgene (and later Perstoff) were called *Grünkreuz* (green cross). *Blaukreuz* (blue cross) gases were vomiting and sneezing agents and *Gelbkreuz* (yellow cross) was all mustard agents.

in covering German retreats. The latter effect was due to the fact that *Buntschiessen* tended to mobilize the previously stagnant front by strategically poisoning swaths of ground for a certain amount of time. Not surprisingly, however, the Allies copied these methods with their own form of mustard gas, which they began using liberally in June 1918.

To counter dense *Buntschiessen* gas attacks, German trench sentries were required to be more vigilant and faster to sound the gas alarm. In 1918, soldiers were nearly constantly required to wear their masks "at-the-ready" or loosely around their necks, while *Entgiftungstruppen* (sanitation troops) were formed for each battalion and tasked with cleaning the persistent mustard gas from trenches and uniforms. German commanders increasingly believed that gas discipline was the key to fighting the chemical war. Given the relative effectiveness of fully functional gas masks, military experts were not entirely wrong in believing that most gas causalities were a result of human error and the fog of war. However, commanders were still unsure how much discipline the army could instill in its soldiers, especially in such a stressful and increasingly hopeless warfare environment.

In the final months of the war, German soldiers were especially fatigued by gas projectiles. Reflecting on this problem, the German poison gas expert Rudolf Hanslian later wrote that "there was no weapon in the war that led to the amount of shirking responsibilities as gas."[71] Gas alerts became more frequent, often straining the nerves of even the most hardened veterans, while most of the men now transferred from the Eastern Front to the western trenches had far less experience with gas warfare and were rarely trained in the correct usage of gas masks. By June of 1918, mustard gas rained down on the weary Germans and the army lacked the number of filters and masks required to cope with the Allied assault. While General Bruchmüller's *Buntschiessen* had made the war more mobile, soldiers were now scurrying between makeshift earthworks seeped in mustard gas.[72]

Alongside harrowing poison gas attacks, German troops were already dispirited in 1918 due to heavy casualties, a general lack of supplies, and the growing influenza epidemic.[73] Nevertheless, General Erich Ludendorff claimed in his memoirs that gas was vital to the German military effort

[71] Hanslian, *Der chemische Krieg*, 202.
[72] Modris Eksteins, *Rites of Spring: The Great War and the Birth of the Modern Age* (Boston: Houghton Mifflin Company, 1989), 161.
[73] By 1918, German soldiers further feared the American use of Lewisite, an odorless gas with effects similar to mustard gas.

even in the final year of the war.[74] While it may have provided the kind of defensive stalling that could drag the war through its last summer, there is no doubt that poison gas took a huge toll on the German soldiers who kept on fighting. Increasingly, these men lost significant faith both in the ability of their gas masks to protect them and in German chemical supremacy more broadly. As Allied gas shell production and weapon technologies surpassed those of the Central Powers, German soldiers began to feel that they had definitively lost the chemical war. Their head start at Ypres, the efficiency of Haber's KWI, and the cooperation of the IG chemical companies produced almost every significant chemical invention up to 1917. But by the war's final year, shortages in materials and the limitations of gas protection technologies gave the Allies a clear advantage. If the chemical war was to be won by the sheer amount of available war materials, then the British, French, and perhaps most clearly, the Americans, now had the upper hand.

Estimating the total number of gas casualties in World War I remains a difficult task. Britain, France, the United States, and Germany did not record gas injuries unless they were the observable cause of sickness or death, and Russian reporting was almost nonexistent. Further still, the inhalation of small amounts of gas was not always noticed until injurious effects appeared later in life. In sum, gas casualty reporting was hazy at best. Interestingly, Fritz Haber's son, the historian L. F. Haber, has done a commendable job in attempting to create an honest statistical account by comparing various postwar national records and considering the technological effectiveness of poison gas in each year of the war.[75] According to L. F. Haber, an estimated 531,000 World War I casualties were caused by gas, including about 17,700 deaths.[76] Without significant knowledge of Russian figures, we can hazard to say that around 74 percent of all gas

[74] Erich Ludendorff, *My War Memories 1914–1918* (London: Hutchinson, 1919), 579, 597.

[75] L. F. Haber, also known as Lutz, was Fritz Haber's second son and his third and youngest child. Lutz spent most of his childhood in Switzerland with his mother Charlotte, later moving to London before the Second World War. In the postwar years, he became a respected economist and economic historian in the UK. His books on chemical warfare were based on deep economic studies of the British, French, and German chemical industries as well as interviews with soldiers who had been gassed during the war. Throughout his life, L. F. Haber asserted that his father was largely responsible for the outbreak of the chemical war and that this responsibility should never be forgotten or ignored.

[76] While the British and the Americans sought to create an accurate account of their gas casualties, the French and Germans inflated or deflated numbers in postwar arguments over the legality and effect of chemical warfare.

Table 2.1 *Estimated gas casualties in World War I*

Nation	1915–1917	1918	Total	1918 as % of total
Germany	37,000	~70,000	107,000	65
Great Britain	72,000	114,000	186,000	61
France	~20,000	110,000	130,000	85
USA	–	73,000	73,000	100
Others	~20,000	~15,000	~35,000	unknown
Totals	~149,000	~382,000	~531,000	~74

Estimated gas casualties in World War I based on the work of L. F. Haber. Haber did not find the numbers for Russia, Italy, or Austro-Hungary to be reliable. In his 1937 book *Chemicals in War*, A. M. Prentiss estimated 1.2 million total gas casualties, with about 400,000 coming from the Russian army; however, there are no reliable sources for this estimate. The total number of men killed by gas is also unknown, although Prentiss hazarded a guess at about 90,000. Interestingly, Prentiss' totals are the most commonly used in online encyclopedias.

casualties came in 1918. The Germans suffered about 107,000 casualties throughout the war, with 65 percent of them coming in the final year. About 4,000 of those casualties were deaths (Table 2.1).[77]

Ultimately, Fritz Haber's initial claim that chlorine cylinders would win the war for the Germans proved incorrect on several levels. His statements and writings from after World War I would suggest that he knew that gas warfare technology was not ready for its unveiling in 1915, but pressure to provide a solution to the standstill on the Western Front influenced his decision-making. Even if this was true, Haber still misjudged the German ability to continue the gas war while he sought a truly decisive technological breakthrough. The KWI's continued development of protective devices such as the gas mask did not buy him enough time. Rather, the development of the mask merely protracted the chemical war, thus significantly contributing to the misery of the German frontline experience. Indeed, the Germans developed the modern military gas mask hand in hand with the new gases and tactics that it was meant to repel. While the mask did save countless lives from weapons that the Germans themselves had unleashed, it also seemingly demanded the escalation of a uniquely horrifying form of warfare that would ultimately help precipitate the final German defeat in

[77] Haber, *The Poisonous Cloud*, 243–244.

1918. Trapped between fatal air and toxic soil, countless German soldiers lost faith in the gas mask's claim to technological salvation. Nevertheless, while face to face with the approximately 12 million gas masks that Germany produced during the war, these men most fully recognized that unmediated life could not continue in a world permeated by deadly gases.

3 The First "Chemical Subjects": Soldier Encounters with the Gas Mask in World War I

When describing the physical environment of the western trenches in 1918, the German infantryman Rudolf Binding wrote:

> Not a blade of green anywhere round. The layer of soil which once covered the loose chalk is now buried underneath it. Thousands of shells have brought the stones to the surface and smothered the earth with its own entrails. There are miles upon miles of flat, empty, broken, and tumbled stone-quarry, utterly purposeless and useless, in the middle of which stand groups of these blackened stumps of dead trees, poisoned oases, killed forever.[1]

While this quote described the totalizing damage caused by industrialized warfare, the "poisoned oases" to which Binding referred were largely a product of the toxic gas that either wafted over or sunk into the ground on the Western Front. Like Binding, other German soldiers regularly described the physical impact of poison gas, often noting the strange greenish-yellow color that seemed to coat everything.[2] This sickly hue, reminiscent of the "pale horse of death" in the Book of Revelations, could tint food, clothing, mounds of earth, and the puddles of poisoned water that slowly seeped into the trenches.[3] Even weapons and other metallic equipment were often recolored by the chlorine gas, turning a deep rust color that added to the strange chemical color pallet that came to make up the soldier's world.

Over the balance of the war, German infantrymen had little real protection against this seeping liquid. In most cases, men only had the gas mask to serve as their primary defense against the poisoned environment. While the mask could filter toxins from the air, sometimes so effectively that soldiers described it as a magic cloak or talisman, it provided

[1] Binding, *A Fatalist at War*, 217.

[2] Oliver Kock, ed., *Das Tagebuch des Leutnants Nilius: Ostern 1916 bis 21.2.1918* (Bayreuth: Scherzers-Militaer Verlag, 2013), 49; Gerhard Scholtz, *Tagebuch einer Batterie* (Potsdam: Rütten und Loening Verlag, 1941), 93.

[3] The color of death's horse in the Book of Revelations is described as *khlōros*. *Khlōros*, the root of the word chlorine, is often translated as the "pale horse," but it can also mean the "greenish-yellow" horse.

little protection against physical contact with liquid poisons. Thus, facing the constant possibility of chemical burns and temporary blinding, the average German soldier often came to question his own agency in the quest for effective World War I chemical protection. This crisis of existential and technological faith became particularly apparent in the final years of the war when the intensification of chemical warfare made the gas mask a regularly described item in soldiers' diaries and letters.

By tracking the personal writings of German soldiers from 1915 to 1918, one can see soldiers' changing psychological and affective responses to poison gas and the gas mask.[4] In particular, the shifting levels of trust and affinity for the gas mask become quite apparent. As the chemical war progressed in 1916 and gas masks were made more effective, soldiers tended to gain trust in their protective devices. They described their masks in human or technological terms that expressed the integration of the mask into the soldier's life world. However, as chemical warfare drastically accelerated in 1918 and German gas masks began to fail, textual descriptions of both poison gas and the gas mask began to take on more supernatural qualities. Gas increasingly became a devilish ghost or a grim reaper, while the mask represented death incarnate. These increasingly popular and macabre descriptors thus reflected soldiers' growing doubts about their ability to make themselves technologically impenetrable to chemical weapons as they concurrently lost trust in their ability to master their environment.

Such a reading of the gas mask seeks what certain historians of materiality have called the "mute power of things" to reshape human consciousness and social relations.[5] Describing this endeavor, the scholar Alan Trachtenberg argued that Wolfgang Schivelbusch's historical treatment of the nineteenth-century railroad sought the subjective experience of newly developing "industrial subjects." Trachtenberg writes:

[4] In an attempt to address issues of time and memorialization in ego documents, this chapter only utilizes diaries and letters that were produced between 1915 and 1918. While some of these texts have been collated or edited at a later date, the selected passages remain sufficiently loyal to the initial writings from the war. The chapter also gathers a broad collection of soldiers' voices by pulling from men who served in different branches of the German military, who espoused different political aims through their later memorialization of the war, and who held different military ranks. The study still skews toward literary-inclined officers, who wrote down their war experiences far more often than enlisted men, but this is mitigated through the inclusion of trench journals aimed at informing and entertaining the average German infantryman.

[5] Tony Bennett and Patrick Joyce, eds., *Material Powers: Cultural Studies, History and the Material Turn* (New York: Routledge, 2010), 10.

> One feature of modernity as it crystallized in the nineteenth century was a radical foregrounding of machinery and of mechanical apparatus within everyday life. The railroad represented the visible presence of modern technology as such. Within the technology lay also forms of social production and their relations. Thus the physical experience of technology mediated consciousness of the emerging social order; it gave a form to a revolutionary rupture with past forms of experience, of social order, of human relations. The products of the new technology produced ... their own subject; they produced capacities appropriate to their own use.[6]

Like the many technologies of the railroad, modern chemical technologies similarly produced their own effects on Europeans living in a rapidly industrializing world. As the German chemical industry expanded at a tremendous pace in the late nineteenth and early twentieth centuries, its products also came to redefine the experience of turn-of-the-century modernity. The most dramatic crystallization of this redefinition was on the battlefields of World War I, where soldiers regularly faced both poison gas and the ancillary devices created to control a chemically volatile world.[7] Admittedly, one could examine historical interaction with poisonous gas itself to reveal fears over the pervasive nature of injury and death in a newly weaponized modernity. Certainly, as weaponized gases were developed and implemented on the World War I battlefield, soldiers developed new relationships to their environment and their own bodies. Nevertheless, gas was a uniquely ephemeral weapon and its physical presence was not always necessary to inspire action, fear, or reflection. On the other hand, as a tool of ostensive resistance and survival, the gas mask stoked the same fears of being gassed while also exposing a soldier's level of faith in the ability to control or defend against this newly modern condition through a combination of technological armoring and individual discipline.

While Richard Willstätter and other scientists and engineers from Fritz Haber's Kaiser Wilhelm Institute for Physical and Electrochemistry (KWI) developed the modern gas mask as a tool to save soldiers from a horrible death by asphyxiation, the mask also tangibly expressed the struggle against death for the soldiers who employed it. As the chemical war progressed and gas discipline became even more stringent, the gas mask more frequently produced anxiety and fear in the increasingly harried German soldiers. In this way, the gas mask was not merely a

[6] Wolfgang Schivelbusch, *The Railway Journey: The Industrialization of Time and Space in the Nineteenth Century* (Berkeley: University of California Press, 2014.), xv.

[7] An affective and corporeal reading of World War I soldiers' texts attempts to bring us closer to these physical experiences. In doing so, one can see the shifting significance of the material things, such as poison gas and the gas mask, that soldiers continually chose to describe. Due to textual and historical mediation, a perfect apprehension of an author's emotive experience may remain impossible, but the hope is for an asymptotic appreciation of subjective feelings in the past.

technological solution derived from necessity, but at the level of actual employment, it was an ambivalent object that revealed potential injury or demise for the average soldier. More broadly, the gas mask forced a direct and visceral confrontation with the rapidly developing technology of the World War I battlefield. Thus, the mask could potentially serve to shape and hone the first "chemical subjects," or the newly envisioned men of the twentieth century who would embrace the now toxic environment as an arena of spiritual, technological, and physical evolution. Through personal discipline and a body augmented and armored by a permanently attached breathing apparatus, this newly imagined "chemical subject" burst forth into a hypothetical future permeated by toxic gas, thus redefining the possibilities for human survival.

Canisters and Cotton Pads: Encountering Early Iterations of Gas Protection

In December of 1914, Fritz Haber's pioneer troops were the first enlisted soldiers to receive artificial respirators. While testing chlorine cylinders at the Wahn artillery range, the pioneers were outfitted with Dräger *Selbstretter* (self-rescuer) respirators. These large devices could supply a soldier with oxygen for up to one hour from canisters held in a satchel bag. Once the oxygen canister was empty, a new one could be attached to the breathing tube that snaked up to the soldier's nose and mouth. The *Selbstretter* did not have an attached face mask, so the pioneer troops were also provided with driving goggles. While this system was effective enough for the pioneer troops to install and test Haber's chlorine cylinders, it significantly hampered mobility and inhibited the ability to fight.

Given that most gas pioneers had backgrounds in scientific research, these troops seem to have had less trouble adjusting to the physical demands of poison gas exposure. While there were poisoning accidents during the early gas trials, the pioneers appear to have felt general confidence in the protective abilities of their devices. The *Selbstretter* was not a new technology, and it had proven its effectiveness against poison gas in numerous mining rescues around the turn of the century. Gas pioneers were provided with enough canisters to last the amount of time that they would be exposed to gas and the regular supply of oxygen inspired a certain confidence in their work. In preparation for gas warfare, the pioneers trained with the *Selbstretter* by performing nonstrenuous activities in a gas-filled chamber for forty-five minutes.[8] The reliability of the

[8] 1914=1918 Dokumente M Wolfen (IZS).

respirator and the scientific backgrounds of the men further created a unique relationship to the chlorine gas that often surrounded them. Most pioneers tended to discuss gas as a mere tool that could be controlled, perhaps even more so than bullets or shrapnel. Furthermore, the pioneers rarely had to witness the suffering of their enemies after releasing gas. Indeed, gas pioneer Otto Hahn later wrote thus:

> The constant handling of these strong toxins had dulled us so much that we had no scruples when we used them at the front ... we front-line observers rarely saw the direct effects of our weapons. Generally all we knew was that the enemy abandoned the positions that had been bombarded with gas shells.[9]

For the average German infantryman, however, the *Selbstretter* was an impractical means of protection. Like at Ypres, infantrymen would have to advance behind a chlorine cloud and clear out enemy resistance, thus requiring full mobility and clear vision. Beyond problems associated with the *Selbstretter*'s unwieldy design, it was an expensive device that could not be quickly manufactured for thousands of German soldiers. Thus, it was only in mid-April 1915 that infantrymen began to receive their own form of protection.[10]

During the delivery of the first iteration of infantry gas protection, the army surgeon Stephen Westman reported that "this most primitive gas-mask was like a surgeon's mask and inside it we laid a pad of cotton wool, soaked in a fluid which was supposed to filter the air one breathed and protect the lungs from the poison. The eyes were left uncovered."[11] Reports from the Ypres front further claimed that many soldiers did not know how to use the cotton pads correctly, often failing to cover both their mouths and noses or applying them to their chest once they felt pain in their lungs.[12] For still more men, the chemical solution dried up or they simply lost their pad, forcing them to use handkerchiefs and other pieces of fabric for protection. In fact, the soldier Wilhelm Hartung claimed that men almost never had the gas-retarding chemicals, let alone any other liquid available at the front. This lack of liquids forced soldiers to urinate on their cotton pads, a stopgap solution that gave Hartung little confidence. He wrote, "I do not think it had any practical value in an emergency."[13]

[9] Hahn, *Mein Leben*, 122.
[10] William Boyd, *With a Field Ambulance at Ypres* (New York: George H. Doran Co, 1916), 67.
[11] Stephen Westman, *Surgeon with the Kaiser's Army* (London: William Kimber, 1968), 92.
[12] Tim Cook, "Creating the Faith: The Canadian Gas Services in the First World War," *Journal of Military History* 62, no. 4 (October 1998): 759.
[13] Wilhelm Hartung, *Großkampf Männer und Granaten!* (Berlin: Verlag Tradition Wilhelm Kolk, 1930), 59.

Not surprisingly, this makeshift outfitting did little to inspire soldier confidence in either their masks or the chlorine cylinders. Many men felt anxious in the presence of the so-called "F-batteries," but they were nevertheless forced to live in close proximity to the entrenched cylinders throughout the March and April of 1915.[14] The fact that the chlorine barrels had leaked several times during installation, asphyxiating upward of twenty men, seemed to justify these early fears.[15] These preattack concerns also help to explain why the German soldiers advanced so slowly and hesitantly behind the gas at Ypres. When these men did finally reach the French trenches, the sight of their enemies served to validate their caution. Soldiers reported that the gas victims were lying on the ground gasping for air. Their faces had turned a dark blue and they coughed a bloody foam.[16] Describing the horror of the scene, the infantryman Rudolf Binding wrote that "the men do not want to hear the word 'gas' mentioned. They are fed up with the stink."[17]

But such was the experience of the chemical war for the German soldier on the Western Front. His greatest predicament was the fact that he had little protection against his own unpredictable weapon. However, it is important to note that German suffering was clearly less intense than that of the Allied soldiers. A Canadian infantryman at Ypres reported, "We did not get the full impact of the gas, but what we got was enough for me, it makes your eyes smart and run, I became violently sick … The next thing I noticed was a horde of Turcos making for our trenches … The poor devils were absolutely paralysed with fear."[18]

Algerian and Moroccan troops in the French army certainly bore the brunt of the first attacks, and the German military chauffeur Anton Fendrich notably remarked on the "bad-smelling Negros" who had been gassed and were now receiving German medical care. Fendrich wrote: "Gently, as if they were babies, the men of the [German] Red Cross carried them to the hospital, where they were given oxygen."[19] Fendrich's pointed description of "bad-smelling Negros" suggests a distinct interest in the Franco-Algerian troops present at Ypres. Interestingly, from its first deployment onward, soldiers attempted to

[14] Trumpener, "The Road to Ypres," 472.
[15] Otto Lummitsch, "Meine Erinnerungen an Geheimrat Prof. Dr. Haber" (MPG) Dept Va, Rep 5, 1480.
[16] Literatur, Kriegserfahrungen, Auslandsnachrichten, 1934–1936 (BA-MA) RH/12/4.
[17] Binding, *A Fatalist at War*, 84.
[18] John Cornwell, *Hitler's Scientists: Science, War, and the Devil's Pact* (New York: Penguin, 2003), 62.
[19] Anton Fendrich, *Mit dem Auto an der Front: Kriegserlebnisse* (Stuttgart: Franckh'sche Verlagshandlung, 1918), 94–96.

make sense of poison gas through categories of identity and belonging. Germans, in particular, were quick to claim that poison gas was an inherently German or Germanic weapon.[20] This suggested to contemporaries that Germans, and perhaps other northern European peoples, were best able to control and withstand the gas. Indeed, the Germans were willing to accept that the Canadians had better resisted their chlorine attacks through a natural fortitude that outclassed the supposedly undisciplined Algerian and Moroccan troops. Not surprisingly, the varying levels of success with gas resistance had, in reality, less to do with intrinsic ethno-national constitution and more to do with a given regiment's preparedness for gas.

In the middle of May 1915, gas attacks on the Western Front abated when the pioneer regiments moved eastwards. This shift in theater gave the Allies time to procure protection for their own soldiers, and on April 25, General Joseph Joffre ordered 100,000 cotton masks impregnated with sodium thiosulphate for the French troops.[21] Unfortunately, the cotton masks could not be immediately produced and French soldiers first learned to dig pits and fill them with water and gas-retarding sodium thiosulphate. The men could then dip handkerchiefs in the pit and wear them as a crude form of protection. Reflecting the lack of available knowledge on chlorine gas, the French also hoped that these pits, in tandem with rifle fire and explosives, might neutralize the large wafting clouds.

The British also provided their troops with cotton pads, which were soon supplemented by the Black Veil Respirator in May 1915. The Black Veil, invented by the Scottish physiologist John Scott Haldane, was a piece of black cotton that could be wrapped around the face to hold a cotton pad in front of the mouth. While the Veil ultimately proved unstable and largely ineffective, it produced a ghastly image when pulled up over the face of a British infantryman. In June of the same year, the Veil was replaced by the British Smoke Hood, also known as the Hypo Helmet. The British military physician Cluny Macpherson designed the Hypo Helmet after supposedly seeing a German soldier place a burlap sack over his head during a gas attack at Ypres. Replicating this soldier's desperate act, Macpherson took a canvas bag, created an eye slit, and coated the entire thing with sodium thiosulphate. Such impromptu British designs inaugurated the regular battlefield presence of oddly disfiguring full-head masks.

[20] Peter Thompson, "The Pale Death: Poison Gas and German Racial Exceptionalism, 1915–1945," *Central European History* 54 (2021): 280–281.

[21] Jonathan Krause, "The Origins of Chemical Warfare in the French Army," *War in History* 20, no. 4 (2013): 549.

Fitting the Mask: A Soldier's Introduction to the Gas Mask

On September 14, 1915, Herbert Sulzbach of the 2nd Bavarian Field Artillery Regiment was first informed about the provision of gas masks. Sulzbach wrote in his memoirs, "We are given talks by medical officers about bits of equipment which are supposed to stop you [from] inhaling poison gas."[22] These "bits of equipment" were most likely the *Linienmaske* and the Willstätter three-layer filter, and this brief passage gives insight into one of the earliest interactions with the first modern military gas mask. In fact, Sulzbach writes days before the first major British gas attack at the Battle of Loos on September 25. His language is dismissive and by calling the masks "bits of equipment" and modifying their purpose with "supposed," Sulzbach reveals his initial indifference for a new and unproven technology. And while these early masks did generally fail at Loos, due to either their inability to protect soldiers from gas or their limited distribution, Sulzbach had little apparent reason to distrust the mask prior to its implementation.

Sulzbach's introduction to gas masks and the technology behind gas warfare was fairly common. Chemical weapons and the technologies associated with them were an integral part of the massive material and tactical changes taking place during World War I. The early use and institutionalization of these technologies represented entirely new, yet unproven, possibilities for industrialized mass killing.[23] Thus, it is perhaps not surprising that when soldiers were first introduced to devices such as the gas mask, gas shells, or the gas itself, they were often confused as to their use and/or value. Some, like Sulzbach, simply did not recognize their purpose, griping about the weight that they added to their kit.[24] Others, reflecting on the inherited rules of war, saw this new technology as morally distasteful or ungentlemanly. Still more found gas masks to be particularly comical or alien in form, referring to them as stinkhoods, mouth drums, pig snouts, muzzles, chloroform hoods, and carnival veils.[25] Yet, regardless of their initial interpretation of the mask in 1915, no German infantryman knew what chemical warfare would eventually entail. Sulzbach and others could afford to view the gas mask with skepticism, disdain, or humor since it was not yet a life-or-death necessity.

[22] Herbert Sulzbach, *With the German Guns* (Hamden, CT: Archon Books, 1981), 67.
[23] William Philpott, *War of Attrition: Fighting the First World War* (New York: Overlook Press, 2014), 4.
[24] Johnson, "A Suffocating Nature," 96.
[25] *Die Gasmaske*. Vol 20 Beilage 2, May 20, 1917.

Soon after experiencing the gas attacks at Loos, however, German soldiers felt increasing respect and attachment to the "bits of equipment" that could save them from asphyxiation. Death by chlorine and phosgene could be a particularly long and painful experience. The infantryman Johannes Martini claimed that "two full breaths suffice to bring about death. Whoever does not die immediately, goes on for a few days until they become lung-sick. The gas eats away all of their internal organs."[26] Martini's inexpert impression was not entirely wrong. Gassed men who did not immediately asphyxiate would first appear pale as their pulse began to race and their breathing became shallow. Usually, such men died an hour later from heart failure. A German military report stated that the "poisoned [men] were usually able to walk back on foot and only complained of pain in the chest, combined with severe coughing. Shortness of breath was absent at first, but it appeared after about an hour, and then increased rapidly, accompanied by violent vomiting. After 2 to 3 hours, death by suffocation occurred."[27] The prolonged struggle of the gassed soldier had profound effects on perceptions of chemical warfare. It was precisely this protracted death that made soldiers particularly anxious about gas inhalation. For many, a bullet to the head represented both a less painful and a more respectable demise. Reflecting on this fear of gas-induced death, one German soldier reported that:

> It is not just a terrible struggle for air, but the struggle to maintain consciousness. One can observe how the dying person pulls himself out of the increasing darkness into consciousness again and again and thereby into fresh agony. I have seen this with some Russians dying of phosgene gas and I am unable to forget it.[28]

As soldiers came to realize the nature of gas death, they increasingly valued their gas masks. Alongside a coat, cigarettes, map, rifle, revolver, bayonet, and steel helmet, the gas mask became part of the soldier's essential belongings in 1916.[29] As the chemical war progressed, men even began to sleep with their gas masks close at hand in case of surprise gas attacks. The infantryman Wilhelm Wittekindt wrote:

> It was less pleasant when we were awakened the next night by the gas alarm. We always slept with gas masks nearby to be prepared for anything, and we put them on immediately ... With the gas mask on, we lay down again and slept on.[30]

[26] Johannes Martini, "Errinerung aus dem Weltkireg 1914–1915" (DTA).
[27] Anlagen-Englischer Gasangriff Juli 1, 1917–September 30, 1918 (LBWH) M 411 Bd. 1276.
[28] Karl Jaspers, *General Psychopathology*, trans. J. Hoenig and Marian W. Hamilton (Baltimore: Johns Hopkins University Press, 1997), 478.
[29] Wilhelm Lustig, *Kriegstagebuch 1915–1918* (LBI) ME 6 MM 2.
[30] Wilhelm Wittekindt, "Meine Kriegserlebnisse 1916–1918" (DTA).

As Wittekindt reveals, by 1916, demands for constant gas preparation had steadily increased. A German gas bulletin from the war's later years ordered that every officer and enlisted man have a gas mask and filter on his person day and night.[31] This included while eating, sleeping, marching, at ease, and on guard duty. The bulletin stated that a second filter should be carried in every man's bag and a third should be kept for reserve at the rear infantry or artillery positions. As demands for preparation increased, Fritz Haber's KWI developed better protective equipment with greater customization. The German gas mask now came in different sizes, and while these sizes did not always perfectly fit each soldier, the process of being measured for a gas mask created a sense of personal ownership and bespoke design. In fact, the trench journal *Die Gasmaske* claimed that the gas mask was so close to the soldier that it might as well be his own cousin, further arguing that "probably no instrument of war has gained such appreciation as the gas mask."[32]

While chemical warfare became a more serious matter in 1917 with the introduction of new weapons such as Clark and mustard gas, German soldiers continued to show a defiant sense of humor and incredulity about the surreal nature of gas attacks. The often-callous Ernst Jünger mocked gas victims, writing: "When I passed the station, there were already a lot of people who had swallowed too much gas. They made a ridiculous impression. Everyone sat around, moaning and groaning, water running out of their eyes."[33] Jünger again found it humorous when one of his comrades forgot his gas mask during an attack. He wrote: "Vogel, who shares my shelter, could not find his gas mask because he had left it somewhere. He danced (I secretly enjoyed this act) under the shouts: 'Oh dear God, my gas mask, my gas mask' from one corner of the shelter to the other."[34] While Jünger's sense of gas humor appears particularly sadistic, plenty of other examples were far more playful. In a trench journal story entitled "The Pioneers of St. Quentin," a gas officer asks a Bavarian soldier if he smells gas. The offended Bavarian soldier replies that it is just his own personal scent.[35] Soldiers further enjoyed posing for photos while performing mundane tasks under the gas mask, whether it was reading a book or drinking wine at the dinner table. Light-hearted subversions of gas discipline were often a soldier's attempt to create a sense of personal control over the irrepressible nature of modern mechanized war. As a

[31] Gorz 18 Hauptmann Ruge Stab 213.J.D., 1918 (BA-MA) PH 8-I/56.
[32] *Die Gasmaske.* Vol 20 Beilage 2, May 20, 1917.
[33] Helmuth Kiesel, ed, *Ernst Jünger Kriegstagebuch 1914–1918* (Stuttgart: Klett-Cotta, 2010), 138.
[34] Ibid., 153.
[35] "Die Pioniere von St. Quentin," *Die Sappe,* March 30, 1918.

coping mechanism, such jokes helped to make these new methods of killing more palatable and to normalize the wartime experience.[36] These regular attempts at gallows humor only began to decrease in frequency once the gas war rapidly accelerated in 1918. By then, the overwhelming sensory experience of chemical warfare had psychologically broken many men, proving that poison gas did indeed do its greatest damage in the realm of the mind.

Ghouls, Goblins, and Beasts of All Kinds: The Sights of Gas Warfare

While cotton mouth coverings and the Black Veil Respirator masked the lower face and created a level of anonymity on the battlefield, the British Hypo Helmet was the first gas mask that fully disguised the soldier's entire head. The historian Tim Cook has described the Hypo Helmet as particularly "nightmarish, especially when soldiers materialized out of the gaseous haze."[37] Gas-ready British soldiers with burlap sacks on their heads now resembled something akin to living scarecrows or a hooded man ready for execution. Certainly, there was something rather dehumanizing and disturbing about the concealment of the face.[38] When full-face gas masks were donned, the trenches were no longer filled with traditional friends and foes, and men could no longer recognize facial displays of emotion, thereby creating a sense of detachment from the quasi-human forms that surrounded them.[39] Seen through the celluloid eyepieces of one's own gas mask, this multilayered mediation between human bodies created a sense of isolation from the exterior world (Figure 3.1).[40]

By the spring of 1916, all the major belligerents had some form of full-face gas mask. The Germans were already using the *Linienmaske*, perhaps the most elegantly designed gas mask due to its small filter size. Yet regardless of the *Linienmaske*'s strengths, German soldiers often claimed that the British maintained the best masks. After abandoning the Hypo Helmet, the British army introduced the highly effective Small Box Respirator in April 1916. French soldiers reportedly had varying qualities

[36] George Mosse, *Fallen Soldiers: Reshaping the Memory of the World Wars* (New York: Oxford University Press, 1990), 155.
[37] Cook, "Creating the Faith," 760.
[38] Historian Modris Eksteins writes that soldiers in gas masks "became figures of fantasy, closer in their angular features to the creations of Picasso and Braque than to soldiers of tradition." Eksteins, *Rites of Spring*, 163.
[39] The 1918 introduction of full-body rubber gas suits only increased battlefield dehumanization. These were given to *Sanitätsoffizieren* (medics) while common infantrymen were given gas blankets, which could be draped over the head.
[40] Encke, *Augenblicke der Gefahr*, 214–215.

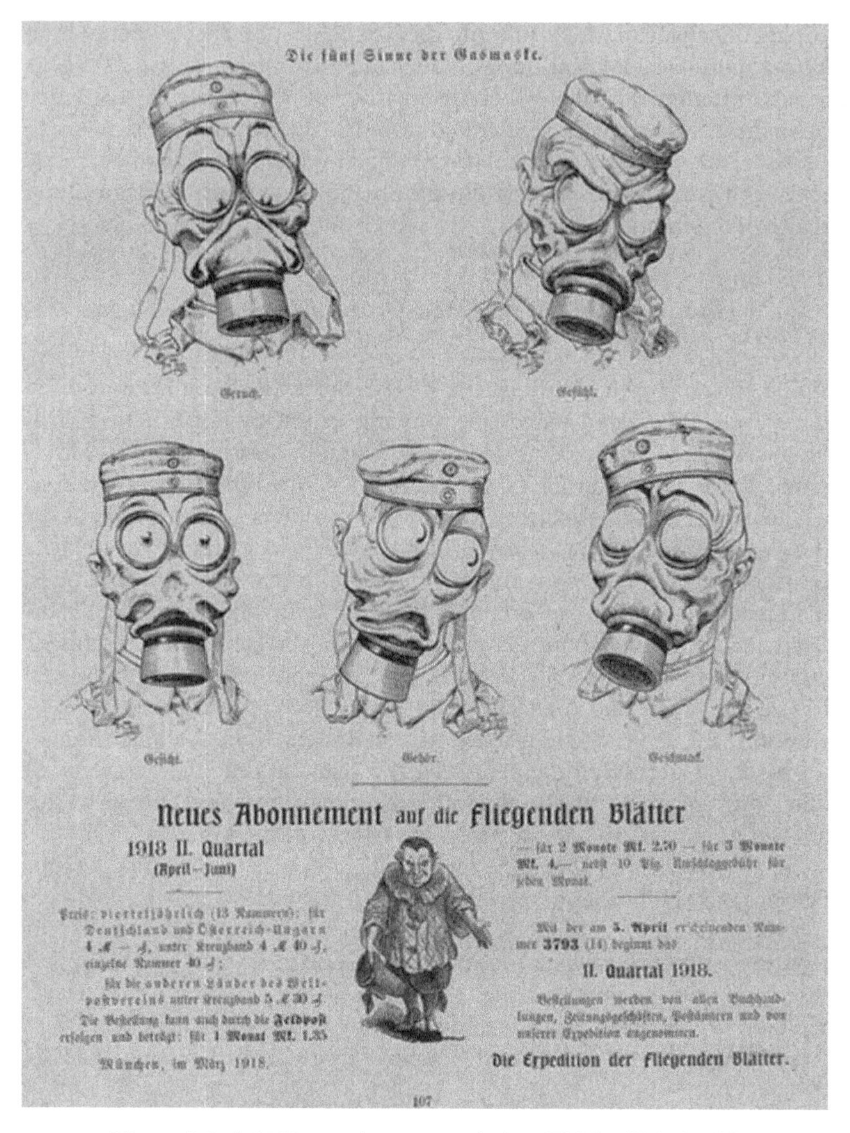

Figure 3.1 A 1918 trench cartoon. Arthur Thiele, "Die fünf Sinne der Gasmaske" 1918 (UH) G 5442-2 Folio RES.

Source. Arthur Thiele, "Die fünf Sinne der Gasmaske" 1918 (UH) G 5442-2 Folio RES

of masks, but the M2 eventually became their standard equipment. Russian soldiers rarely received a gas mask and those they did receive were poorly constructed.[41] German scientists at the KWI tested all of these Allied masks in the process of further developing gases that could penetrate their filters.[42]

As the *Linienmaske,* Small Box Respirator, and M2 became visually pervasive, soldiers began to describe fellow combatants with the tropes, analogies, and metaphors of magical realism, often detailing an outlandish landscape populated by ghosts, goblins, and insects of various kinds.[43] For instance, soldier Wilhelm Hartung described these new masked men as a "horde of the underworld," while Ludwig Renn thought they all looked like outlandish monkeys.[44] Ernst Jünger claimed that men now had "ghost-eyes" and "fantastic beaks," while Richard Dehmel portrayed the battlefield as a "fantasy of modern technology and industry, filled with insects."[45] Infantryman Otto Ahrends wrote that under the gas mask, "one becomes unrecognizable to the other, and that the gas mask becomes a form of protection much like a magic hood."[46] And finally, Lieutenant Gregor Wörsching's diary featured the most striking evidence of the gas mask's process of dehumanization. When reminiscing about his fallen comrades, Wörsching recounted their lost gas masks, which now served as each soldier's horrifying death mask (Figure 3.2).[47]

Beyond recounting the outlandish image of the newly masked soldier, men frequently described both the color of chlorine gas clouds and their slow haphazard movement across the ground. Infantryman Cornelius Breuninger wrote that "with a strong hiss, the gas escaped and rolled into a widening wedge shape against a small protective trench, a yellow-white cloud, which then melted into the dugout."[48] The men also

[41] Steven J. Main, "Gas on the Eastern Front during the First World War (1915–1917)," *Slavic Military Studies* 28, no. 1 (2015): 109.

[42] Anweisungen und Verfugungen uber die Gasschutzlager, die Ausbildung, den Bedarf von Gasschutzmitteln und den Gaskampf 1915–1917 (LBWK) 465 F 3, 620.

[43] The uncanny visual surrealism of masked combat was further compounded by animal gas masks, which were developed for dogs, horses, and other pack animals.

[44] Hartung, *Großkampf Männer und Granaten!*, 233; Ludwig Renn, *Krieg* (Leipzig: Philipp Reclam, 1979), 201.

[45] Ernst Jünger, *Feuer und Blut* (Magdeburg: Frundsberg-Verlag, 1926), 102. While *Feuer und Blut* is considered one of Jünger's war novels, the text draws substantially on his war diaries; Richard Dehmel, *Zwischen Volk und Menschheit: Kriegstagebuch von Richard Dehmel* (Berlin: S. Fischer Verlag, 1919), 71.

[46] Otto Ahrends, *Mit dem Regiment "Hamburg" in Frankreich 1914–1916* (München: Verlag von Ernst Reinhardt, 1929), 2014.

[47] Robert Sauter, ed., *Ich setze mich nieder und schreibe. Gregor: Eine Lebensgeschichte zum Ersten Weltkrieg* (Plaidt: Cardamina Verlag, 2014), 215.

[48] Frieder Riedel, ed, *Cornelius Breuninger: Kriegstagebuch 1914–1918* (Leinfelden-Echterdingen: Numea Verlag, 2007), Selection from March 1, 1916.

Figure 3.2 German infantrymen wearing their *Linienmasken* in the trenches at Neuve Chapelle, July 1916.
Source. Linienmasken Neuve Chapelle (IZS) Nr 106. Neg. Nr 60/30 aus N: Leibfried

commented on the sheer density of these clouds once they reached their trenches since both the chlorine gas and the construction of the gas mask obstructed and confused their vision. Gas clouds regularly collected in topographic depressions and trench dugouts, and Dieter Finzen wrote of a fog so thick that "we can't see our own hands."[49] Alfred Hermann was better positioned on higher ground, but he still complained of only being able to see ten feet in front of his trench.[50] Because gas attacks were often conducted at night, soldiers further struggled with low levels of light.[51] Figures would materialize and melt away into the dark clouds, thereby emphasizing the spectral quality of the chemically mediated battlefield.

[49] Dieter Finzen, "Kriegstagebuch: Das Tagebuch von Dieter Finzen Aus Beiden Weltkriegen," *The War Diaries of Dieter Finzen* (blog), June 26, 2017, https://dieter-finzen.blogspot.com/search?q=gas, 21 März 1918.
[50] Alfred Hermann Fried, *Mein Kriegstagebuch* (Bremen: Donat Verlag, 2005), 94.
[51] Ludwig Steinmetz Kriegstagebuch 1918 (DTA).

The gas mask only exacerbated this poor visibility. The celluloid eyepieces did not always sit directly over the eyes and they fogged regularly, thus obscuring a soldier's field of vision. Furthermore, mask eyepieces were set on an angle, making depth perception and weapon accuracy incredibly difficult during battle.[52] Given that soldiers also had to wear gas masks in expectation of attacks, its visual hindrances could also impact reading, writing, and other daily tasks.[53]

The unique killing method of weaponized gas added a grisly spectacle to the visual field of the German soldier. Infantryman Hartung described gas-sick men as "miserable bundles of people, laboriously dragging themselves along with crumpled, green-gray faces, vomiting where they went and dying."[54] Ernst Jünger wrote that "some of the bystanders begin to writhe in convulsions, and at last with a jerk, pull off the protective masks to vomit."[55] The retching of gas-sick men enhanced this strange sense of sudden jerking movement that was often present in the soldier's field of vision during gas attacks.

A Deadly Bouquet: The Smells of Gas Warfare

From April 1915 to March 1918, German soldiers increasingly relied on their gas masks for protection against chlorine, phosgene, and various tear gases. While the mask could prove cumbersome and tiring to wear, a properly functioning mask did generally succeed in preventing the inhalation of poison. Up until the very end of the war, most soldier gassings occurred when gas masks were lost, forgotten, or improperly applied. For this reason, veterans with previous chemical warfare experience kept their masks close at hand. These men would disinfect their masks regularly to clean out trace chemicals, thus protecting the rubberized fabric as well as the head straps. During this cleaning process, soldiers applied soap to the eyepieces to prevent eyeglass fogging.[56] They would also make sure that they were packed in their cases in such a manner that they could be quickly removed and applied with the filter on. Even more cautious soldiers would simply wear their gas masks around their necks during any potential enemy engagement. Reflecting on the importance of the gas mask in 1916, one trench journal read, "the soldier is inseparable

[52] Konrad Steuernagel, *Kriegstagebuch 1914/1918 des II. Batls. Ref.=Inf.=Regts. Nr. 221* (Worms am Rhein, 1937), From May 23, 1916.

[53] Wolfgang Foerster, ed., *Wir Kämpfer im Weltkrieg: Feldzugsbriefe und Kriegstagebücher von Franktkämpfern aus dem Material des Reichsarchivs* (Berlin: Neufeld & Henius Verlag, 1929), 495.

[54] Hartung, *Großkampf Männer und Granaten!*, 59.

[55] Jünger, *Feuer und Blut*, 132.

[56] Ibid., 102.

from [the gas mask], and so, too, he leads it into the quiet divide ... the rifle was his original bride, but he has elevated the gas mask to his cousin. This, too, expresses a very intimate relationship, for the soldier usually attends to his cousin as his second bride."[57]

Soldier writings sometimes reported on the successful use of the gas mask and the feelings of relief that accompanied the survival of a gas attack. Infantryman Hans Burr reported the excellent design and construction of the German gas mask on multiple occasions.[58] Furthermore, in a trench journal dialogue entitled "Do you have your gas mask ready to hand," two soldiers named Fritz and Karl expound on the importance of always keeping your gas mask nearby. Karl claims, "I will never be separated from [the gas mask], whether I am in the trench or the grave."[59]

With the increasing diversity of chemical weapons in 1916 and 1917, the scientists at Haber's KWI encouraged soldiers to use all of their senses in detecting gas. Troops were provided with gas sample cases that could be used for detection training.[60] These were large suitcases that contained vials of various chemical weapons and reactive materials that could be used to alert soldiers to the presence of gas in the air. The soldiers would attempt to identify each gas based on its color, its effect on a small patch of skin, and most importantly, its smell.[61] A soldier's sense of smell was particularly important when phosgene was employed because the gas was not visible to the naked eye. The gas sample case claimed that phosgene smelled like musty hay, cut grass, or green corn, although this scent was known to be faint, especially in diphosgene's liquid form.[62] While chlorine clouds could usually be seen, they could be further detected by their strong bleach odor. Several tear gases such as ethyl bromoacetate and xylyl bromide smelled fruity while chloropicrin had the distinct scent of fly paper. Vomiting and sneezing gases were supposed to be odorless but cyanide-based gases smelled of bitter almonds and Lewisite gave off the faint smell of geraniums.[63] Mustard gas was the easiest chemical to detect

[57] *Die Gasmaske: Feldzeitung der Armee Abteilung C.* vol 20 (2) May 20, 1917, p. 1.
[58] Hans Burr, *Das Württemburgische Infatrie-Regiment Nr. 475 im Weltkrieg* (Stuttgart: C. Belsersche Verlagsbuchhandlung, 1921), 25, 54.
[59] "Habt euere Gasmaske stets zur Hand!" *Kriegszeitung der 1. Armee,* 148, April 8, 1918, p. 4.
[60] Dehmel, *Zwischen Volk und Menschheit*, 426.
[61] Carl Zuckmayer, *Second Wind* (New York: Doubleday, Doran and Co, 1940), 86–87.
[62] Denis Winter, *Death's Men: Soldiers of the Great War* (London: Allen Lane, 1978), 122.
[63] Otto Borggräfe, Mein Tagebuch 1914–1915 (IZS) N10.3-7.

via smell since it emitted a strong garlic or horseradish stench once it was aerosolized.[64]

By 1917, these various chemical smells became a regular part of the trenches on the Western Front. The pungent gas aromas were further accompanied by the unfamiliar odors of protective agents such as lime.[65] When enemy gas attacks arrived, soldiers received a fresh burst of such strong smells, and once they had donned their gas masks, these aromas were replaced with a distinctly rubberized scent that could equally turn a man's stomach.[66] This constant discomfort led many soldiers to appreciate the relatively unpleasant odor of a battlefield that was free from poison gas. Fresh air remained a rarer and even more desired luxury.[67] At the same time, a willingness to embrace the smells of chemical warfare could save a soldier's life. While taking deep whiffs of the chemically saturated battlefield was assuredly unpleasant, a discerning sense of smell was a useful ally for survival. Reflecting on this newly desired skill, the soldier Cornelius Breuninger wryly wrote that in 1916, "suddenly everyone has a wonderfully developed sense of smell."[68]

A Cacophony of Silence: The Sounds of Gas Warfare

The sounds of gas warfare were most distinctive in their stark contrasts. In preparation for a gas attack, men would remain as still and silent as possible, listening for the hiss of gas as it was released from cylinders or shells.[69] This pregnant silence heightened the feeling of anxiety over the possibility of an attack by emphasizing the near-silent and insidious nature of chemical weapons. Soldiers certainly struggled to detect the gas and, at both the local and corporeal levels of experience, men feared gas' ability to quietly infiltrate and destroy from the inside out. In his diaries, Ernst Jünger described this sense of impending doom when writing, "In the distance, the ghostly cry of gas echoes through the darkness a few times, but then it becomes very quiet again, and now it almost seems too quiet."[70]

[64] Lilac was also a concerning smell on the battlefield because it was used to cover the odor of mustard gas. Andrew J. Rotter, *Hiroshima: The World's Bomb* (New York: Oxford University Press, 2009), 19.

[65] Carl Werner Müller, *Verzicht auf Revanche: Das Kriegstagebuch 1914/18 des Divisionspfarrers der Landauer Gernison Dr. Anton Foohs* (Speyer: Pfälizische Gesellschaft zur Förderung der Wissenschaften, 2010), 187.

[66] Rudolf Hoffmann, ed., *Der deutsche Soldat: Briefe aus dem Weltkrieg* (München: Albert Langen/Georg Müller, 1937), 435.

[67] Jünger, *Feuer und Blut*, 102.

[68] Riedel, *Cornelius Breuninger*. From January 10, 1916.

[69] Ibid. From March 1, 1916.

[70] Jünger, *Feuer und Blut*, 99.

The association between a pregnant silence and gas could become so strong that the infantryman Paul Zech reversed Jünger's metaphorical description. He wrote, "And suddenly there was silence … a second-long stillness, which completely upset us like a new, fast, and thoroughly deadly gas."[71] On the other hand, the intense silence of creeping gas was often juxtaposed against the chaos of gas detection. The German soldier Albert Sagewitz described the explosive moment of detection, writing: "The gas seems to come out of the ground. It gets closer and closer, and we can hardly see each other anymore. My eyes begin to hurt and tear up. Something swirls in your nose. Gas! – One roars in the other's ears – GAS!"[72] Another artilleryman similarly attempted to recreate the horrible tear in the silence:

> At 4 o'clock in the morning we were suddenly awakened with the cry: Gas! Gas Mask! And then there was the stupid crash of the grenades that flew against our walls. All light in the trenches was immediately doused. You stood in the dark and tried to find your gas mask. In the room, everyone was calling or shouting for fear of death. One man had no mask, the other had no filter, and one more was looking for his wife and children. Everything was chaos.[73]

The cacophony of gas alerts increased in volume and frequency as the war progressed and the German army formalized its gas drills. While some trenches eventually received sirens that would alert soldiers to the presence of gas, most relied on a gas sentry to bang on pots, pans, or other pieces of metal.[74] Hans Zoeberlein wrote: "Gas! Gas! I barked and pulled the mask down over my face – it was already coming in … [the gas sentry] hammered on gongs and bells as if he were mad."[75] Even with all of this sudden noise, some men managed to sleep through what became nearly nightly gas alarms, thereby risking asphyxiation in their sleep.[76]

Once a soldier had applied his gas mask, his auditory sensations became even more foreign. It was difficult to hear noise outside of the mask because the rubberized cotton covered the ears. Instead, soldiers could hear their labored breathing and/or their heart beating at an amplified volume.[77] This sensation could become deafening, thereby increasing the soldiers' sense of isolation and vulnerability.[78] Making matters

[71] Paul Zech, *Von der Maas bis die Marne* (Leipzig: Griefenverlag zu Rudolstadt, 1986), 119.
[72] Hoffmann, *Der deutsche Soldat*, 435. [73] Foerster, *Wir Kämpfer im Weltkrieg*, 272.
[74] Dehmel, *Zwischen Volk und Menschheit*, 422.
[75] Zoeberlein, *Der Glaube an Deutschland*, 199.
[76] Die feindlichen Gasangriffe in der Somme-Schlacht 1916 (BA-MA) PH 5-II/19.
[77] Hoffmann, *Der deutsche Soldat*, 435.
[78] The feeling of auditory isolation under the gas mask is still a concern for contemporary militaries. When commenting on the use of gas masks in the Gulf War, the historian of military strategy Rick Gabriel wrote, "Clothes and hoods that save a man from lethal gas can make him more vulnerable to psychological breakdown. Once you don a chemical suit you increase the isolation of the soldiers; he cannot hear very well, the lenses get

worse, soldiers could not easily hear their officer's orders when coordinated action was essential for collective protection. In an attempt to solve this problem, the KWI created prototypes of gas masks with internal speakers and microphones.[79] These masks were never made widely available and engineers never found an effective solution to the internal beating that reverberated in the soldier's head.

A Burning Sensation: The Feel of Gas Warfare

Soldiers most often associated gas attacks with a sensation of burning or uncomfortable warmth. As a powerful vesicant, mustard gas notoriously created a burning sensation that could lead to inflamed skin, blistering, and blinding.[80] When mustard gas was first used against German troops in late 1917, many soldiers were so confused by the burning sensation on their skin that they assumed that their gas masks had failed and that they were now experiencing an effect of internal poisoning.[81] This same burning feeling was also regularly felt in the eyes, especially when gas masks were slowly applied or unavailable.[82] The then-corporal Adolf Hitler later wrote of his mustard gas exposure at Passchendaele, "Towards morning I began to feel pain. It increased with every quarter of an hour; and at seven o'clock my eyes were scorching as I staggered back and delivered the last dispatch that I was destined to carry in this war. A few hours later my eyes were like glowing coals and all was darkness around me."[83]

It is important to note, however, that the scorching sensation of chemical warfare was not only felt by those exposed to mustard gas. Other gases also produced a burning tightness, primarily located in the chest and throat. After exposure to phosgene, Otto Mallebrein reported feeling "chest pains when breathing, so I could only breathe superficially. The slightest effort to breath faster caused bad palpitations."[84] Chlorine and phosgene also created a heavy sensation in the chest as fluid drained into the lungs.[85] Rudolf Binding described this feeling, writing that

fogged, he cannot see, and he cannot feel his body." Anthony Babington, *Shell-Shock: A History of the Changing Attitudes to War Neurosis* (London: Leo Cooper, 1997), 176.

[79] Armeeoberkommando 6 1917–1918 (BHK) Armeeoberkommando 6, 2427.

[80] Cook, "Creating the Faith," 771.

[81] Anlagen zum Kriegstagabuch Juli 1916–August 1916 (LBWK) 456 F 26, 22.

[82] Reinhold Maier, ed., *Feldpostbriefe aus dem Ersten Weltkrieg 1914–1918* (Stuttgart: W. Kohlhammer Verlag, 1966), 161.

[83] Adolf Hitler, *Mein Kampf*, trans. Marco Roberto (Independent, 2017), 151–152.

[84] Otto Mallebrein, Mein Kriegstagebuch 1914–1918 (DTA). See entry for August 4, 1918.

[85] Cook, "Creating the Faith," 762.

"respiration becomes more and more difficult as the lungs fill with liquid instead of with air, and [soldiers] eventually drowned, so to speak. It is horrible to watch people who have apparently escaped death succumb to this gradual suffocation."[86]

Fears over the painful, burning effects of gas inhalation inspired a distinct soldier appreciation for the basic composition of the atmosphere and for the physical sensations of wind, water, and solar heat. While the German meteorological services and the artillery developed a deep knowledge of pressure fronts and wind speeds, infantrymen also began to watch the weather with increased interest. Up until the first use of gas shells in 1916, static air, wild winds, or a breeze blowing away from the German trenches meant that an enemy gas attack was essentially impossible. Rain, excessive heat, and cold weather could further prevent attacks. However, a slight breeze blowing toward the German trench necessitated more gas sentries and raised general levels of anxiety. Thus, the basic sensations of wind in the hair or water on the skin had far more critical implications for the German soldier between 1915 and 1917, as these atmospheric perceptions could be the difference between life and death.[87]

In the poem "Dulce et Decorum Est," perhaps the most well-known literary depiction of gas warfare, the British poet Wilfred Owen described a gas attack as "an ecstasy of fumbling."[88] In their own writings, German soldiers concurred with Owen's sense of physical clumsiness. They stressed the panicked searching about for their gas masks and the eventual gripping of the rubberized fabric.[89] These were moments of possible physical salvation, decided by the presence of a redemptive object. But the gas mask could similarly demonstrate an ineptitude that exacerbated a man's "ecstasy of fumbling," thus turning feelings of safety and deliverance into a uniquely horrifying experience. The gas mask required a tight fit with the face in order to keep gas out of the space between the wearer's skin and the rubberized material. To achieve this, the German mask had several interior bands that encircled the skull and held the chin. Soldiers were instructed to shave their entire face below the upper lip so that no facial hair would break the mask's seal.[90] This airtight seal

[86] Binding, *A Fatalist at War*, 175.

[87] Andreas Meyer, *It's Not the Fatherland's Fault!? – Keep the Letters for Later: Dr. Otto Meyer, from Gunner to Lieutenant, a German-Jewish War Career* (Andreas Meyer, 2013), 102.

[88] Wilfred Owen, "Dulce et Decorum Est." in *Poems* (New York: Viking Press, 1992).

[89] Jünger, *Feuer und Blut*, 102.

[90] For this reason, the highly stylized beards that previously served as a mark of the German soldier largely disappear from photos taken after 1916. Abwehr von Stinkgeschossen, etc, 1916– (BA) 201-008-001.

created a significant amount of pressure on the wearer's face, particularly around the temples and the bridge of the nose where circulation was often impeded.[91] When the mask had to be worn for hours, these pressure points could become unbearable, forcing soldiers to remove their masks at their own risk.[92] One Albert Sagewitz described the gas mask as a "deceitful instrument," further writing that "all activities become more difficult with the mask. It is true that breathing is heavier and it puts pressure on the chest."[93] Still more soldiers reported frequent headaches while wearing the gas mask (Figure 3.3).[94]

Gas masks could also prove treacherous when they slipped out of position. German military reports claimed that the *Stahlhelm* (German steel helmet) could easily push the gas mask out of position and break the airtight seal.[95] Still more, the mask tended to trap body heat against the head, thereby raising the soldier's body temperature and making the wearer's face sweat. This sweat, which was further elicited by the uniquely stressful nature of gas attacks, could also break the mask's seal. The former soldier Edlef Köppen described this problem, writing, "God, this gas-mask! It is soaking wet from the condensation of my breathing. It rubs against my chin while running and sometimes the water sloshes in my mouth. Ugh! The devil! And then the eyepieces fog over and it looks like everything is covered in mist."[96] Such difficulties could be so bothersome that soldiers would be forced to remove the gas mask in order to carry out strenuous tasks. Ernst Jünger described one such instance:

> I had scarcely got the mask out before I quickly pulled it down off my face. I had run so hard that the air supply through the gas mask was not enough, and I sweated so much that the eye glasses were fogged and completely opaque, so I just tore the mask off. There was a violent stinging in my chest.[97]

Over the course of the nearly three years of intensifying chemical warfare, German soldiers needed to increasingly rely on their gas masks despite these numerous and serious limitations. But in such instances of technological failure, the gas mask appeared rather untrustworthy. Soldiers could never

[91] Geratetechnisches, Konstruktion, Erfindungen, Versuche, Formveranderungen usw. 1933–1936 (BA-MA) RH/12/4.
[92] Bericht der Kaiser Wilhelm Institut fur Arbeitsphysiologie, 1934 (BA-MA) RH/12/4; PH 2 Preussisches Kriegsministerium, 1918 (BA-MA) PH2 282–302.
[93] Hoffmann, *Der deutsche Soldat*, 435.
[94] Albrecht Dorn, eds., *Kriegstagebuch Der 6. Kompagnie 18. Rgl. Sächs.Inf.Regt. Nr. 192* (Leipzig: Buchhandlung Sebr. Fändrich, 1930), 31.
[95] Ibid., 31.
[96] Edlef Köppen, *Heeresbericht* (München: Deutsche Verlags-Anstalt, 2004), 338.
[97] Kiesel, *Ernst Jünger Kriegstagebuch 1914–1918*, 135.

Figure 3.3 A German soldier's lined face after wearing the gas mask. The photograph is from a set of interwar military studies on the effectiveness of the gas mask.
Source. Geratetechnisches, Konstruktion, Erfindungen, Versuche, Formveranderungen usw. 1933–1936 (BA-MA) RH/12/4

assume that they were fully protected from gas; for it was precisely such a naïve technological faith that produced the most horrible deaths. Captain Hermann Geyer described this predicament: "Mask leaks were often difficult for the layman to recognize. Leaky masks, however, were more dangerous than having no mask at all, because if the defect had not been fixed, the mask provided a false sense of security."[98] If the gas mask slipped or the airtight seal was broken, then the mask would begin to fill with gas and asphyxiate the soldier in what had essentially become an individual gas chamber. Thus, the painful pressure that the gas mask put on the soldier's face served as a lasting reminder that while the mask might prove his only salvation, its reliability always remained in doubt.

[98] Hermann Geyer, "Der Gaskrieg," in *Der Weltkampf um Ehre und Recht: Der Seekrieg, der Krieg um die Kolonien, die Kampfhandlungen in der Türkei, der Gaskrieg, der Luftkrieg. Band 4* (Leipzig: Alleinvertrieb durch Ernst Finking, 1922), 507.

Synthesizing the Chemical Subject: Discipline versus Anxiety

In 1918, the German soldier's already fraught relationship to the gas mask changed further as technological failures became more frequent and gas tactics evolved. Since the Second Battle of Ypres, German infantrymen had received instructions on their expected conduct during gas attacks. According to these directives, the men were to remain calm and hold their ground whether the gas was blowing toward or away from them. They were further instructed to don their gas masks quickly and confidently. By 1916, these orders were delivered by the regimental gas officers who received formal training in week-long gas courses held at military training sites back in Germany. In these "gas schools," gas officers studied topics such as the chemical nature of the various weaponized gases, correct tactics for troops under gas, and how to drill soldiers to better resist gas attacks.[99] When they had returned to their companies, gas officers would then conduct monthly lectures regarding new developments in chemical offensives or new tactical strategies for resistance. Ernst Jünger served as one such gas officer, finding great satisfaction in calculating the speeds of gas clouds and surprising his men with gas drills.[100] In these drills, the gas officer would enter a trench dugout and scream "Gas!," at which point all of the men were expected to drop what they were doing and put on their gas masks. They were supposed to be able to don their masks in about fifteen seconds while holding their breath.[101] The gas officer would time them and then proclaim slow soldiers dead.

Not surprisingly, regular drilling did not make gas officers particularly popular with enlisted men. This divide was compounded by the amount of work required of gas officers, leading some to express dissatisfaction with their position. For instance, the sergeant Reinhardt Lewald wrote, "The post brings a lot of work with it, as you have to check the regiment's gas protection equipment daily. You have to ride and run around a lot and then record all of the results."[102] Hermann Boeddinghaus further added, "Today I've checked the fit of the gas mask for all the men in the regiment. This is now my new job. This gas service is quite stupid, but it's just a so-called necessary job."[103] One of the more loathsome duties was the

[99] Hermann Knaüer Tagebuch (DTA) 1221.1.
[100] Kiesel, *Ernst Jünger Kriegstagebuch 1914–1918*, 49.
[101] Anhaltspunkte fur die Prufung des Standes der Ausbildung in der Gasabwehr bei Besichtigungen Ubungen usw (BA-MA) RH/11/IV.
[102] Reinhardt Lewald Kriegstagebuch 1914–1918 (DTA).
[103] Ingrid von der Dollen, ed. *Briefe an die Eltern von Hermann Boeddinghaus 1898–1941* (Bad Honnef: KAT-Verlag, 2009), from August 11, 1918.

requisition or building of gas chambers. The gas officers would either create small airtight shacks or demand the temporary use of troop dugouts in which they could disperse tear gas for training purposes. When requisitioning a dugout, the gas officers often met strong resistance from the men who preferred to keep their quarters gas-free.[104] Once the gas officer had found a suitable gas chamber, however, he would order the men to don their gas masks and enter the room. When inside, the men would then perform mild exercise for ten minutes to test the effectiveness of their masks.[105] Comically expressing their discomfort with this drill, soldiers call it the *Stinkkurs* (stink course), while the gas chamber was the *Stinkbude* (stink shack).[106] Infantryman Richard Dehmel caustically wrote:

> Around noon, the gas officer … came by and confused us brave soldiers with highly distressing questions. Of course, he found fault with all sorts of things, otherwise his arrival would have been pointless. Finally, the "holy man" issued a harsh command that from now on gas drills had to be performed regularly on the front lines, even if accidents might occur.[107]

The increased frequency of drills was a result of the KWI's belief that the most disciplined army would best be able to resist chemical weapons. Once it was clear that the chlorine cylinders would not win the war for Germany, Fritz Haber began to argue that he simply needed more time to produce a definitive chemical weapon. An extension of the chemical war required that German soldiers successfully resist cylinder attacks through personal determination and the correct use of the gas mask. Thus, while commanders and scientists seemed to believe that the gas mask provided ample protection against chlorine and phosgene gas, the survival of the German army still required compliant men. As the gas mask manufacturer Bernhard Dräger put it, "Half-understood or unorganized respiratory protection is worth less than none; it is incredibly dangerous. The best respirator cannot change that because its proper application depends on the human being."[108]

The demand for gas discipline was also part of a larger change in German tactics that followed the massive loss of life at Verdun and the Somme. In late 1916, German military tacticians wrote both the "Principles of Field Construction" and the "Principles of Command in the Defensive Battle in

[104] Wilhelm Wittekindt, "Meine Kriegserlebnisse 1916–1918" (DTA) 1229.
[105] Die Kriegstagebücher des Sebastian Heinlein, 11 Juli 1916 (IZS) N04.3-6.
[106] Theodor Imme, *Die deutsche Soldatensprache der Gegenwart und ihr Humor* (Dortmund: Fr. Wilhelm Ruhfus, 1917), 117.
[107] Dehmel, *Zwischen Volk und Menschheit*, 424.
[108] *Dräger-Vorträge: Atemschutz Wiederbelebung*, 7.

Positional Warfare."[109] As part of these larger tactical manuals, the "Infantry Training Guidelines" stressed the need to quickly train frontline soldiers in the new ways of industrialized warfare. The guidelines favored what they called "combat training," meaning that soldiers would receive their instruction through exposure to frontline battle. German tacticians argued that this would instill personal discipline and force soldiers to continue fighting even under extreme fatigue.[110] When put in mortal danger, soldiers would have no choice but to continue their struggle even without an officer's orders, relying solely on themselves and their comrades. Then, after a certain amount of time at the front, soldiers could be cycled back behind the lines for more formalized drilling. Historian Michael Geyer described this as "the development of a military practice in which the conscripted soldier is remade into an instinctive warrior."[111]

These new tactical manuals further stressed the remobilization of the Western Front. To achieve this, they argued that German units should no longer hold a line but rather control the battlefield through small counterstriking encounters that aggressively employed new military technologies such as poison gas.[112] This, again, put great responsibility on small groups of stormtroopers, who would now act much more independently on the battlefield. Commander resistance to this loss of hierarchical control delayed the adoption of these new tactics, but with dwindling manpower and supplies in 1918, the Germans began to implement these measures with surprising success. While the Germans could not match Allied offensive firepower in 1918, they could put up ferocious resistance in retreat. At the same time, however, stranding men in warzones created extreme amounts of mental stress, which was amplified by harassing warfare technologies such as mustard gas shells. Continuous gas alarms stretched soldiers' nerves to the breaking point and even the slightest whiff of gas could lead to panic.[113] Most men could not psychologically handle this form of fighting, and desertion and trauma cases spiked.[114] Those few soldiers, like Ernst Jünger, who could

[109] Michael Geyer, "German Strategy in the Age of Machine Warfare, 1914–1945," in *Makers of Modern Strategy* (Princeton: Princeton University Press, 1986), 539–340.

[110] Michael Geyer, "How the Germans Learned to Wage War: On the Question of Killing in the First and Second World Wars," in *Between Mass Death and Individual Loss: The Place of the Dead in Twentieth-Century Germany* (New York: Berghahn Books, 2008), 32.

[111] Geyer, "How the Germans Learned to Wage War," 33.

[112] Matthias Strohn, *The German Army and the Defense of the Reich: Military Doctrine and the Conduct of the Defensive Battle 1918–1939* (Cambridge: Cambridge University Press, 2011), 60–61.

[113] Ben Shephard, *A War of Nerves: Soldiers and Psychiatrists in the Twentieth Century* (Cambridge, MA: Harvard University Press, 2000), 64–65.

[114] Geyer, "How the Germans Learned to Wage War," 37.

withstand and sometimes even thrive in this style of warfare became the first of a new type of mobile infantryman.

Physical and psychological resistance to chemical weapons was a necessary requirement for this newly envisioned soldier. Changes in gas shelling tactics and a remobilized battlefield meant that men could be living amidst gas for hours, days, or weeks on end.[115] For instance, infantryman Konstantin Kraatz wrote on July 20, 1918, "We are bombarded with gas grenades every day. Nevertheless, we have no losses. There is no sleep, food is not available. From this terrible situation, we come out in the evening at 23 o'clock."[116] This life of regular gassings required that men either be continuously prepared for a gas attack or that they permanently wear their gas masks. Reflecting on this situation, Carl Zuckmayer wrote:

> A man in the front-line trenches has a peculiar way of sleeping ... he sleeps with 'all his senses,' like an animal whose complete relaxation is not ruffled by reflexes from his blacked-out consciousness ... A thin, shrill whistle rapidly swelling to an unbearable pitch, suddenly followed by the dull blast of an explosion. The man springs with both feet from his bunk; his left hand darts for his gas mask, while his right hand snatches his rifle from the wall.[117]

As Zuckmayer makes clear, by 1918, the gas mask was no longer just a part of every soldier's immediate tools. In many ways, it had become so omnipresent that it was now part of the new soldier's metaphorical body.

On the other hand, soldiers who failed to meet the demands for constant preparedness often began to reevaluate the value of their gas masks. One Fritz Ebeling described a scene in which men lost faith in the mask's ability to provide technological salvation, writing: "The units break apart and the scattered divisions are not found again in the darkness. The failure of the officers and NCOs adds to the confusion. The onward march is delayed since advancing with gas masks is cumbersome. Some men take off the mask and get sick on the gas."[118] Indeed, as Ebeling suggested, it was not entirely uncommon for men to intentionally breathe small amounts of gas or to wipe mustard agent in their eyes rather than continue fighting. The gas itself now became a potential ally in attempts to earn several days in a field hospital.[119]

[115] "Die Schlacht an der Somme im Monat Juli: Erster Teil," *Militär Wochenblatt* no. 39 (1916): 866.
[116] Konstantin Kraatz, Kriegstagebuch 1916–1919 (DTA).
[117] Zuckmayer, *Second Wind*, 45.
[118] Fritz Ebeling, *Geschichte Des Infanterie Regiments Herzog Friedrich Wilhelm von Braunschweig Nr 78 Im Weltkriege* (Berlin: Gerhard Stalling Verlag, 1924), 161.
[119] "Literatur, Kriegserfahrungen, Auslandsnachrichten" (BA-MA) RH/12/4; Sauter, ed., *Ich setze mich nieder und schreibe*, 266.

While gas-disciplined veterans often survived gas attacks in 1916 and 1917, the attacks of 1918 claimed men regardless of battlefield experience.[120] It was not general battlefield knowledge that could save a man in 1918 but rather a subjective acceptance of a world in which fresh air could no longer be taken for granted. On this new battlefield, every place and moment was potentially steeped with deadly poison. As one German gas officer now complained, "The troops were trained to breath under the mask for a short time, but they were not trained to 'live' under the mask, as the war required."[121] The men who could survive best in this environment embraced the gas mask as part of their physical body, thus seemingly exemplifying the first envisioned "chemical subjects" and accepting the chemically precarious nature of life amidst poison gas. It is not surprising that in 1918 these new soldiering bodies were often described as supple, lean, and most importantly, machine-like.[122] The German gas scientist Rudolf Hanslian triumphantly described the culling process that helped to identify such soldiers:

> Chemical weapons placed the highest requirement on the morale and intelligence of the soldier ... it separated the wheat from the chafe, the men that maintained gas discipline and accomplished their combat mission from the inferior masses that collapsed mentally and gave up. Thus, the gas narrowed the circle of people who could survive in a modern struggle into a very special group.[123]

As Hanslian's quote would suggest, German tacticians began to lionize those who could survive the desperate fighting of 1918. This combat was envisioned as a crucible in which the most battle-hardened men were molded. Similarly, in the context of chemical warfare, poison gas served as a collective physical and psychological test, and those who survived it were described as the best and most fully formed modern soldiers.

The Pale Horse and the Death's Head: The Symbolic Importance of Gas Warfare

Monographs surveying World War I often discuss poison gas in small sections that deal with the broader technological changes that occurred throughout the war. This is often justified by the fact that under 5 percent

[120] Fritz Rider, Mit 18 als Freiwilliger an die Front 1916–1918 (DTA); Karl Minning Kriegstagebuch von 1916 (DTA) 1.
[121] Kriegserfahrungen im Gasschutz 1919 (BA-MA) PH 14/230.
[122] Joanna Bourke, "The Body in Modern Warfare: Myth and Meaning 1914–1945," in *What History Tells: George L. Mosse and the Culture of Modern Europe* (Madison: University of Wisconsin Press, 2004), 203–204.
[123] Hanslian, *Der chemische Krieg*, 200.

of war casualties were caused by gas. However, some historians concede that gas should not be evaluated by its military effectiveness but rather by its psychological impact.[124] This is certainly true, but the psychological (let alone embodied) changes that gas produced throughout the war cannot be given justice in several paragraphs.[125] Given that soldiers often claimed to fear artillery and gas more than any other weapons, it is clear that gas was unique from other new military technologies.[126] Further highlighting the difference between traditional artillery fire and gas, historian Tim Cook asserts that soldiers could bring themselves to sleep through explosive shell bombardments but not through gas attacks. He writes, "Gas destroyed the idea of chance, and that was its terrifying nature. There is a considerable difference between crawling on a trench floor or advancing under fire and praying that a shell or bullet does not get you, as opposed to simply waiting for a death cloud to pass with certainty over you."[127]

Beyond the absence of individual agency in gas attacks, the weapon also inspired a deep sense of personal injustice. When the German military employed gas, most soldiers were able to distance themselves from the decision-making and implementation processes carried out by commanders, engineer corps, and artillerymen. Soldiers in the trenches could justify their army's use of gas based either on the idea that it was retribution for Allied violence or merely the grim reality of war. However, when the Allies employed gas against the Germans, soldiers often lost all conception of their own army's actions, viewing themselves solely as helpless victims.

Finally, as philosopher Peter Sloterdijk has pointed out, chemical weapons had the unique aim of poisoning the enemy's entire environment – air, soil, and water.[128] Gas could theoretically usurp life-preserving air and turn it into inescapable death, creeping undetected into the most intimate spaces.[129] Thus, gas threw all previous conceptions of the rules of war into question. The all-encompassing and unruly

[124] Jörn Leonhard, *Pandora's Box: A History of the First World War*, trans. Patrick Camiller (Cambridge, MA: Harvard University Press, 2018), 265–266; Alexander Watson, *Enduring the Great War: Combat, Morale and Collapse in the German and British Armies, 1914–1918* (Cambridge: Cambridge University Press, 2008), 33.

[125] Joanna Bourke, *Fear: A Cultural History* (London: Virago Press, 2005), 210.

[126] Winter, *Death's Men*, 121–125. [127] Cook, "Creating the Faith," 779.

[128] John Keegan, *The Face of Battle: A Study of Agincourt, Waterloo, and the Somme* (London: Pimlico, 2004), 307–308.

[129] The German soldier Georg Bruchmüller writes of men who attempted to seal the cracks in the walls of their trench to keep out gas. The gas pushed through the sealed cracks and penetrated their gas masks, asphyxiating them in their quarters. Georg Bruchmüller, *Die Artillerie beim Angriff im Stellungskrieg* (Berlin: Offene Worte, 1926), 166.

nature of the weapon again forced soldiers to carefully reconsider their own agency in warfare.[130] Literary scholar Samuel Hynes writes:

> The minds of the fighting men changed ... Soldiers came to war believing that individual wills would have a role there, that what a man did – his decisions, his actions – would affect whether he lived or died. [Gas] made death accidental. It wasn't the violence of the power or the cruelty of those weapons that made the war different; it was the vast randomness and anonymity of their ways of killing.[131]

The lack of individual control over gas warfare most clearly manifested itself in the metaphors that soldiers chose to describe their fear and anxiety.[132] Scholar Mary R. Habeck has argued that soldiers discussed warfare technologies using a combination of supernatural (monstrous or demonic), subhuman (machine-like, or force of nature), or human (bodily, anthropomorphic, or animalistic) metaphors. Supernatural metaphors were most commonly used for technologies over which men had no control, while subhuman metaphors expressed some level of control, and human metaphors near total control.[133] As the memoirs, diaries, and letters in this chapter have shown, gas was often described in supernatural terms. By 1918, gas was best personified by death on his horse, come to reap wild and aimless as one of the four horsemen of the apocalypse. Using such spectral language, the soldier Franz Schauwecker wrote that, "gas in its eerie silence and ghostly cloudiness smothered our courage like cloth."[134] Contrasting passages most frequently came from soldiers in technical specialties (e.g., gas services, artillery, engineering corps) who could appreciate these deadly technologies for their own sake.[135] For instance, the gas officer Ernst Jünger often thought in objective terms, writing, "Now I got my first mouth full of gas. Aha! Chlorine! One breath was enough to bring back memories of the physics lessons of yesteryear, so mask on!"[136]

[130] W. M. K. Pfeiler, *War and the German Mind: The Testimony of Men of Fiction Who Fought at the Front* (New York: Columbia University Press, 1941), 79; Santanu Das, "'An Ecstasy of Fumbling': Gas Warfare, 1914–18 and the Uses of Affect," in *The Edinburgh Companion to Twentieth-Century British and American War Literature* (Edinburgh: Edinburgh University Press, 2012), 398.

[131] Samuel Hynes, *The Soldiers' Tale: Bearing Witness to Modern War* (New York: Penguin Press, 1997), 56.

[132] Poison gas could blur fear and anxiety because, while it created a distinctive and objective threat in the moment of attack, the anticipation of gas was often more oppressive. Bourke, *Fear: A Cultural History*, 188–189.

[133] Mary R. Habeck, "Technology in the First World War: The View from Below," in *The Great War and the Twentieth Century* (New Haven: Yale University Press, 2000), 102.

[134] Franz Schauwecker, *So war der Krieg* (Berlin: Frundsberg Verlag, 1929), 25.

[135] Habeck, "Technology in the First World War: The View from Below," 118.

[136] Kiesel, *Ernst Jünger Kriegstagebuch 1914–1918*, 135.

Varying levels of technological trust are still better demonstrated in passages detailing the gas mask. Even closer to the soldier's body, the mask was often more vividly described through sensorial language. Reflecting on the importance of the gas mask to soldier interpretations of chemical warfare, the gas specialist Rudolf Hanslian wrote that "the soldier's emotional impressions of gas were much less due to the novelty or sensation of the gas itself, than to the feeling of a lack of adequate gas protection."[137] Among others, the historian Eric Leed and the literary scholar Santanu Das have expressed the importance of embodied language for understanding soldier writings. Leed writes, "The impact of the war upon the human sensorium of combatants is the point where one must begin to understand the necessity of illusion, fantasy, and myth."[138] In this case, mediated language expressed the changing bodily experience with the gas mask from 1915 to 1918 and revealed the ways in which the mask seemingly conspired to discipline the soldier's body.

While the mask was essential to survival, and it did in fact save many lives, it was also a constant physical and mental burden. Always at a soldier's side or neck, it was incessantly hugging his body and reminding him of an ever-present danger.[139] For the men who could endure this, the mask either retained human metaphors and became part of their own bodies or it was described in strictly technological terms.[140] The far greater number of men who could not withstand the strict discipline of the mask turned again to supernatural language. As the German gas mask increasingly failed due to the intensification of chemical shelling and the military's material shortages in 1918, the mask was less frequently described as a technological device; nor was it a pig snout, monkey face, muzzle, carnival veil, or bug eyes. At war's end, these soldiers rather perceived the gas mask as a death's head, an omen of what was inevitably to come.[141] As an article from the military journal *Protar* would later point out, "gas masks emerged as horrible, scary-looking devices that appeared to many as the symbol of all the horrors of the modern age."[142]

[137] Hanslian, *Der chemische Krieg*, 200.

[138] Eric J. Leed, *No Man's Land: Combat and Identity in World War I* (Cambridge: Cambridge University Press, 1979), 117; Santanu Das, *Touch and Intimacy in First World War Literature* (Cambridge: Cambridge University Press, 2005), 23.

[139] For a contemporary study of this phenomenon, see Kenneth T. MacLeisch, "Armor and Anesthesia: Exposure, Feeling, and the Soldier's Body," *Medical Anthropology Quarterly* 26, no 1 (March 2012): 49–68.

[140] Susan Buck-Morss describes these highly disciplined and unfeeling bodies as "anesthetic" subjects. Susan Buck-Morss, "Aesthetics and Anaesthetics: Walter Benjamin's Artwork Essay Reconsidered," *October* 62 (1992): 33.

[141] Kiesel, *Ernst Jünger Kriegstagebuch 1914–1918*, 271.

[142] J. P., "Die Erste Gasmaske," *Protar* 11, no. 3 (1936): 200.

4 The Limits of Sympathy: The Medical Treatment of Poison Gas during and after World War I

After dropping out of medical school in 1913, the expressionist poet Johannes Becher joined several communist organizations and actively spoke out against the stated German war aims of World War I. Thirteen years later, as a continuation of his antinationalist and anti-capitalist activism, Becher published the novel (CHCI = CH)3 (*Lewisite*). Utilizing a collage of modernist literary techniques, the book condemned the saber rattling of World War I through a series of loose vignettes, fictional newspaper reports, and other media of the modern cityscape. Like many fellow communists, Becher saw war as a product of modern capitalism, claiming that the conflict's technological arms race only served to line the pockets of German industrialists. As such, he inculpated all those associated with German applied science and industry in the oppression of the working classes. This included German doctors, who he saw as the anesthetists of the modern affliction. In describing the role of the military hospital, Becher wrote:

> In the soldier's language, the military hospital was called the "poison scales" and it mainly housed gas-sick men. There was also a separate lunatic asylum, but the patients in the two wards did not differ significantly, especially the paralytics and the stutterers. It was the devastation of their nervous system during the war that triggered the paralytic attack. Remissions rarely occurred due to the adverse nature of the condition and the inept treatment. Thus, they flowed into the graves like waterfalls.[1]

Inspired by his own personal struggles with drug use, Becher further wrote that the doctors carelessly "splashed" their patients with morphine and cocaine to both dull their pain and turn them into more pliable subjects.[2]

At the time of publication, Becher's novel was largely considered unsuccessful due to its lack of popularity and its often ham-fisted literary

[1] Johannes Becher *(CHCl=CH)3As (Levisite) oder Der einzig gerechte Krieg* (Berlin: Aufbau-Verlag, 1968), 45.

[2] Ibid., 146–147.

techniques. However, the book's prosaic disappointments were the least of Becher's concerns, as its publication resulted in his 1926 indictment for "literary high treason" against the Weimar Republic. While Becher's novel remains an unapproachable and highly stilted read, its reproach of the medical profession in World War I is rather distinctive. In particular, the novel's focus on both the treatment of gas-sick men and the handling of neuroses speaks to an understudied history of the medical treatment of gassed World War I soldiers and veterans.

As a central feature of the new *Materialschlacht* (war of materials), poison gas led to around 107,000 German casualties over the course of the war, thus presenting medical professionals with previously unseen injuries and deaths.[3] Beyond the fact that gas poisoning was a fairly new injury on the battlefield, continually developing battle gases led to different health problems that required differing treatments.[4] Furthermore, as an often invisible inhalant, gas did not always produce very clear external symptoms, leading to particular difficulty in diagnosis. With little prior knowledge of Fritz Haber's new chemical weapons, overtaxed doctors and nurses could do very little to treat the countless different symptoms that presented themselves.

Consequently, army hospitals often became sites of medical experimentation and testing, potentially exciting those doctors who saw gas warfare as a new frontier for medical fields, such as pulmonology, cardiology, and toxicology. It is quite clear that, as the historian Wolfgang Eckart has pointed out, the development of poison gas was always closely tied to the attempts to heal or protect against its effects.[5] Beyond the physiological symptoms of gas exposure, gassed soldiers also increasingly displayed what was termed "gas neurosis." As a subgroup of war neurotics, men experiencing "gas neurosis" faced similar censure for their inability to cope with the mental stresses of modern warfare. However, the uniqueness of gas gave a distinctive quality to the medical discussion surrounding these cases. According to authoritative German doctors, the cause of gas neurosis was not just weak nerves or a fragile psychological makeup but also the incapacity to internalize gas discipline. Due to their regular inability to treat new gas symptoms, doctors increasingly relied on the effectiveness of the gas mask as a disclaimer for their own medical insufficiencies. Medical insistence on individual composure and internalized

[3] German field hospitals treated approximately 69,212 gas cases. Literatur, Kriegserfahrungen, Auslandsnachrichten, 1934–1936 (BA-MA) RH/12/4.

[4] Heather R. Perry, *Recycling the Disabled: Army, Medicine, and Modernity in WWI Germany* (Manchester: Manchester University Press, 2014), 1.

[5] Wolfgang Eckart, *Medizin und Krieg: Deutschland 1914–1924* (Paderborn: Ferdinand Schöningh, 2014), 77.

self-control created a distinct discourse surrounding poison gas that then entered wider Weimar society as a normalized condemnation of the constitutional failures of gassed veterans. Furthermore, such claims served as a reassertion of the importance of collective gas discipline for a national public that was increasingly concerned with the possibility of large-scale chemical attacks on civilian population centers.

Overburdened and Underprepared: Treating Gassed Soldiers in World War I

Prior to the outbreak of war, the German scientific world did maintain significant knowledge of the pharmacological effects of chlorine and phosgene gas. Much of this knowledge was derived from the study of industrial accidents at the major production facilities of chemical companies. These studies were supplemented by animal experiments conducted in the late nineteenth century by firms such as Bayer and BASF.[6] Through such testing, scientists were able to record both the lethal doses for such gases and their physiological effects. However, much of this knowledge remained strictly within the realm of the chemical industry, since there was no assumption at the time that large numbers of people would ever be exposed to chlorine or phosgene. Consequently, most German physicians knew little about the medical research of chemical companies and had no familiarity with treating poison gas exposure.

While information on poison gas was eventually funneled through Fritz Haber's KWI, the 8,986 German medical doctors initially mobilized for military service had scarce knowledge of what they were about to face.[7] After the Second Battle of Ypres, frontline doctors now scrambled to treat hundreds of men with blue faces, shortness of breath, chest pain, headache, and bloody expectoration.[8] In April and May of 1915, the field hospitals treated about 200 Germans who complained of gas sickness.[9] These men were given oxygen treatment from Dräger Pulmotors, or suitcase-sized boxes that could pump air into a small breathing mask.[10] Unfortunately, military doctors did not have enough Pulmotors or oxygen for all of the gas-sick in the spring of 1915.[11] This was partly due to the fact that severely afflicted soldiers would often be treated with Pulmotors for hours at a time.

[6] Kranken Arztenwesen Korperbeschadigte, 1915– (BA) 231.012.
[7] Niko Amend, "Die Sanitätsdienstliche Versorgung der gasversehrten Deutschen Soldaten des Ersten Weltkrieges," Dissertation (2014), 106.
[8] Dehmel, *Zwischen Volk und Menschheit*, 424–426.
[9] Literatur, Kriegserfahrungen, Auslandsnachrichten, 1934–1936 (BA-MA) RH/12/4.
[10] Generalstabes des Heeres on Gas, 1915 (BA-MA) PH3/252.
[11] Verwaltungsunterlagen, 1917–1918 (LBWK) 456 F 114, 111.

Pulmotor shortages forced some of the moderately ill to share the devices, while others were simply placed in a well-ventilated area and told to breathe deeply. Beyond supplying men with available oxygen, doctors had little knowledge of subsequent treatment. In desperation, they often gave the men adrenaline or other asthma medications, assuming that the gas induced bronchial muscle cramping.[12] Other doctors administered tetanus shots and/or intravenous salicylic acid in the hope that these treatments would thin the mucus and reduce fever.[13] Some even attempted to bleed the soldiers in an effort to improve oxygen circulation by draining the veins and/or lymph nodes of the poison.[14] In fact, many frontline doctors initially doubted that the Pulmotors' provision of oxygen helped gassed men and they tended to prefer bloodletting and ammonia treatments.[15] The medic Theodor Zuhöne wrote of the desperate situation: "Among our gas patients, we have a lot of serious cases. That's why I have a lot of work to do: oxygen inhalations, intravenous injections, bloodletting."[16]

Further complicating any form of treatment, gas-sick soldiers tended to be difficult to diagnose upon first arrival in the field hospital. Afflicted men presented varying internal symptoms, including irritation of the mucous membrane, blood thickening, infection, and lung pain. These differing symptoms could reflect a given soldier's cardiac health, or the different amounts of gas inhaled and the varying time since exposure. Some men who had been lightly gassed could remain on the battlefield for a few hours to a few days, later coming into the hospital complaining of dizziness, shallow breathing, and/or headaches. On arrival, they were required to give an account of their gassing that could sufficiently convince doctors that theirs was not a case of malingering.[17] If their story was believed, then these men would be given a fresh bed and a hot water heater if available. They would remain in the hospital for a few days while being encouraged to breathe fresh air. They might also receive warm liquids or schnapps to fight off any coughing attacks and to "boost the spirits."[18]

[12] Zur Kenntnis und Behandlung der Gasvergiftungen, 1917 (LBWH) M 635/1 Nr 963.

[13] Tetanus shots and salicylic acid treatments actually tended to produce favorable results. Verwaltungsunterlagen, 1917–1918 (LBWK) 456 F 114, 111.

[14] Gasschutzwesen, Allg.u.Besonderes, 1915–1918 (BHK).

[15] Mark Harrison, *The Medical War: British Military Medicine in the First World War* (Oxford: Oxford University Press, 2010), 108.

[16] Jürgen Kessel, ed., *"Jetzt gehts in die Männer mordende Schlacht..." Das Kriegstagebuch von Theodor Zuhöne 1914–1918* (Damme: Heimatvereins, 2002), entry for July 31, 1918.

[17] Harrison, *The Medical War*, 109.

[18] Walter Schmidt, *Das 7. Badische Infanterie-Regiment Nr. 14 im Weltkrieg 1914/18* (Freiburg im Breisgau: Herder Verlag, 1927), 100.

While this was the standard care for most lightly gassed men, some soldiers exhibited more serious symptoms and received differing levels of treatment. For instance, after inhaling a small amount of gas while sleeping, the infantryman Otto Mallebrein wrote:

> The following morning, I had chest pains when I breathed … After about two hours, the condition had not improved, so I moved with slow, small steps … to the medical dressing area. The doctor there explained that I was seriously ill and he forbade me from moving any further. However, then he allowed me to crawl back into my trench dugout … I was left to lie about for three days, gripping my gas mask in case of a new gas attack.[19]

Although Mallebrein's gas inhalation appears quite serious, it paled in comparison to soldiers who breathed in significant amounts of poison gas. Typically, these men immediately began to choke and writhe on the ground. In instances where these men survived the first minutes of exposure, stretcher bearers were ordered to swiftly remove them from the field so that they could be provided oxygen through an artificial respirator. Once in the field hospital, seriously ill soldiers were made to inhale alcohol to fight potentially deadly coughing attacks and given caffeine to stimulate cardiac activity before they were allowed to rest in the base hospital for three to four weeks.[20] A hospital report from 1916 stated that "All [gas] patients were initially conscious. Increasing weakness, little and irregular pulse, stinging on the chest, anxiety, severe coughing, spitting blood."[21] Regardless of any initial relief of symptoms, these seriously ill soldiers who made it to the medical tent frequently died two to three hours later as their lungs slowly dissolved and filled with liquid. If they did survive initial exposure, they then often experienced nausea, vomiting, general weakness, dizziness, and sometimes diarrhea. Thus, they required near-constant medical attention from available nurses. Two to three days later, the survivors again faced mortal danger, as pneumonia often set in.

Soldiers who breathed nearly pure concentrations of chlorine or phosgene gas tended to immediately asphyxiate in the field.[22] Even if they survived for up to an hour, they often subsequently succumbed to cardiac arrest. Based on first-hand medical reports, these men were difficult to behold, as their breathing could increase to 40, 60, or 70 breaths and their pulse to 160–180 beats per minute. Their face and body would

[19] Otto Mallebrein Mein Kriegstagebuch, 1914–1918 (DTA) Sig 3814.
[20] Zur Kenntnis und Behandlung der Gasvergiftungen, 1917 (LBWH) M 635/1 Nr 963.
[21] Analgen zum Kriegstagebucher, April 1916–Juni 1916 (LBWK) 456 F 50, 145.
[22] Anlagen-Englischer Gasangriff, Juli 1, 1917–September 30, 1918 (LBWH) M 411 Bd. 1276.

swell and slowly turn a mixture of blue and red, while their thorax and diaphragm would convulse as their lungs expanded with foamy yellowish fluid.[23] As fellow soldiers transformed into "slimy creatures," witnesses confronted the horrible sights of "white eyes writhing in [the] face" and "blood ... gargling from the froth-corrupted lungs."[24] The doctor Alfred Schroth described such cases as a "sight of the greatest horror," in which the afflicted men began to violently cough, spit up blood, and then die of a pulmonary edema while "almost fully conscious."[25]

Further reflecting the lack of medical knowledge at the start of the chemical war, doctors only began to record gas sickness in January of 1916. In military medical reports and monthly hospital records, medical practitioners recorded gassed soldiers under No. 39 "Other Diseases." About 2,000 German men fell into this category in the first year of gas warfare.[26] As the war progressed, categorization continued to be a problem, as new gases presented vastly different symptoms, and other bronchial ailments caused by the common cold and influenza proved difficult to distinguish. After receiving gas training and treating gas-sick soldiers at the front, the physician Hermann Büscher wrote of his frustration, "When I was a battalion doctor on the Western Front ... I helped the soldiers as best as I could. I was very uncertain about gas diseases, so I often breathed a sigh of relief when the gas patients were taken away."[27]

The most difficult diagnoses were those soldiers who were exposed to gas and reported to the field hospital with no physical symptoms. Doctors had little time to fully inspect these men, leaving nurses to observe them for twenty-four to thirty-six hours. Even the worst cases could potentially take time to develop, and gas-exposed men could begin to asphyxiate at any moment. On the other hand, the psychological fear of being gassed led some soldiers to overestimate their level of gas exposure, and there were cases in which men never experienced any symptoms while under supervision.[28] Given the difficulty of categorization and the importance of timely treatment, the coordination of medical

[23] Vorschriften uber Gasschutz, 1931–1933 (BA-MA) RH/12/4.

[24] Shephard, *A War of Nerves*, 63. [25] Eckart, *Medizin und Krieg*, 80.

[26] Carbon monoxide poisoning was also common during artillery bombardments, further complicating this medical categorization. Literatur, Kriegserfahrungen, Auslandsnachrichten, 1934–1936 (BA-MA) RH/12/4.

[27] Hermann Büscher, *Grün- und Gelbkreuz: Spezielle Pathologie und Therapie der Körperschädigungen durch die chemischen Kampfstoffe der Grünkreuz- (Phosgen und Perchlorameisensäuremethylester [Perstoff]) und der Gelbkreuz-Gruppe (Dichloraethylsulfid und β-Chlorvinylarsindichlorid [Lewisit])* (Johann Ambrosius Barth Verlag: Leipzig, 1932), 8.

[28] Vorschriften uber Gasschutz, 1931–1933 (BA-MA) RH/12/4.

care was essential in gas cases.[29] Even stretcher bearers and ambulance drivers were provided with some level of medical knowledge about gassed soldiers. By 1917, every company maintained a gas rescue squad comprised of one noncommissioned officer, two ambulance workers, and three substitutes. The NCO was further required to attend a gas course and familiarize himself with gas rescue procedures.[30]

Even when doctors finally created a series of best practices for treating chlorine gas exposure, they were then confronted with the effects of phosgene, mustard gas, Clark, and a score of other battle gases. Phosgene gas produced effects that were most similar to chlorine, although it required less exposure to create serious injuries. Conjecture swirled around the nature of these more devastating pulmonary gas symptoms, leading doctors to believe that phosgene and chlorine exposure regularly brought on colds, influenza, or even latent tuberculosis.[31] In fact, some medical professionals hypothesized that the influenza outbreak of 1917–1918 was a product of mass gas exposure, rather than poor hygienic conditions. The irritating effects of Blue Cross gases such as Clark I and II were temporary and nonlethal, but there were also fairly isolated reports of paralysis and the loss of "mental control."[32]

Mustard gas required yet another entirely different set of treatments. The effects of the gas were usually felt two to three hours after exposure when men would normally begin to sneeze. Subsequently, their eyelids would swell, mucus would start to drip from their noses, and their throat and skin began to burn. Mustard gas patients could further experience severe headaches, an accelerated pulse, a rise in temperature, and the formation of red patches on the skin. Over the course of twenty-four hours, these red patches would eventually bubble up into blisters and any exposure to the eyes could lead to permanent blinding.[33] If and when these initial symptoms subsided, many soldiers subsequently developed pneumonia.

Medics and stretcher bearers were provided with rubberized suits in order to immediately treat men who had been exposed to mustard gas in the field.[34] The medics would undress an afflicted soldier before carrying him back to the field hospital because any residual mustard agent brought a severe risk of contamination. Upon arrival, orderlies and nurses would then shear off the soldier's hair and bathe him with hot

[29] Akademisches Wissenschaftliches Arbeitsamt, *A.W.A. Gasschutz-Lehrgang* (Berlin: Akademisches Wissenschaftliches Arbeitsamt e.V, 1932), 48.
[30] Literatur, Kriegserfahrungen, Auslandsnachrichten, 1934–1936 (BA-MA) RH/12/4.
[31] Knack, "Kampfgasvergiftungen," *Deutsche Medizinische Wochenschrifft* 43 (1917): 1246.
[32] Jones, "Terror Weapons," 6. [33] Winter, *Death's Men*, 122.
[34] Rubberized suits did not offer complete protection because sufficient quantities of mustard gas could dissolve both leather and rubber.

soap and water for thirty minutes in makeshift showers or tubs, an excruciatingly painful experience for men who had already formed skin blisters.[35] Hospital workers then washed any stripped clothing in boiling water, often reissuing it to healthy men due to material shortages.[36] If the soldier's eyes were exposed to mustard agent, then they would receive an alkaline eye ointment and paraffin wax to keep their eyelids from bonding together.[37] Not much could be done for afflicted lungs, but antiseptics helped ease coughing and menthol soothed the throat. All of this treatment required a significant amount of work from the field nurses assigned to gas injuries. Perhaps not surprisingly, nurses often described gas wounds as the most difficult to both treat and witness.[38]

All members of the German medical services received gas masks as basic protection against errant chemical attacks or secondary contamination. However, some medical pamphlets encouraged doctors not to put on their gas mask too quickly when doing rounds or treating patients because the sight of the mask tended to frighten patients.[39] As part of these protective measures, medical professionals were further encouraged to regularly powder their hands with lime, change their clothes regularly, and thoroughly wash their bodies. Even when all these guidelines were carefully followed, hospital contamination remained a serious risk. In fact, the smallest spot of liquid mustard gas on a soldier's uniform could taint an entire field hospital. This became a particularly pressing issue in 1918, when both the Germans and the Allies began to fire thousands of mustard gas shells at each other's trenches, leading to a chemically saturated battlefield and field hospital overcrowding. The sheer number of lightly gassed men led to the creation of subsidiary medical centers that specialized in gas treatment.[40]

Further Complications: Gas Neurosis

Alongside the wide range of physical gas injuries, soldiers also described symptoms that were most likely psychosomatic. One German military doctor wrote of men who complained of pain in their legs fourteen days after breathing chlorine gas. Another soldier was reportedly unconscious

[35] Fitzgerald, "Chemical Warfare and Medical Response during World War I," 618.
[36] Literatur, Kriegserfahrungen, Auslandsnachrichten, 1934–1936 (BA-MA) RH/12/4.
[37] Fiona Reid, *Medicine in First World War Europe: Soldiers, Medics, Pacifists* (London: Bloomsbury, 2017), 78–79.
[38] Christine Hallett, *Containing Trauma: Nursing work in the First World War* (Manchester: Manchester University Press, 2009), 59.
[39] Zur Kenntnis und Behandlung der Gasvergiftungen, 1917 (LBWH) M 635/1 Nr 963.
[40] Shephard, *A War of Nerves*, 64.

for three days and when he awoke, he claimed that one of his legs was paralyzed and that he felt severe arm pains.[41] Men further complained of nasal or lung irritation, even when they had avoided gas exposure. These kinds of cases were difficult to separate from both those men who had been gassed but had not yet felt symptoms and soldiers who developed persistent nervous disorders.[42] The gas officer Reinhard Lewald described the troubling psychological effects of gas after an enemy artillery attack destroyed some nearby German gas cylinders. He wrote, "Forty officers and six-hundred-forty-five men were killed. Hundreds wounded. The few who escaped have all suffered severe emotional shock from the indescribable confusion."[43] Such gas-shocked men could crowd the field hospitals, presenting symptoms that required psychological specialization that most frontline doctors did not possess.[44]

Over the course of the chemical war, the impact of poison gas on the mind was equally as confounding as any physical injuries. In 1915, psychiatrists reached little consensus on how to treat the increasing number of soldiers who displayed mental disorders after exposure to poison gas. Any previous medical experience with similar neuroses stemmed from treating those who had suffered industrial accidents prior to the war, but no medical professionals were prepared for the sheer number of psychiatric issues that weapons such as artillery shells, tanks, and poison gas would ultimately create. As historian Paul Lerner has revealed, it was only after a 1916 German Association for Psychiatry conference in Munich that the psychiatric field took an ostensibly unified position on the diagnosis and treatment of war neuroses. In the wake of this conference, psychiatrists formally categorized war neuroses as a form of hysteria and began to prescribe a wide range of "active treatments" in their therapeutic endeavors.[45] By the end of the war, over 300,000 men (roughly 4 percent of all German casualties) were treated for various nervous disorders.

Dr. Robert Gaupp, a professor of psychiatry at the University of Tübingen, was one of the leading psychiatrists recruited to treat the increasing number of German soldiers who displayed some form of war neurosis. Gaupp wrote of the troubling situation, "The immense advances in war machinery, the dreadful destructive force of modern artillery, the rolling barrages, the gas shells, the bombs from the air, the

[41] Knack, "Kampfgasvergiftungen," 1246.
[42] Herstellung und Lieferung von Geschoss-Fullungen, 1914– (BA) 201-005-002.
[43] Reinhard Lewald, Kriegstagebuch, 1914–1918 (DTA) Sig 3502.1.
[44] Gaupp Akten, 1918 (TU) 308/89.
[45] Paul Lerner, "Psychiatry and the Casualties of War in Germany, 1914–18," *Journal of Contemporary History* 35, no 1 (January 2000): 20.

flame-throwers … have accounted for a greater accumulation of violent symptoms of shock than has any other war."[46] Through his clinical work with these "gas-shocked" men, Gaupp noticed that their neuroses often compelled them to either continually claim that they were swallowing imaginary gas or to physically reenact the convulsions of gas poisoning.[47] Medical professionals subsequently termed these specific psychiatric responses as "gas neurosis" or "gas hysteria."[48]

When compared with traditional artillery shells and bullets, chemical weapons seemed to produce disproportionately high numbers of psychiatric issues due to the unique form of terror that they induced. In the wake of the war, soldiers routinely claimed that gas, by penetrating all aspects of the soldier's environment and preventing escape or retaliation, destroyed the idea of legitimate chance in wartime survival. Furthermore, the apparently primal fear of asphyxiation routinely led to severe mental stress and even panic.[49] While shrapnel was statistically far more dangerous throughout the war, soldiers felt that its dangers were more tangible; it could be seen entering a man's body and, thus, it could theoretically be avoided or removed. For many men, the horrific flesh wounds of World War I were relatively consistent with previous expectations of war. Thus, many at least claimed that they would rather face the poor survival chances involved in crossing No Man's Land rather than sit in a dugout that may or may not eventually be filled with gas. It was precisely this experience of tense and passive waiting that made men feel helplessly trapped. Furthermore, the insidious nature of gas led to regular visions of inhalation and painfully slow death.[50] Even though traditional artillery shells had a much greater presence on the World War I battlefield until the final year of the war, it was precisely the psychological burdens of gas warfare that impelled one soldier to insist that "gas shock was as frequent as shell shock."[51]

[46] Wolfgang Eckart, "War, Emotional Stress, and German Medicine," in Chickering and Förster, *Great War, Total War*, 140.

[47] Wolfgang Eckart, "The Soldier's Body in Gas Warfare: Trauma, Illness, *Rentennot*, 1915–1933," in Friedrich et al., *One Hundred Years of Chemical Warfare*, 220.

[48] Harold S. Hulbert, "Gas Neurosis Syndrome," *American Journal of Insanity* LXXVII, no. 2 (October 1920): 213–216. English language sources also refer to this condition as "gas fright."

[49] Rebecca Ayako Bennette, *Diagnosing Dissent: Hysterics, Deserters, and Conscientious Objectors in Germany during World War One* (Ithaca: Cornell University Press, 2020), 92.

[50] Reid, *Medicine in First World War Europe*, 81.

[51] Jeremy Paxman and Robert Harris, *A Higher Form of Killing: The Secret Story of Chemical and Biological Warfare* (New York: Hill and Wang, 1982), 17.

Soldiers often displayed signs of gas neurosis soon after they realized or assumed that gas was present on the battlefield. Physical contact with gas was not necessary for experiencing symptoms; merely the anticipation of the weapon could elicit a wide variety of physical signs such as spasmodic dilation of the nostrils or opening of the mouth, convulsive movements of the throat muscles, gasping for breath, painful paralysis, grasping at the throat or face as if to remove a gas mask, attacks of delirium, permanent facial expressions of horror or anxiety, full-body convulsions, and even unconsciousness. Once gas-shocked soldiers were brought to the field hospital, doctors further observed stuttering, shallow breathing, repeated sighing, and compulsive moving of the hands or feet. Over the course of their time spent in the field hospitals, gas neurotics regularly experienced insomnia and/or nightmares involving poison gas.[52] Psychiatrists noted one gas-shocked man who habitually woke in the middle of the night screaming "Gas!"[53]

Most men who displayed the signs of gas neurosis remained in field hospitals near the front lines.[54] After a few days of rest, they would then be sent back to the trenches, often still displaying various neurotic tics.[55] A German psychiatric report from 1918 admitted that most frontline medics could not afford to play the role of psychiatrist and that the best form of treatment was to give these men a "short vacation" where they might calm their nerves under close supervision.[56] Even when neurotic symptoms did fully subside in the field hospitals, they would often resume when the soldier returned to the front. This recurring mental trauma put a great strain on frontline medical professionals and medical resources.

To combat the overburdening of German field hospitals, psychiatrists and neurologists continually advocated for a speedy diagnosis and processing of patients. If a soldier was deemed unfit for further military service, then he was usually sent to a nerve clinic that attempted to quickly restore the man's mental health through industrial service.[57] This centralized system of corrective rehabilitation reflected the accepted medical understanding of wartime neurosis. By diagnosing war neuroses as "hysteria," a term that had traditionally been reserved for female anxiety attacks, German psychiatric consensus denied the importance of the traumatic event in the formation of neuroses. Rather, psychiatrists

[52] Gaupp Akten, 1918 (TU) 308/89. [53] Franz Emmel, 1918 (TU) 669/30128.

[54] The British and French also kept their gas neurotics in field hospitals due to limited treatment options. From July to October 1916, the French sent 91 percent of their war neurotics back to the front. Babington, *Shell-Shock*, 98.

[55] Fitzgerald, "Chemical Warfare and Medical Response during World War I," 617.

[56] Verwaltungsunterlagen, 1917–1918 (LBWK) 456 F 114, 111.

[57] Lerner, "Psychiatry and the Casualties of War in Germany, 1914–18," 19.

such as Gaupp claimed that neuroses were a product of an individual predisposition to mental weakness. Like other mentally ill soldiers, gas neurotics were thus explicitly deemed constitutionally weak and implicitly cast as effeminate. As such, some psychiatrists advocated harsh therapies that aimed to mentally fortify their patients and allow them to be quickly reintroduced into society as capable German subjects. Such treatments included hypnosis, electrotherapy, solitary confinement, mock operations under ether, and the insertion of a laryngo bougie.[58]

Robert Gaupp's clinic tended to prescribe months of rest alongside medicines that aided sleep and eased any physical pain. And gas neurosis certainly did have its physical manifestations. For example, one of Gaupp's patients reported headaches, severe insomnia, and bodily shaking that became so violent that he eventually had to walk with a cane.[59] A German surgeon described the miserable condition of still more men who suffered from gas neurosis, writing:

> The patients were kept in large wards, but the unruly cases were often put in padded cells, where they stood in an agitated state ... Others had hallucinations; they were still living through the horrors of an artillery bombardment or a gas attack. They covered their faces with their hands, so as to protect them from the shell splinters. Others cried out for their gasmasks, which they could not find, and still others heard voices under their pillows or under their bed covers, threatening them with death.[60]

While these conditions elicited concern from many medical professionals, a greater distrust of malingering could override calls for sympathy.[61] Given that most German psychiatrists claimed war neuroses as products of a weak mental constitution, they regularly assumed that many neurotics preferred their condition to frontline service. Admittedly, some soldiers did attempt to fake neurotic symptoms to escape the trenches, but medical professionals tended to exaggerate the frequency of this ploy. In fact, the societal stigma against war neurotics and the harsh treatment that they often received was likely enough to prevent most men from attempting to imitate neurotic symptoms.

Poor treatment of war neurotics was, of course, not limited to the German context. On the whole, the British and French similarly viewed soldier expressions of psychosis as a form of malingering, thereby calling

[58] Wolfgang U. Eckart, "Aesculap in the Trenches: Aspects of German Medicine in the First World War," in *War, Violence and the Modern Condition*, ed. Bernd Hüppauf (New York: Walter de Gruyter, 1997), 188.

[59] Bennette, *Diagnosing Dissent*, 56. [60] Westman, *Surgeon with the Kaiser's Army*, 163.

[61] Stephanie Neuner, *Politik und Psychiatrie: Die Staatliche Versorgung Psychisch Kriegsbeschädigter in Deutschland 1920–1939* (Göttingen: Vandenhoeck & Ruprecht, 2011), 24–25.

into question both the masculinity and self-discipline of the afflicted.[62] Nevertheless, the German case was unique due to the sheer intensity and politicization of debates over war neurosis. After the loss of the war, the fragility of the German nation made the prospect of mass military pensions nearly unthinkable for many doctors.[63] Additionally, the experience of defeat encouraged many Germans to conceptualize war neurosis through a distinctly national lens. Unlike many French and British doctors, who viewed war neuroses as products of an individual's fitness, their German counterparts tended to see neuroses as symptoms of the national state's larger inability to successfully confront or navigate modernity.[64] Thus, the psychiatrist Alfred Hoche wrote, "It is not the case, as was assumed at the beginning, that it is a matter of simulation, of intentional faking of symptoms that are not there. The individuals are in fact sick, but they would be well, strangely enough, if the [German pension] law did not exist."[65] For Hoche, it was merely the discussion of neuroses and pensions on a national level that weakened and sickened the German mind. The apparent solution to this so-called *Rentenneurose* (pension neurosis) was a refashioning of the national spirit through labor and productivity.[66] Cures could be found either through a soldier's reentry into war, which would supposedly counter the weaknesses of life on the home front, or placement in the many work programs that the German medical services subsequently set up.

If these methods did not produce satisfactory results, then harsh medical treatments appeared fully justified in the name of the German nation. While political opposition to this treatment arose in the 1920s, it never led to the kind of normative legislation that came about in Britain. Furthermore, pensions for German war neurotics were gradually

[62] Tracey Loughran, "Shell-Shock and Psychological Medicine in First World War Britain," *Social History of Medicine* 22, no. 1 (2009): 87; Julien Bogousslavsky and Laurent Tatu, "French Neuropsychiatry in the Great War: Between Moral Support and Electricity," *Journal of the History of the Neurosciences* 22 (2013): 150.

[63] Historian Deborah Cohen points out that the Weimar state actually did more for its veterans than the British state. The German government provided veterans with significant educational, occupational, and financial opportunities, while the British handed over much of this work to private charitable organizations. But in taking on this massive endeavor, Weimar state services also absorbed any and all criticism of veterans' programs, eventually leading to major political disagreement over such programs during the Great Depression. Deborah Cohen, *The War Come Home: Disabled Veterans in Britain and Germany, 1914–1939* (Los Angeles: University of California Press, 2001), 63.

[64] Fiona Reid, *Broken Men: Shell Shock, Treatment and Recovery in Britain 1914–30* (New York: Continuum, 2010), 69.

[65] Lerner, "Psychiatry and Casualties of War in Germany, 1914–18," 15.

[66] Bruno Cabanes, *The Great War and the Origins of Humanitarianism, 1918–1924* (Cambridge: Cambridge University Press, 2014), 117.

reduced throughout the 1920s. Of the 300,000 German soldiers diagnosed with war neuroses during the war, only about 20,000 received pensions by the onset of the Great Depression.[67] It remains practically impossible to say how many of these soldiers, in addition to those who never received a diagnosis, suffered from "gas neurosis."

Prevention versus Treatment: The Medical Value of Gas Discipline

One of the main points of concern in the medical treatment of gas poisoning was the extrication of gassed soldiers from the battlefield, especially given the fact that protecting injured soldiers from subsequent gas exposure remained difficult. During extraction, medics and stretcher bearers were ordered to leave any injured soldiers in their gas masks and to put new masks on any unprotected men.[68] However, affixing gas masks to soldiers who had sustained significant head trauma presented an obvious practical problem. For these cases, the medical services eventually received gas protection hoods, or large bags that could be inflated with oxygen around an injured soldier's head.[69] The distribution of gas protection hoods was part of the medical service's initial responsibility to distribute gas masks to all soldiers and to prepare them for chemical warfare. However, as the German medical services struggled to keep up with the manifold demands of the ever-changing chemical war, Fritz Haber's KWI increasingly took over the bureaucratic duties of gas mask distribution and instruction in gas protection.[70]

After the onset of the chemical war, information on the treatment of gassed soldiers gradually trickled down from academic institutes such as the KWI to frontline doctors. In early 1916, the pathologist Ludwig Aschoff and the senior military doctor Oskar Minkowski published the manual "Zur Kenntnis und Behandlung der Gasvergiftung" ("The Recognition and Treatment of Gas Poisoning").[71] The chief of the medical services, Dr.

[67] Jason Crouthamel, "Nervous Nazis: War Neurosis, National Socialism and the Memory of the First World War," *War & Society* 21, no. 2 (2013): 57.
[68] Therapie der Kampfgas-vergiftungen, 1918 (MPG) Va 5, 524.
[69] Kriegsmedizin, 1918 (BA-MA) PH 2/653.
[70] Kriegstagebuch Band III, 23 Dezember 1915 bis 12 November 1916 (LBWH) M 660/032 Nr 23.
[71] Gasubungen Sanitätsamt, 1918 (BHK) Stv GenKdo I. AK SanA 136. At the onset of war, Ludwig Aschoff (1866–1942) was a respected professor of pathology at the University of Freiburg. He was most known for his work on the pathology and pathophysiology of the heart. Oskar Minkowski (1858–1931) was known for his diabetes research at the University of Strasbourg, but he was also a leading expert on poison gas, leading to an appointment as head of field sanitation.

Otto von Schjerning, then distributed this manual to all German field hospitals. The booklet detailed how to treat soldiers who had been exposed to chlorine gas, exhorting doctors to administer oxygen quickly and diligently and to provide analgesics such as codeine, morphine, and cocaine.[72]

Over the following three years, the German military released four different versions of "Zur Kenntnis und Behandlung der Gasvergiftung." Toward the end of 1916, Aschoff and Minkowski expanded the pharmacology section and gave suggestions on when to end oxygen treatment. In June 1917, they added a quantitative clinical testing section for chlorine, phosgene, and tear gases. And finally, in January 1918, the doctors included a comprehensive section on the treatment of mustard gas.[73] While field hospital doctors received copies of the manual, sanitary officers and stretcher bearers learned these gas treatment guidelines through the same gas protection schools that trained gas officers. Initially, men in the medical services took week-long trips to Berlin in order to attend such courses, but as the war progressed, two more formalized gas schools were opened in Leverkusen and Greppin.[74]

While Aschoff and Minkowski's manual served as the official guidelines for gas treatment, the medical services also gained access to an expanding body of supplementary literature on poison gas.[75] Not surprisingly, this information was at its most detailed toward the end of the war, as frontline doctors reported their gas findings and treatments back to the central medical services office and the KWI.[76] Nevertheless, even with greater toxicological knowledge, alleviating gas-related illnesses remained difficult and a speedy extraction from the battlefield was still the most beneficial action. The toxicologist Ferdinand Flury wrote:

> The first and most important task in gas poisoning is the rescue of the casualty. Since the measures taken in this case usually decide on the fate and life of the victim, every person should know and master the basic concept of first aid in the event of injuries. In gas poisoning, the removal from the area of danger … is the most urgent requirement.[77]

Even in the final year of the war, one-third of all seriously gassed soldiers were dead on arrival at field hospitals and a large number of the remaining

[72] Only so much of a strong analgesic could be administered to soldiers who were exposed to chlorine or phosgene gas because high doses of a drug such as morphine could dangerously depress the respiratory system.

[73] Amend, "Die Sanitätsdienstliche Versorgung der Gasversehrten," 24.

[74] Literatur, Kriegserfahrungen, Auslandsnachrichten, 1934–1936 (BA-MA) RH/12/4.

[75] Gasschutz Abteilungen CUF, 1918 (MPG) Va 5, 526.

[76] Armee Oberkommando der 6. Armee Gasschutz, 1915–1918 (BHK).

[77] Ferdinand Flury and Franz Zernik, *Schädliche Gase: Dämpfe, Nebel, Rauch- und Staubarte* (Berlin: Springer-Verlag, 1931), 565.

two-thirds died in the next seventy-two hours.[78] Given the general inability of doctors to save severely gassed men, resources and manuals strayed from the treatment of symptoms. Focusing instead on prevention, official guidelines pointed to the gas mask and "gas discipline" as crucial prophylactics in the fight against gas poisoning. Ludwig Aschoff noted:

> Doctors' powerlessness against the gravest [gas poisoning] cases ... makes it all the more necessary to be mindful of the proper use and maintenance of preventive measures. Ever since troops were equipped with the latest models of the protective mask, only a very small number of severe gas poisoning events have occurred – mostly in special cases where crews did not have their masks at hand, lost them in battle, or put them on too slowly.[79]

By stressing the supreme importance of the gas mask, Aschoff placed the responsibility for a soldier's well-being and survival on his ability to use gas protective devices. As early as spring 1916, calls for "gas discipline" now appeared alongside published guidelines for medical treatment. In one such leaflet, the first point of emphasis urged men to "Trust your mask, it protects you when it's well taken care of (in good condition, without holes, cracks, etc.) and you know how to use it safely and quickly."[80] Another flyer claimed that gas attacks were only dangerous to soldiers who were not prepared and that nothing could happen to a man who properly maintained his gas mask, gas tarpaulin, skin detox ointment, and eye drops.[81]

By the fall of 1916, German medical reports took care to relate the specific ways in which men had been gassed. Without exception, these reports put the greatest stress on failures to correctly apply gas masks.[82] For instance, one such report mentioned a soldier who took off his mask because he had run too strenuously, another in which a soldier fell and knocked his mask off his head. Still others described men caught between two trenches without masks or men who had forgotten their masks somewhere in the trenches. A final instance described a soldier as *blöde* (stupid) because he did not notice that his mask had major leaks.[83] Fully revealing the growing medical role assigned to the mask, a 1917 report estimated that 30 percent of all cases of gas poisoning were instances in which men were too surprised to put on their masks. An additional 50 percent of all

[78] Generalkommando I. Armee-Korps (BHK) 1276.
[79] Ludwig Aschoff, *Über anatomische und histologische Befunde bei "Gas"- Vergiftungen* (Berlin: Reichsdruckerei, 1916), 18.
[80] 1914=1918 Dokumente M Wolfen (IZS). [81] RH/12/9 Sig 85, 1916 (BA-MA).
[82] Literatur, Kriegserfahrungen, Auslandsnachrichten, 1934–1936 (BA-MA) RH/12/4.
[83] Gas als Kampfstoff, August 12, 1916–April 27, 1917 (LBWH) M 33/2 Bü 402.

cases were due to the mask not being nearby and 5 percent were due to the mask being ripped off by explosions or physical activity.[84]

Failed gas discipline and incorrect use of the gas mask quickly became catch-all explanations for instances of gas poisoning, simultaneously eliding both the technical difficulties associated with the mask and the inability to effectively treat gas exposure. Certainly, the gas mask was the best method of protecting against gas, but it was not infallible even if perfect gas discipline was possible. This is not to say that the military medical services did not attempt to find more effective remedies for gas poisoning, nor does it mean that they never found moments of success in treatment, but it does appear that appeals to personal protection provided a disclaimer for any who would blame medical shortcomings in the treatment of gas. Furthermore, the doctrinal importance assigned to gas discipline served to normalize methods of gas protection and to posit poison gas as an acceptable form of warfare. If every soldier could protect themselves against gas through supposedly simple preventative measures, then gas really was a more humane weapon than artillery shells or bullets.

This argument was particularly important for the Germans, who felt the need to justify their use of gas to the international community. In fact, toward the end of the war, Fritz Haber penned several lectures on the supposed "humane-ness" of poison gas. Apparently, even the seemingly unrepentant Haber felt the need to avoid postwar censure for the grisly medical outcomes of many gassed soldiers. Furthermore, the concept of gas discipline fit nicely within the German medical community's broad attempt to refashion the German nation according to the hard realities of the modern battlefield. Following the logic of gas discipline, gassed men had no one to blame but themselves since a disciplined soldier and a true German man could supposedly resist gas through the correct operation of the gas mask. In this way, gas discipline served to pathologize the soldiers who failed to survive and/or embrace chemical warfare.

Clinical Trials and Tribulations: World War I as a Great Experiment

In a 1937 text that attempted to narrate the realities of World War I gas warfare, the medical doctor Hermann Büscher wrote:

> ...when people came up with the idea of using poisonous gases to kill in the last war, only then did they foretell and feel all the terrible aspects of poison gas. Since then, the poisonous gases are oppressing us like a nightmare. Toxic gases have

[84] Gasschutzwesen, Allg.u.Besonderes, 1915–1918 (BHK).

become a problem, and the war has stamped them into our imagination with that unique power before which we find ourselves almost motionless, impotent, helpless. Are we truly helpless and unprotected against poison gas? Are we no longer masters of the gas? Do we have to bow to the poison gases as to a fate that cannot be avoided? No! We simply must face gas with clear eyes and courage, in order to be able to again master them.[85]

Büscher's bold assertion is representative of two overriding and persistent themes in the broader German medical community's conceptualization of gas warfare during and after World War I. First, by positing gas as a problem that could be scientifically solved, Büscher envisioned the chemical war as a grand scientific/medical experiment. With the application of the scientific method to fundamentally solvable problems, humans could develop effective protection against poison gases. In doing so, Büscher explicitly claimed that mankind could again master its changing physical environment. Certainly not all medical professionals described gas warfare through the thematic lenses of experimentation and mastery, but a substantial number of influential doctors, medical researchers, and poison gas specialists evoked these themes in their writings during and after the war. Interestingly, many of these men were involved in the medical specialty services such as toxicology that grew in importance as the chemical war developed.[86]

Undeniably, the war presented twentieth-century medical professionals with new forms of bodily destruction. Prior to 1915, few doctors had seen men blown apart, burned beyond recognition, or asphyxiated by poison gas. While these were undoubtedly distressing sights that created their own form of psychological trauma for those in the medical services, they also presented army doctors with nearly limitless opportunities to advance medical knowledge. Physicians in numerous medical fields could now study the effects of modern explosives and chemical substances on the human body, while also testing various drugs, prophylactics, and therapies on injured soldiers without significant oversight. It was within this context that many doctors believed that they were bound to invent a distinctly humane gas weapon – one that would merely put soldiers to sleep and spare future bloodshed.

Countering the claim that World War I served as a grand in vivo medical experiment, one could note that the German medical services generally lacked the time, manpower, and resources to conduct rigorous

[85] Hermann Büscher, *Giftgas und wir?: Die Welt der Giftgase Wesen und Wirkung, Hilfe und Heilung* (Leipzig: Johann Ambrosius Barth, 1937), 7.
[86] Harrison, *Medicine and Victory*, 279.

medical experiments in the field.[87] However, while rigorous testing of new treatments was rare, desperate experimental treatment (especially in the case of gas poisoning) was not uncommon. The medical professionals who often worked tirelessly and without much guidance in the field hospitals saw these unproven medical applications as their only option.[88] Furthermore, much of the information gathered from field hospitals and wartime studies did advance the therapeutic knowledge and abilities of medical specialists in fields such as cardiology, dermatology, and internal medicine.[89] If nothing else, the desperate wartime situation of German military medicine shaped many doctors' conception of the war as a new and untamed medical frontier.[90] The German specialist in hygiene and tropical medicine, Karl Mense, summed up the situation, writing:

> ... right behind the lines and back home, each doctor – even the one who does not practice any more, but who is peacefully doing research – is now fervently trying to heal the wounds of the war ... Directly in front of us, an experiment, so extensive that it could hardly derive from imagination is taking place ... The people of the globe are taking part in an epidemiological experiment so huge, scientists would have never dreamed of it.[91]

Alongside the development of fields such as epidemiology, this grand medical experiment was certainly central to the treatment of poison gas. As early as June 1915, Dr. Hans von Pezold, who would later become a senior physician at the Karlsruhe Municipal Hospital, expressed excitement when treating men who had been exposed to chlorine:

> In the morning, I first visited our *most interesting* patient, the pioneer Herzog, who had been poisoned by his own gas bomb at Ypres. He came to us with shortness of breath and constant vomiting, even though he was poisoned fourteen days ago. Now he is doing much better.[92]

Dr. Ferdinand Flury also reflected on the experimental nature of treating gas during the war, writing: "With the daily growing danger of gas hazards, the experience of the war ... has also been a great teacher in this field."[93]

[87] Derek Linton, "The Obscure Object of Knowledge: German Military Medicine Confronts Gas Gangrene during World War I," *Bulletin of the History of Medicine* 74, no. 2 (2000): 316.
[88] Erkrankungen durch Kampfgas, 1916 (LBWK) 456 F 115, 67.
[89] Stefanos Geroulanos and Todd Meyers, *The Human Body in the Age of Catastrophe: Brittleness, Integration, Science, and the Great War* (Chicago: University of Chicago Press, 2018), 37.
[90] Eckart, *Medizin und Krieg*, 64.
[91] Karl Mense, "Zum neuen Jahre," *Archiv für Schiffs und Tropenhygiene* 19 (1915): 1.
[92] Kriegstagebuch Band II- 11. August bis 22. Dezember 1915 (LBWH) M 660/032 Nr 22.
[93] Flury and Zernik, *Schädliche Gase*, 595.

Ferdinand Flury (1877–1947) studied pharmacology, chemistry, and biology at the University of Erlangen from 1892 to 1901. After his year of required military service, he received his doctorate in pharmacy in 1902. For the next three years, Flury worked as an *Assistent* in the Chemical Laboratory of the University of Erlangen, passing his exam in food chemistry in 1904. The following year, he returned to the military and served as a staff pharmacist for a garrison hospital while simultaneously studying medicine at the University of Würzburg. After receiving his medical doctorate and producing a habilitation in pharmacology, he obtained a position as a professor at Würzburg. At the outbreak of the war, Flury returned to military service as a staff pharmacist. This position lasted until 1916, when Fritz Haber hand-selected him to serve as the head of his KWI's Department D, focusing on toxicology, animal testing, and industrial hygiene.

Once in Berlin, Flury began to collect much of the prewar industrial knowledge on poison gases and to combine it with new experimental data obtained primarily from animal testing.[94] Through clinical trials, chiefly involving dogs but sometimes humans, Flury was able to record blood pressure, body temperature, and pulse rate under various levels and types of gas exposure. These tests were augmented by information arriving from the front, thereby allowing Flury to draft large tables in which he would record the various ways in which soldiers were gassed, their first recorded symptoms, and any later developing complications.[95] Flury concurrently studied the impact of the gas mask, recording the length of time that filters lasted and the influence of the mask on bodily fatigue. All this research was crucial to the development of what was eventually termed the "Haber Constant," or a rough mathematical relationship between concentration and exposure to poison gas and the amount of time to produce death.

Viewing his work as crucial to the war effort, Flury described scientific and medical knowledge as the only avenues toward the mastery of chemical weapons. He wrote, "Matters surrounding gas warfare would be much enhanced if the knowledge of the use of the new weapon penetrated deeper into the army during major actions. The overwhelming majority of the troops … had no idea of the extent of gas warfare."[96] While Flury was particularly influential, he was not the only scientist or medical doctor undertaking such research. The major German chemical companies, gas mask manufacturers, and various pharmacological and

[94] Amend, "Die Sanitätsdienstliche Versorgung der Gasversehrten," 12.

[95] The KWI's Friedel Pick also conducted such studies. Friedel Pick, "Über Erkrankungen Durch Kampfgase," *Zentralblatt Für Innere Medizin* 39, no. 20 (1918): 305–310.

[96] Bericht uber Beobachtungen an Gaskranken an der Westfront, 1917 (BA-MA) PH/2/.

toxicological laboratories conducted similar clinical trials. Repeating many of Flury's animal tests, the Elberfelder Laboratory was particularly precise in recording changes in oxygen levels, thickening of blood, and time of death after exposure to all of the major battle gases.[97]

By repeatedly arguing that knowledge was the best way to combat gas exposure, Flury was one of the many doctors and scientists who insisted on a mental mastery of the chemical construction of the environment. According to Flury, the men who could best understand the physical characteristics of poison gases could also best survive and control them. Given that treatments for extended gas exposure were still mostly ineffective by the end of the war, this call for scientific and medical knowledge extended to a systematic understanding of protective technologies such as the gas mask and a familiarity with their use. This growing body of knowledge was, of course, an important part of the newly mechanized and mobile military tactics that were imposed on German soldiers in 1918. However, Flury was not just concerned with *surviving* poison gas exposure; his desire for chemical mastery also focused on the effective offensive deployment of the most fatal gases. For scientists such as Flury and his supervisor Fritz Haber, control of the chemical environment implied the harnessing of aggressive power and the achievement of battlefield domination.

In the postwar world, many German doctors and scientists continued to view gas warfare as an experimental branch of toxicology that sought total mastery over a chemically uncertain environment. Flury continued to work in the field and to foster this view, writing in 1931 that:

> The use of gases and gas protection gained extraordinary importance in the last war – no branch of war technology has met such interest as militarized gas, nothing has filled army leaders with such concern, nothing so much preoccupies scientists' thinking and research. Nothing has so influenced the public imagination as the gas battle.[98]

Other doctors and medical scientists echoed Flury's claims, molding gas protection research into both a formalized branch of applied science and a broad public concern. Joining the post-war military theorists who claimed that all future wars would feature aero-chemical bombings, these toxicologists and related medical professionals insisted that the study of poison gases would better prepare the German nation for supposedly imminent attacks. By the mid-1920s, such doctors were writing in

[97] Brief an den Feldsantitätschef, August 3, 1916 (BA-MA) PH/2/.
[98] Flury and Zernik, *Schädliche Gase*, 17.

everything from professional bulletins to military journals in order to call for nationwide preparedness against poison gas.[99]

Many of these medical professionals who became distinctly involved in poison gas protection disseminated their ideas through small, interconnected networks that persisted throughout the 1920s and 1930s. For instance, Ferdinand Flury and Oskar Minkowski continued to publish on gas protection while Flury furthered his animal testing with poison gas as a professor of pharmacology at the University of Würzburg. In 1924, Flury also became one of the German military's primary contacts for questions related to chemical weapons.[100] Before writing several tracts on poison gas, Hermann Büscher elected to remain in the postwar medical services, treating gassed industrial workers when a train carrying gas shells exploded outside of Hugo Stoltzenberg's chemical weapons plant in 1919.[101] Younger medical professionals such as Rudolf Hanslian would later join these newly titled "gas specialists" in their interwar endeavors.

Medical studies on the effects of gas masks also continued after the war, often under the supervision of the Auer and Dräger gas mask companies.[102] One such study argued, "In our time, when more and more people are forced to wear a gas mask – be it for professional reasons or to work in air defense – it is of utmost importance to know what influences the gas mask has on cardiac activity."[103] Much like interwar studies of poison gas, this research made normative appeals for a mastery of gas knowledge and technical protection. Furthermore, such work on all of the various facets of chemical warfare retained the nationalist overtones that were born in the war. Thus, this small yet distinctive group of "gas specialists" appealed directly to their fellow Germans in their calls for chemical mastery. While admitting the dangers of any future aero-chemical war, Dr. Hermann Büscher was already proclaiming a future German victory, writing:

[99] W. Schweisheimer, "Die Medizinischen Grundlagen Des Giftgaskrieges," *Militär Wochenblatt* 17, no. 110 (1925): 577–582.

[100] Helmut Maier, *Chemiker Im "Dritten Reich" Die Deutsche Chemische Gesellschaft Und Der Verein Deutscher Chemiker Im NS-Herrschaftsapparat* (Weinheim: Wiley-VCH, 2015), 263.

[101] Büscher, *Giftgas und wir?*, vii.

[102] Harald Schuster, *Die Aufnahmeleistung von Atemfiltern und ihre Bedingtheiten* (Lübeck: Heinr. & Bernh. Draeger, 1936), 4.

[103] Karl Graus aus Schmechten, *Elektrokardiographie Untersuchungen bei Belastung älterer Personen unter der Gasmaske* (Höxter: Huxaria Druckerei und Verlagsanstalt, 1939), 5. See also Karl Schneider, *Blutgase und Kreislauf bei Arbeit under der Gasmaske* (Würzburg: Universitätsdruckerei H. Stürtz A.G. Würzburg, 1939); "Wochenschrift für die gesamte Heilkunde," *Vierteljahrsschrift für gerichtliche Medicin und öffentliches Sanitätswesen* 59, no. 1 (January 1920).

And if iron fate again imposes a new war against our will, then we will accept it. But we, as a race of men and of heroes who know how to make world history on battlefields, will be vigorous and victorious and go through this difficult journey for Germany's legacy and greatness. Perhaps millions of us will be destroyed. Our faith remains unshakeable: Germany will not perish. Germany will always exist. Germany will be eternal. "Long live Germany!"[104]

Excising Pain: The Medical Translation of Wartime Suffering into Interwar Fear

The third section of Hermann Broch's celebrated 1930 novel *The Sleepwalkers* used the interactions between injured soldiers and medical professionals to reveal what Broch saw as the physical and mental decay of German society during World War I. In the novel, a small-town clinic in the Moselle valley is transformed into a military hospital as injured men stream back from the front. One particularly poignant scene contrasts the cultural militarization of medicine with its lack of preparation for the modern methods of war. Broch writes, "Dr. Friedrich Flurschütz was making his round of the wards. He was wearing a military cap with his doctor's white overalls; a combination which Lieutenant Jaretski characterized as absurd." Approaching a wounded soldier, the doctor tells him that he will need to have his arm amputated. Casually, the doctor says, "Don't worry about it, it's only the left," to which the soldier replies, "Yes, all you want to do is cut things off." Flurschütz then shrugs his shoulders and retorts, "Can't be helped; this century has been devoted to surgery and rewarded by a world-war with guns … now we're beginning to find out about glands, and by the time the next war comes along we'll be able to do wonders with these damned gas-poisonings … but for the present the only thing we can do is cut."[105]

With a conspicuous reference to poison gas, Broch used this brief scene to express both the changing nature of warfare and medicine's inability to keep pace. But even more importantly, Flurschütz's shrug conveys a certain callousness in the doctor's relationship to his patient. Historians such as Paul Lerner and others have already shown the moments in which an objective sense of national medical duty outweighed personal compassion in studies of German psychiatric treatment, surgical operations, disability rehabilitation, and pension allocation during and after World War I. Certainly, in these instances, many soldiers and veterans viewed military doctors as unsympathetic or

[104] Büscher, *Giftgas und wir?*, 219–220.
[105] Hermann Broch, *The Sleepwalkers* (New York: Vintage, 1996), 354–355.

even malevolent. This was sometimes evident in the appellations used for medical spaces and tools. For instance, operating rooms were called "bone mills" or "torture chambers," while the chloroform hood was the "gas mask" and the use of anesthesia was a "gas attack."[106]

In the case of military pensions, Lerner has shown how psychiatrists used the diagnosis of hysteria to deny benefits to certain veterans with psychiatric disorders.[107] This refusal to acknowledge the psychological damages of mechanized war was justified as a necessary step in the salvation of German masculinity and citizenship as well as the preservation of an overtaxed pension system.[108] Indeed, throughout the war, some psychiatrists were far more concerned with exposing malingerers than developing effective psychiatric therapies. Their so-called Kaufmann-Kur comprised a series of horrifying tests, aimed at forcing fraudulent neurotic claimants to confess their deception. This included taking X-rays in dark rooms, making patients swallow their own vomit, and pretending to conduct unnecessary surgical operations.[109] Unsympathetic doctors regularly rationalized these tests in the terms of Social Darwinism, opining the possibility that the mentally strong would be wiped out, while the deceitful, frightened, weak, and truly neurotic would instead survive.[110]

Many field hospital doctors harbored similar skepticism for gassed soldiers both during and after the war. Given the unseen, internal manifestations of both mental illness and gas poisoning, soldiers most commonly used these two afflictions to avoid frontline service. Since it was difficult for medical professionals to determine if a soldier had, in fact, inhaled small amounts of gas, doctors and company commanders were forced to either observe the men who claimed to be ill for several days or send them back to the front with the hope that they were lying. This is apparent in a 1917 letter in which one infantryman Birzer claimed that he had been gassed. When he reported this to his company commander, however, he was denied medical leave. Birzer wrote, "Instead he locked me up. Here you go and see how they act. He would be happy if I were *kaputt*. Then he told me off like no one did before. He called me a coward and a malingerer and did not let me go and see the doctor."[111]

[106] Imme, *Die deutsche Soldatensprache der Gegenwart und ihr Humor*, 132.
[107] Eckart, "The Soldier's Body in Gas Warfare," 224.
[108] Paul Lerner, *Hysterical Men: War, Psychiatry, and the Politics of Trauma in Germany, 1890–1930* (Ithaca: Cornell University Press, 2003), 234.
[109] Eckart, "Aesculap in the Trenches," 188. [110] Ibid., 186.
[111] Berndt Ulrich and Benjamin Ziemann, eds., *German Soldiers in the Great War: Letters and Eyewitness Accounts*, trans. Christine Brocks (Barnsley: Pen and Sword, 2010), 83.

In the immediate postwar years, a similar skepticism confronted the steadily increasing pension claims for gas exposure. Veterans asserted a variety of respiratory diseases including bronchitis, emphysema, and asthma.[112] Others complained of regular dizziness and sleeplessness that they believed stemmed from gas exposure, even if the symptoms appeared years after the war. The overwhelming majority of these cases were denied war pensions after medical authorities claimed that there was no proof of causation. Furthermore, several German research studies from the 1910s and 1920s took great pains to deny any connection between gas exposure and subsequent tuberculosis.[113]

The reliance on gas discipline and the gas mask as the best form of gas protection encouraged this often-harsh view of veterans who claimed to have been gassed. According to the logic of gas discipline, gassed men were those who could not master themselves or their environment. Through some personal deficiency, they failed in a rote task such as cleaning their gas mask, hanging it from their neck, or sleeping with it affixed. Resonating with the ideas of staunch German nationalists and Social Darwinists, gas effectively weeded out those who could not exist in a modern world that would now be permeated by atmospheric poison. These were victims of their own insufficiencies; men who were no longer needed or desired in an imagined Germany comprised of both individually and collectively disciplined subjects who were ready to enter the imminent, violent struggle for national survival.

Of course, not all German medical professionals were so callous toward their gassed patients. As historian Christine Hallett points out, "Some of the most powerfully emotive writings of World War I nurses were those in which they described their experiences of combating the effects of innovative weapons of war. All seemed to agree that the worst of these were the poisonous gases."[114] Such expressions of sympathy eventually spurred on real political action. Indeed, it was the medical community's first-hand experience with gas victims that inspired the International Committee of the Red Cross to call for a battlefield ban of poison gas in 1918. This appeal was met with some enthusiasm from the Allies, but the German government rejected a complete ban, reiterating their claim that their use of gas was only a retaliation against Allied tear gases. Nevertheless, this international dispute sparked a February

[112] Vorschriften uber Gasschutz, 1931–1933 (BA-MA) RH/12/4.
[113] Günter Hänsel, *Die Spätfolgen von Kampfgasvergiftungen an den Lungen und ihre versorgungsrechtliche Bewertung* (Lippstadt: Thiele Lippstadt/Westf., 1934), 3–4.
[114] Hallett, *Containing Trauma*, 59.

1918 debate in the Reichstag over the use of gas in a preplanned German offensive for later that year. Only the Independent Social Democratic Party (USPD) supported ending Germany's military use of gas, and the German public remained unaware of such political discussions due to strict government censorship.[115]

While many doctors were appalled by gas warfare and deeply committed to their patients' recovery, condemnations of gas from the German medical community were largely censored or ignored. Historian Joanna Bourke has argued that over the course of the nineteenth century, the development of new medical technologies and the increasing disdain for the importance of affect in healing established the appreciation for "detached concern" as an idealized mode of medical comportment. According to many doctors, this normative relationship to a patient would allow medical professionals to balance emotional engagement with a scientific demeanor that "encouraged the confidence of patients."[116] But in the case of gas poisoning, it proved extremely difficult for doctors and nurses to ignore this unique form of physical anguish. The unseen nature of gas and gas poisoning often made medical staff feel particularly helpless in treating the men who had inhaled gas. Coughing and retching until they died, these men proved that medical treatment did not always keep pace with the scientific and technological advances of war.[117]

But as both Bourke and literary scholar Elaine Scarry have pointed out, there is a great difficulty, if not an impossibility, in narrativizing pain.[118] Whether or not the phenomenon of pain is a deconstruction of language, as Scarry argues, it was quite conceivably impossible for World War I medical professionals to translate their patients' pain into an interwar politics that could fully express the dangers of poison gas.[119] It is also symbolically and tragically fitting that gassed men, who often had burned the lining of their throats, were usually unable to speak after gas exposure.[120] On the other hand, German medical professionals were able to craft their own lessons of chemical weapons from World War I by

[115] Eckart, "The Soldier's Body in Gas Warfare," 222–224.
[116] Joanna Bourke, "Pain, Sympathy and the Medical Encounter between the Mid Eighteenth and the Mid Twentieth Centuries," *Historical Research* 85, no. 229 (2012): 450.
[117] Leo van Bergen, "The Poison Gas Debate in the Inter-War Years," *Medicine, Conflict and Survival* 24, no. 3 (2008): 175.
[118] Joanna Bourke, "Bodily Pain, Combat, and the Politics of Memoirs: Between the American Civil War and the War in Vietnam," *Histoire Sociale* 46, no. 91 (2013): 46.
[119] Elaine Scarry, *The Body in Pain: The Making and Unmaking of the World* (New York: Oxford University Press, 1985), 72.
[120] Vorschriften uber Gasschutz, 1931–1933 (BA-MA) RH/12/4.

either lobbying for the relative "humane-ness" of an imperfect and underutilized weapon or arguing against the use of a weapon for which there was little known medical treatment. This second claim was frequently tied to pacifistic and left-leaning political positions that frequently argued for greater distribution of war pensions, especially to war neurotics.[121]

However, in the end, most medical attempts to convey the effects of gas warfare inspired greater interwar fear, rather than politically meaningful empathy for soldier pain. The growth of chemical fear throughout the 1920s and 1930s would then have major implications for subsequent debates over the stockpiling and use of poison gas in Germany. Ultimately, these increasingly wild fears forced all political deliberations onto the rhetorical ground of national protection. Interwar gas politics often ignored the real capabilities of poison gas, instead relying on poison gas specialists' invocation of German national discipline or near-hysterical prophecies of doom and destruction to frame their arguments. In a moment of real introspection, the gas specialist Dr. Otto Muntsch attempted to describe the problem of expressing wartime experience in the postwar world: "Many a word has been said about the humanity of gas warfare: Anyone who has ever seen a gas-sick person at the height of lung damage must, if he still has a spark of humanity, fall silent."[122] It was precisely this problem of representation and a subsequent dearth of empathy in the rhetorical framing of arguments that would set the parameters for the interwar debates surrounding chemical warfare.

[121] Jason Crouthamel, "War Neurosis versus Savings Psychosis: Working-Class Politics and Psychological Trauma in Weimar Germany," *Journal of Contemporary History* 37, no. 2 (2002): 165.

[122] Muntsch, *Leitfaden der Pathologie und Therapie der Kampfstofferkrankenungen*, 43.

5 Atmos(fears): The Poison Gas Debates in the Weimar Republic

In his 1930 essay "Die totale Mobilmachung" ("Total Mobilization"), the then established German writer Ernst Jünger reflected on the changes in warfare tactics and militarized industrial production that had taken place since World War I.[1] Meditating on the mobilization of the German home front during the war, Jünger argued that the gradual transformation of the entire nation into a collective military force was an inevitable process of technological modernity. Jünger further recognized that while the collective arming of Germany offered the nation unforeseen military power, it also changed the face of warfare. In this new age of total mobilization, battles would no longer be won by the bravery of individual soldiers but through an industrial production and a mass-produced destruction that required physical and mental fortitude from every citizen. Consequently, each and every man, woman, and child now became a soldier, contributing to the national war effort through various forms of organized production and violence.

According to Jünger, the militarization of civilians further changed the means and rules of warfare. If all civilians were now contributing to a national war effort, then modern wartime violence would target both soldiers and civilians alike, attempting to cripple national production and destroy the will of a people. Aerial bombing and aero-chemical attacks were naturally suited to this new style of fighting since they could conceivably terrorize civilians from afar.[2] As engineers created military airplanes that could fly further distances at higher altitudes, the formerly safe interior of every nation now became a potential battlefront. Simultaneously, research into gas shell technology allowed for a denser diffusion of chemical weapons at greater distances. Reflecting on the

[1] Ernst Jünger, *Sämtliche Werke*, Band 7 (Stuttgart: Klett-Cotta, 1980), 126.

[2] For the long-term accuracy of Jünger's predictions, see John Armitage, "On Ernst Jünger's 'Total Mobilization': A Re-evaluation in the Era of the War on Terrorism," *Body and Society* 9, no. 4 (2003): 191–213.

seemingly inevitable combination of interwar airpower and chemical weapons, Jünger wrote:

The warrior who commands night-time bombardment knows no difference between combatants and non-combatants, and the deadly gas cloud moves like an element above all living things. But the possibility of such a threat is still only partially realized. It presupposes a total mobilization, which, in its own right, frightens even the child in its cradle.[3]

Jünger's writings from the early 1930s serve as an ideal entry point into the larger interwar preoccupation with the sociopolitical alterations that stemmed from the Great War. Discussions of new warfare tactics, technologies, and existential threats expressed the widespread sense of a distinct break from the past. As historian Peter Fritzsche has argued, Weimar-era political discourse was deeply impacted by a broad attempt to toss aside what appeared, in the wake of World War I, as an invalidated historical tradition.[4] This would then seemingly allow for a German national reconstruction that could trouble all preconceptions of "contingency, possibility, and necessity," including, but not limited to, national defense against gas attacks.[5] As Fritzsche reminds us, in the interwar period:

the proliferate images of gas war and the gas mask, which civilians, particularly women and children, were shown using, demonstrated the reach of this militarized new time into the household goods and private lives of contemporaries, just as had the descriptions of "the new life" (das neue Leben) and "the new times" (die neue Zeit).[6]

Certainly, thinkers and politicians across Europe reconceptualized the most intimate aspects of life following the cataclysmic reverberations of World War I.[7] But in Germany, this process took on a greater sense of urgency, primarily driven by a recurrent sense of imminent contingency and risk stemming from the loss of World War I.[8] With a significantly weakened air force and a geographic position that made aerial attack possible from any side, Germany sat expectant for an aero-chemical attack on a major population center. Reflecting on this situation in

[3] Ernst Jünger, *Blätter und Steine* (Hamburg: Hanseatische Verlagsanstalt, 1941), 134.
[4] Peter Fritzsche, "The Economy of Experience in Weimar Germany," in *Weimar Publics/Weimar Subjects: Rethinking the Political Culture of Germany in the 1920s*, eds. Kathleen Canning, Kerstin Barndt, and Kristin McGuire (New York: Berghahn Books, 2010), 362.
[5] Ibid., 363. [6] Ibid., 375.
[7] Michele Haapamäki, *The Coming of the Aerial War: Culture and the Fear of Airborne Attack in Inter-War Britain* (New York: I. B. Taurus, 2014), 57.
[8] Peter Fritzsche, "Did Weimar Fail?," *The Journal of Modern History* 68, no. 3 (1996): 655.

1931, the gas specialist Ferdinand Flury wrote, "No state today may deny the possibility that in future wars the militarily uninvolved population will be threatened by gases, especially in air raids."[9] Furthermore, such expert claims inspired various reactions from regular Germans, including both futile hand-wringing and significant attempts at social and political reform.[10] The sheer amount of political debate surrounding aero-chemical warfare would seem to suggest that it was one of the most serious and pressing emergencies in the Weimar period. For, as historian Jonathan Wright has pointed out, "fear of war – in extreme cases even psychosis – was a recurrent theme [in interwar Germany], particularly given the preparations everywhere to be seen not simply in the build-up of the armed forces but in classes on air defense and gas warfare."[11]

While men such as Jünger seemed to either implicitly or explicitly embrace the new realities of totally mobilized warfare, a variety of pacifist critiques of chemical warfare and industrialized armament arose amidst these interwar fears of aero-chemical attack. Criticisms of Germany's chemical warfare program in World War I stimulated both attempts to reform the state through effective international peace and disarmament treaties as well as to dismantle the state through various forms of revolution. These distinct programs envisioned vastly different ends, but strictly defined pacifists and communists alike realized the need to not only rid Germany of chemical weapons but also expose the impossibility of national gas protection in a future war. This meant concurrently critiquing both fervent militarists and those nationalists who claimed to provide defense against chemical weapons through protective technologies. And while pacifists of all stripes aimed to reduce the possibility of aero-chemical attack by tearing the links between science, industry, and the military that were painstakingly forged by men such as Fritz Haber, they did so on rhetorical grounds that presupposed a future of war and violence. Repeatedly relying on apocalyptic visions and xenophobic fears, interwar pacifists rarely offered a form of chemical defense that appeared viable and lasting to the German public.

On the other hand, many of the scientists and engineers involved in chemical warfare during World War I offered seemingly instantaneous protection. Now increasingly referring to themselves as scientific specialists in the field of gas protection, or "gas specialists," men such as Ferdinand Flury and Rudolf Hanslian applied their wartime knowledge of poison gases to interwar political problems. Just as they had previously

[9] Flury and Zernik, *Schädliche Gase*, 565. [10] Fritzsche, "Did Weimar Fail?," 653.

[11] Jonathan Wright, *Germany and the Origins of the Second World War* (Basingstoke: Palgrave Macmillan, 2007), 71.

called for a mix of technological armoring and gas discipline in the trenches, they now argued that German civilians could survive an aero-chemical attack if they were supplied with both the necessary technology and knowledge. Recognizing that available gas bomb technology still failed to create a dense enough cloud to easily kill from the air, the gas specialists advocated for a variety of techniques for survival alongside a broad acknowledgment of the importance of national security against aero-chemical attack. They further assured the public that the rearmament measures necessary for national security were the best way of providing a long-lasting international peace, thereby positing themselves as the true pacifists.

According to many of the gas specialists, the measures necessary for national protection were not just related to the rebuilding of national defenses. In this era of assumed "total mobilization," certain aspects of aero-chemical protection needed to be drilled into every German citizen. Thus, like the infantrymen of 1918, civilians would now need to internalize gas discipline, always prepared to don a gas mask or run for a gas shelter. However, it remained unclear to the gas specialists whether the oftentimes unruly civilian subject would be able to act as an idealized soldier. For, as the pacifist Waldus Nestler pointed out in 1930:

> To wear a gasmask requires extraordinary discipline; people put them on with teeth clenched. If discipline is difficult for trained soldiers, how much more so with civilians. A mother could not endure to hear her child crying under its mask. Women and children will certainly not be able to make full use of a protective apparatus; every gas attack would cause a panic.[12]

With these problems in mind, the gas specialists remained at odds with both various pacificists and each other over the viability of gas protection technologies. Through various political and scientific publications, individuals and interest groups vied for dominance over the future of German chemical policy and national rearmament. In the early and mid-1920s, international treaties and general disarmament, largely supported by the pacifists, gained great traction as the preferred solution for avoiding a future chemical war. However, in 1928, a major poison gas leak in Hamburg seemed to justify mounting calls for both increased gas preparedness among civilians and a stronger national defense against aero-chemical attack.

[12] Waldus Nestler, "Collective and Individual Protection," in *Chemical Warfare: An Abridged Report of Papers Read at an International Conference at Frankfurt am Main* (London: Williams & Norgate, 1930), 77.

Poison Gas Research and Production after the War: The Travails of Hugo Stoltzenberg and His Competitors

The Treaty of Versailles, signed on June 28, 1919, to formally end World War I, maintained several articles that explicitly dealt with chemical warfare. For instance, Article 171 reiterated the ban on poison weapons from the earlier Hague Conventions, while simultaneously blaming the Germans for the breaking of those conventions during the war. The article mandated that the use, manufacture, and importation of "asphyxiating, poisonous or other gases and all analogous liquids, materials or devices" was prohibited in Germany.[13] Article 172 intended to ensure that the Germans would abide by these restrictions, while also attempting to requisition sophisticated German chemical production processes for Allied industries. The article read:

> Within a period of three months from the coming into force of the present Treaty, the German Government will disclose to the Governments of the Principal Allied and Associated Powers the nature and mode of manufacture of all explosives, toxic substances or other like chemical preparations used by them in the war or prepared by them for the purpose of being so used.[14]

The treaty further prohibited the manufacture or purchase of all German air protection technologies such as airplanes, antiaircraft guns, and searchlights.[15]

The treaty went into effect on January 10, 1920, and in that same year, the Military Inter-Allied Control Commission (MICC) under the French Marshal Ferdinand Foch began to formally assess the German chemical industry. The MICC relied heavily on the Hartley Report, a 1919 attempt by the British chemist Sir Harold Hartley to document the wartime activities of Fritz Haber and the major German chemical companies in the British sector of occupation. The 1921 publication of French chemical liaison officer Victor Lefebure's supplementary study, *The Riddle of the Rhine*, further sparked international debate over the military contributions and relative strength of the German chemical industry. Lefebure wrote, "The IG is, if you will, a huge and cleverly disguised mine in the troubled waves of world peace ... The existence of

[13] The Treaty of Versailles, Article 171, p. 119, www.loc.gov/law/help/us-treaties/bevans/m-ust000002-0043.pdf.

[14] The Treaty of Versailles, Article 172, page 119, www.loc.gov/law/help/us-treaties/bevans/m-ust000002-0043.pdf.

[15] Erich Hampe, *Der Zivile Luftschutz im Zweiten Weltkrieg: Dokumentation und Erfahrungsberichte über Aufbau und Einsatz* (Frankfurt am Main: Bernard & Graefe Verlag, 1963), 9.

this gigantic monopoly raises vital military and economic questions that can truly be called 'The Riddle of the Rhine.'"[16] And indeed, at war's end, the major chemical companies such as Bayer, BASF, and Hoechst did control a near total monopoly on chemical production in Germany.[17]

While the Allies intended to shut down German chemical factories that produced military-related products during the war, the MICC ultimately struggled to differentiate between military and civilian applications in factories that were intentionally set up to transition their chemical production seamlessly between both markets.[18] Furthermore, now fully recognizing the importance of industrial chemistry in the wake of the war, the German government was apt to turn a blind eye as German chemical companies and the Kaiser Wilhelm Institute attempted to hide much of their production and research activities before 1920.[19] While the demilitarization of the Rhineland did allow French officials to harass the major IG companies such as BASF, Hoechst, and Bayer, the Allies failed to significantly hamstring German chemical production or to learn substantial proprietary production processes. The MICC ceased its inspections in 1924 and finally left Germany in 1927.

The failed attempt to extradite Fritz Haber was yet another part of the Allies' lofty plans to dismantle the German chemical industry. Yet in allowing Haber to return to Germany to take up both his original post as the head of the Kaiser Wilhelm Institute for Physical and Electrochemistry as well as a new role as the driving force behind the Emergency Committee for German Science, the Allies also allowed him to rebuild the web of personal and professional connections that so successfully joined German academic science, industrial production, and the military. In fact, the main purpose of the Emergency Committee for German Science was to solicit funding from the federal and state governments as well as private industry in order to jumpstart scientific projects that had laid dormant following the war. Haber remained particularly adept in this task and the

[16] Victor Lefebure, *The Riddle of the Rhine*, 1921 (BA) 201-043. The title of Lefebure's study is a play on Erskine Childers' *The Riddle of the Sands* (1903), a spy novel that dealt with a secret German plan to invade England.

[17] Hew Strachan, *Financing the First World War* (Oxford: Oxford University Press, 2004), 104–105.

[18] Chemische Abteilung II, 1927 (BA-MA) RH/12/4. See also Jeffrey A. Johnson and Roy Macleod, "The War the Victors Lost: The Dilemmas of Chemical Disarmament 1919–1926," in *Frontline and Factory: Comparative Perspectives on the Chemical Industry at War, 1914–1924* (Dordrecht: Springer, 2006), 230–238.

[19] Roy M. MacLeod, "Chemistry for King and Kaiser: Revisiting Chemical Enterprise and the European War," in *Determinants in the Evolution of the European Chemical Industry, 1900–1939*, eds. Anthony S. Travis, Harm G. Schröter, Ernst Homburg, and Peter J. T. Morris (Dordrecht: Kluwer Academic, 1998), 47.

majority of this money ultimately came from private donors. Reflecting on his skill in fundraising, Haber's godson wrote, "[Fritz Haber] had contacts everywhere – with officialdom, with politicians ... and with people of rank in the private world."[20]

At his KWI, Haber claimed that he had completely pivoted his research program away from chemical weapons. Instead, he turned his attention to pesticides, applying his knowledge of poison gases to the extermination of the head lice, fleas, mosquitoes, and mites that had plagued soldiers throughout World War I. Since the onset of the war, the German army had operated large sanitation centers at railway junctions where louse-ridden soldiers could bathe and have their clothing fumigated with sulfur dioxide.[21] However, the Germans rarely had a sufficient supply of sulfur dioxide and the delousing treatment often damaged scarce military uniforms. Seeking a solution to these problems, the military turned to the Degussa company, whose 1916 clinical trials had shown that hydrocyanic poison gas was effective for killing insects.[22] Hydrogen cyanide, or prussic acid, was a byproduct of the dye industry that proved too weak an irritant to serve as a battlefield gas. However, in enclosed spaces, the gas was far more deadly to insects and humans alike. With the support of Fritz Haber and military contracts in hand, Degussa began to successfully treat lice-infested military buildings with this new gas.[23]

As early as February 1917, Haber had already begun to stress the importance of his chemical weapons research for both military and civilian pest control. He claimed that agricultural pest control represented the most promising future for the application of his chemical technologies. However, given Haber's general desire to continue his poison gas research after the war, there is reason to believe that this claim was at least partly a calculated attempt to establish poison gas as an integral part of both scientific progress and the broader progress of human civilization.[24] This could be further seen in Haber's mid-war attempt to create a Kaiser Wilhelm Institute that would exclusively continue research on civilian applications for poison gas, a proposal that was ultimately rejected by the Kaiser Wilhelm Society. However, Haber

[20] Stern, *Einstein's German World*, 134. [21] Szöllösi-Janze, "Pesticides and War," 100.

[22] Degussa gained its first recognition through its work with the Society for Applied Entomology. In 1916 Degussa conducted gassing tests with members of the Agricultural College of Berlin, hoping to set up an institute for applied entomology in Berlin or Frankfurt. While this institute was never realized, chemists at the Berlin Hygiene Institute introduced Degussa's work to the German military and to Fritz Haber.

[23] Degussa also worked with Ferdinand Flury to develop a gas mask filter for hydrocyanic fumes. Paul Weindling, *Epidemics and Genocide in Eastern Europe, 1890–1945* (Oxford: Oxford University Press, 2000), 93.

[24] Szöllösi-Janze, "Pesticides and War," 102.

was able to organize a formal conference in which various research institutes and Degussa presented on the effectiveness of pesticides. This conference then led to the formation of the Technical Committee for Pest Control (TASCH) in February 1917. The committee, led by Haber, was comprised of volunteers who advocated for pest control research under the control of public law.[25] In that same year, in cooperation with Haber and Degussa, the entomologist Albrecht Hase (1882–1962) first erected a major hydrocyanic delousing site outside of Katowice. Over the final two years of the war, several more centers were brought into operation.[26]

Just like these military sites, civilian mills and storehouses also demonstrated a need for pest control during the war. In response, Haber staffed pest control committees with former gas pioneers and sent them on regional fumigation assignments.[27] This helped to establish pest control as both a lucrative consumer market and an applied scientific discipline in the postwar years. But by 1919, the Weimar government had outlawed the civilian use of the highly volatile and deadly hydrocyanic gas.[28] Accordingly, Haber then transformed his pest control committees into the Deutsche Gesellschaft für Schädlingsbekämpfung (Degesch), a unique company that was allowed to continue its experimental application of hydrocyanic acid under public law.[29] This experimentation eventually led to the development of a highly effective type of hydrocyanic acid called Zyklon, but the newly synthesized compound remained unstable and dangerously undetectable. During this same immediately postwar period, Degesch became a private company and moved its laboratories from Berlin to Frankfurt. In Frankfurt, Degesch built ties to Degussa, which would eventually buy Degesch in 1922.[30]

[25] The Technical Committee for Pest Control was a branch of the Chemical Department of the Prussian War Ministry. While several entomologists were part of the TASCH committee, TASCH and Haber engaged in several political battles with entomologists at the Society for Applied Biology who sought sole legal control over the use of hydrocyanic gas. Weindling, *Epidemics and Genocide in Eastern Europe, 1890–1945*, 95.

[26] Szöllösi-Janze, "Pesticides and War," 103.

[27] Jürgen Kalthoff and Martin Werner, *Die Händler des Zyklon B Tesch & Stabenow Eine Firmengeschichte zwischen Hamburg und Auschwitz* (Hamburg: VSA-Verlag, 1998), 17–18.

[28] Even during the war, Degussa primarily conducted its delousing in the Oberost since the Ministry of Agriculture maintained restrictions on the use of hydrocyanic gas in Germany proper. Weindling, *Epidemics and Genocide in Eastern Europe, 1890–1945*, 94.

[29] Peter Hayes, *From Cooperation to Complicity: Degussa in the Third Reich* (Cambridge: Cambridge University Press, 2004), 7.

[30] Paul Weindling, "The Uses and Abuses of Biological Technologies: Zyklon B and Gas Disinfestation between the First World War and the Holocaust," *History and Technology* 11 (1994): 293.

Fritz Haber also advanced postwar poison gas research by setting up a KWI pharmacological department under Ferdinand Flury and Albrecht Hase.[31] This department clandestinely continued Flury's wartime toxicological studies, using code words for various war gases to skirt the Allies' armament restrictions.[32] Flury and Hase tested mustard gas on cockroaches, created arsenic bombs, and even acquired airplanes and chemical shells for pesticide dispersion studies.[33] But by 1920, when the Military Inter-Allied Control Commission began to enforce the chemical restrictions of the Treaty of Versailles, the KWI had dismantled all chemical warfare research including Flury and Hase's pharmacological department. This was not the end, however, and Fritz Haber was able to move much of this research program to the newly restructured Reich Institute of Biology.[34] By 1926, Haber had raised funds from the German military to staff this renewed pharmacological laboratory with four chemists and four assistants.[35] And while there is no definitive evidence of the laboratory's interwar activities, it is generally suspected that it worked on both civilian and military poison gas applications throughout the Weimar era. Furthermore, in 1924, Ferdinand Flury became the central contact for the German military in "coordinating weapons research."[36] Albrecht Hase, on the other hand, continued to work on stabilizing Zyklon with supplementary funds from Degesch. In 1924, in Hase's laboratory, the scientists Walter Heerdt and Bruno Tesch succeeded in producing Zyklon B, a stabilized pellet form of Zyklon that was more easily detected and could be safely packed in sealed canisters.[37]

Between his collaboration with Degesch and his assistance for Flury and Hase, Fritz Haber remained quite active in poison gas research through the 1920s.[38] Historian Fritz Stern has even speculated that Haber may have been involved in several military discussions about the future wartime applications of poison gas throughout the interwar

[31] Kampfstoff Forschung Beiheft: Sitzungberichte, 1925–1931 (BA-MA) RH/12/4.

[32] Scott Christianson, *The Last Gasp: The Rise and Fall of the American Gas Chamber* (Berkeley: University of California Press, 2010), 96.

[33] For Flury and Hase's other interwar research, see Versuche mit Nebelgerat Ausfuhr vom Nebelgerat, 1930–1931 (BA-MA) RH/12/4; Sarah Jansen, "Histoire d'un transfert de technologie: De l'étude des insectes à la mise au point du Zyklon B," *La Recherche* 340 (2001).

[34] Sarah Jansen, *"Schädlinge": Geschichte eines wissenschaftlichen und politischen Konstrukts, 1840–1920* (Frankfurt: Campus Verlag, 2003), 353–354.

[35] Szöllösi-Janze, "Pesticides and War," 107.

[36] Maier, *Chemiker im "Dritten Reich"*, 263.

[37] Rh/12/9 Sig 26, 1926 (BA-MA).

[38] Thomas Hager, *The Alchemy of Air: A Jewish Genius, a Doomed Tycoon, and the Scientific Discovery That Fed the World but Fueled the Rise of Hitler* (New York: Harmony, 2008), 186.

period.[39] Haber maintained still more connections to the chemical weapons industry through his former gas pioneers, who now served as technical experts in various public and private chemistry labs.[40] One of the more important of these former employees was Hugo Stoltzenberg.

At the start of World War I, Stoltzenberg served as an infantry lieutenant on both the Western and Eastern fronts. After suffering five different war wounds, he was selected by Haber to serve in the pioneer regiments due to his completion of a prewar chemistry dissertation. While installing a gas cylinder during one of Haber's chlorine test trials, the chlorine drum exploded and blinded Stoltzenberg in the left eye. After participating in the first chlorine attack at the Second Battle of Ypres, this eye injury then prevented Stoltzenberg from serving in combat positions throughout the remainder of the war. Transferred to the KWI laboratories, Stoltzenberg worked on gas mask technologies in Department A before serving as the director of the newly built Adlershof shell-filling plant and then the Klopper Works shell-filling plant at Breloh.[41] Stoltzenberg left Breloh in August 1918 and returned to the KWI, where he worked in Department D under Professor Heinrich Wieland to synthesize new forms of mustard gas.[42]

At the end of the war, with the German army's chemical production capacity vastly reduced, Stoltzenberg's future as a chemical weapons specialist appeared uncertain. Of course, Stoltzenberg's job was not the only military career in jeopardy. The stipulations of the Treaty of Versailles had reduced the size of the German military to 100,000 men and disbanded the General Staff. To oversee these changes, General Hans von Seeckt was

[39] Stern, *Einstein's German World*, 134.

[40] The status of German industrial chemists changed dramatically during World War I and the early 1920s. While, prior to the war, industrial chemists were often poorly compensated relative to their education, wartime chemical applications revealed the broader societal importance of their work. In the 1920s, these chemists began to carve out new professional roles that continued to advance the field's status and power. Jeffrey A. Johnson and Roy M. MacLeod, "War Work and Scientific Self-Image: Pursuing Comparative Perspectives on German and Allied Scientists in the Great War," in *Wissenschaften und Wissenschaftspolitik Bestandsaufnahmen zu Formationen, Brüchen und Kontinuitäten im Deutschland des 20. Jahrhunderts* (Stuttgart: Franz Steiner Verlag, 2002), 176; Jeffrey A. Johnson, "Academic, Proletarian … Professional? Shaping Professionalization for German Industrial Chemists, 1887–1920", in *German Professions, 1800–1950*, eds. Konrad Jaraush and Geoffrey Cocks (New York: Oxford University Press, 1990), 137.

[41] Kalthoff and Werner, *Die Händler des Zyklon B*, 22.

[42] Shortly after the final German defeat in 1918, a train carrying gas shells exploded outside of the Breloh gas plant, effectively destroying much of the facility. German authorities reported this explosion as an accident, but it was more likely a method to prevent the Allied appropriation of the chemical facility. Rolf-Dieter Müller, "Die deutschen Gaskriegsvorbereitungen 1919–1945. Mit Giftgas zur Weltmacht?," *Militärgeschichtliche Zeitschrift* 27, no. 1 (1980): 26.

appointed the Chairman of the Committee for the Organization of the Peacetime Army.[43] Seeckt had served under General Mackensen during World War I and was central in planning Mackensen's victories on the Eastern Front. Several of these victories involved the use of Haber's gas cylinders, and Seeckt vocally supported innovative offensive military tactics that took advantage of new technologies.

While Seeckt was obligated to reduce the size of the German military, he remained deeply committed to nationalist and monarchist politics, ensuring that the army retained connections to its imperial past. As such, he expressed the military's broadly shared anger over the terms of the Treaty of Versailles and rejected the idea of Germany joining the League of Nations.[44] As early as 1920, Seeckt had created a special Reichswehr working group called Sondergruppe R that sought to secretly maintain the German armament industry in Soviet Russia. After the 1922 signing of the Treaty of Rapallo, which established diplomatic relations between Germany and the Soviets, the Reichswehr's discussions on the outsourcing of military industries to Russia gained significant traction. Derived from his general disregard for the terms of Versailles and his overwhelming concern for interwar national security, Seeckt pushed for the secret continuation of German chemical weapons research and production.[45] A German military report from 1919 revealed the strategic importance of continued training with chemical weapons:

> It can be said that the training has been enough, but a uniform management of gas protection is necessary for the whole empire in the event of a war on German soil. Even if we do not want to employ chemical weapons in the future, the need to outfit troops with gas shell projectors and train them in gas protection still exists since there is no certainty that our future opponents would refrain from using gas.[46]

43 James S. Corum, *The Roots of the Blitzkrieg: Hans von Seeckt and German Military Reform* (Lawrence: University of Kansas Press, 1992), 93.

44 Wolfram Wette, *The Wehrmacht: History, Myth, Reality* (Cambridge, MA: Harvard University Press, 2006), 144.

45 In 1923, Seeckt created the *Kommission für chemische Frage* (the Commission for Chemical Questions) in the Reichswehr Ministry. This military commission, in coordination with Ferdinand Flury, attempted to envision future chemical warfare methods, spending significant time and effort on the development of *Kunstnebel* or "artificial fog" during the interwar years. Artificial fog was often a code term for poison gas, although commanders also hoped that particulate fog could create large smoke screens for hiding troops or ships. Versuche mit Nebelgerat usw Ausfuhr von Nebelgerat usw, 1923–1928 (BA-MA) RH/12/4.

46 Allgemeine Erfahrungen im Gaskampf, 1919 (BA-MA) PH 14/226.

Other German commanders who shared these sentiments assumed that chemical warfare would surely be part of any future European war.[47] Thus, many continued to have their soldiers train with gas shells and gas masks throughout the 1920s.[48] The sanitary services carefully documented and stored old rubber and leather gas masks from World War I and gas training schools remained covertly operational.[49] At these schools, men were expected to practice World War I gas mask training procedures at least twice per year.[50]

Seeckt was not only intent on organizing the Reichswehr's chemical protection measures; he also wished to restart the production and stockpiling of poison gases for offensive warfare. A 1924 report from first lieutenant Joachim von Stülpnagel expressed the broadly perceived importance of producing offensive chemical weapons: "Only when we are capable of covering enemy cities with the same amount of gas, will the enemy reconsider using chemical weapons … In a plethora of thoughts and questions, confusing and oppressive, strikes the historical responsibility for the liberation of our Land."[51]

In order to jumpstart a chemical weapons program, Seeckt turned again to Fritz Haber, who suggested contacting Hugo Stoltzenberg in 1920.[52] With the assent of Seeckt, Stoltzenberg first began to sell off chemicals from the now-defunct Breloh plant to foreign countries such as the United States and Sweden. The Allies allowed this as a necessary postwar "clean-up process" that also saw Stoltzenberg dumping hazardous chemicals in the North and Baltic Seas as well as the peat bogs near Münster.[53] Between 1920 and 1925, Stoltzenberg's new Hamburg-based chemical factory, the Chemische Fabrik Doktor Ingenieur Hugo Stoltzenberg, served as the clearing house for still more chlorine, mustard gas, and ethylene.[54] Stoltzenberg's company began setting up offices in Berlin, Madrid, Moscow, New York, and Constantinople in order to sell German chemical weapons or chemical production systems to

[47] Hans von Seeckt, Vortrag von Gaskrieg (BA-MA) RH 2/2207, File HL 27 1923 1A.

[48] Gasschutzgerät, 1919–1920 (BHK) Reichswehrbrigade 23 127.

[49] Bestimmungen Ausstattung des Heeres mit Gasschutzgerat, 1920–1924 (BA-MA) RH/12/4.

[50] Befehle zum Schutz gegen Gas, 1925 (BA-MA) RH/12/4.

[51] Kalthoff and Werner, *Die Händler des Zyklon B*, 22.

[52] Haber suggested Stoltzenberg because the Allies still carefully monitored the IG Farben factories in the demilitarized Ruhr zone.

[53] Benjamin Garrett, "Hugo Stoltzenberg and Chemical Weapons Proliferation," *The Monitor* 1, no. 2 (Spring 1995): 11.

[54] In 1923, Stoltzenberg's factory held about one tenth of the chemical weapons that Germany produced in 1918. Hans Günter Brauch and Rolf-Dieter Müller, *Chemische Kriegführung–Chemische Abrüstung: Dokumente und Kommentare* (Berlin: Berlin Verlag, 1985), 27.

foreign militaries.[55] The large-scale movement between German chemical weapons depots, Stoltzenberg's seemingly legitimate chemical factory, and international buyers also allowed for a certain level of clandestine work. For instance, in 1924, Stoltzenberg established a second, larger factory at Gräfenhainichen, a town north of Leipzig, where he began to illegally produce mustard gas and chlorine for the German army.[56]

In 1921, Stoltzenberg also began serving as the chief engineer at Spain's new La Marañosa mustard gas plant.[57] The Spanish military used Stoltzenberg's mustard gas between 1922 and 1927 to subdue Abd el-Krim's Rif rebellion in Morocco.[58] The Spanish crown viewed chemical weapons as an ideal tool for the suppression of colonial uprisings, and the aerial deployment of mustard gas against Moroccan rebels went largely unnoticed by the rest of the world until the late 1920s.[59] After manufacturing new gas masks, designing new gas bombs, and helping to draft the strategy for the first aerial gas attacks, Stoltzenberg was even granted Spanish citizenship in 1923. That same year, he received another contract from the Soviet Union to modernize their chemical industry and produce 35 million Reichsmark worth of chemical weapons. This contract was directly tied to Seeckt's desire to outsource German armaments production to Russia.[60] With both Fritz Haber and Colonel Max Bauer serving as the primary liaisons for the endeavor, Stoltzenberg set about creating a joint German-Russian chemical factory at the Volga River town of Saratov.[61] Named Bersol AG, the collective project aimed at turning an

[55] Hugo Stoltzenberg, *Anleitung zur Herstellung von Ultra Giften* (Hamburg: Norwi-Druck, 1930), 3–8.

[56] During this period, Stoltzenberg repeatedly tried to obtain government approval to begin the production of hydrocyanic acids like Zyklon A and B. Weindling, *Epidemics and Genocide in Eastern Europe 1890–1945*, 134.

[57] Michael Freemantle, *The Chemists' War 1914–1918* (Cambridge: The Royal Society of Chemistry, 2015), 185.

[58] In 1924 alone, the Spanish ordered thirty tons of mustard gas from Stoltzenberg's factory. Sebastian Balfour, *Deadly Embrace: Morocco and the Road to the Spanish Civil War* (Oxford: Oxford University Press, 2002), 147.

[59] Between May and September 1924, the Spanish dropped 24,104 bombs on the Moroccan rebels. Over the course of twenty-one days in May 1925, the Spanish dropped 3,000 mustard gas bombs on the town of Anjera alone. Rudibert Kunz and Rolf-Dieter Müller, *Giftgas Gegen Abd El Krim: Deutschland, Spanien und der Gaskrieg in Spanisch-marokko, 1922–1927* (Freiburg: Verlag Rombach Freiburg, 1990), 72.

[60] Vasilis Vourkoutiotis, *Making Common Cause: German-Soviet Secret Relations, 1919–22* (New York: Palgrave Macmillan, 2007), 172–173.

[61] The Reichskabinett gave the military 60 million Reichsmark for the project in Russia. This was part of the larger GEFU (Gesellschaft für Förderung gewerblicher Unternehmungen) deal, which attempted to create German industrial sites for the manufacture of airplanes and chemicals in Russia. Rolf-Dieter Müller, *Das Tor zur*

old Russian chlorine plant into a modern mustard gas facility.[62] The Reichswehr subsequently sent German engineers and chemists to Russia in order to undertake government-directed chemical weapons research, with special emphasis on the aerial delivery of chemical weapons. At the same time, these technicians were instructed to impart their specialized knowledge to their Russian counterparts.[63]

The Bersol factory succeeded in producing mustard gas, but Stoltzenberg's role in the endeavor ended in 1927 when the Soviets expressed a desire to take over all aspects of production. After four years of joint work, both the Germans and Russians soured on the deal, describing each other as unreliable partners. This fallout was further precipitated by the IG Farben concern, who leaked the Bersol project to the foreign press in order to thwart Stoltzenberg's competition in the chemical industry. *The Manchester Guardian* reported the project on December 3, 1926, thus forcing the Germans to scale down their involvement in order to avoid international confrontation.[64] In the following year, the Reichswehr removed Stoltzenberg from his position in Bersol AG, liquidated the project, and dismantled Stoltzenberg's Gräfenhainichen factory. Fritz Haber cut ties with his former protégé after this major professional setback, and Stoltzenberg was left in financial ruin. For the remainder of his career in the chemical industry, Stoltzenberg was compelled to seek independent armament contracts from foreign countries.[65]

Even after the Bersol failure, the German military continued other chemical weapons projects in the Soviet Union. This included the Tomka chemical test school, located to the northeast of Saratov. The school was first erected in 1927, and by 1929, thirty-three expert German chemists lived and worked in Russia.[66] Up until 1933, these men tested everything from the chemical resistance of rubber clothing to the feasibility of aero-chemical attacks.[67] While the Reichswehr would have

Weltmacht: Die Bedeutung der Sowjetunion für die deutsche Wirschafts- und Rüstungspolitik zwischen den Weltkriegen (Boppard am Rhein: Harald Boldt Verlag, 1984), 141.

[62] Garrett, "Hugo Stoltzenberg and Chemical Weapons Proliferation," 11.

[63] Ian Ona Johnson, *Faustian Bargain: The Soviet-German Partnership and the Origins of the Second World War* (Oxford: Oxford University Press, 2021), 59–60.

[64] Henning Schweer, *Die Geschichte der Chemischen Fabrik Stoltzenburg bis zum Ende des Zweiten Weltkrieges* (Diepholz: Verlag für Geschichte der Naturwissenschaften und der Technik, 2008), 79.

[65] Stoltzenberg received chemical weapons contracts from Yugoslavia and Brazil in the late 1920s and 1930s. Zahlungsschwierigkeiten der chemischen Frabrik Dr Hugo Stoltzenberg, 1926 (BArch-B) R 43-I/420.

[66] Tomka, 1930–1931 (BA-MA) RH/12/4.

[67] Bericht V, 1931–1933 (BA-MA) RH/12/4.

undoubtedly preferred to conduct these tests on German soil, Allied supervision continued to prevent direct cooperation with the major German chemical companies. Separated from government contracts in the wake of World War I, these chemical concerns sought protection and stabilization on the reopened world market.

Escaping mostly unscathed from the Military Inter-Allied Control Commission, Bayer, BASF, and Hoechst remained giants in the world of chemical production. Nevertheless, in 1925, they chose to further consolidate their regional alliances into the IG Farben conglomerate. As part of the merger, BASF, Bayer, and Hoechst now each made up 27.4 percent of the interest group's holdings.[68] Primarily under the leadership of BASF's Carl Bosch and Bayer's Carl Duisberg, the IG generated two-thirds of all German chemical industry profits from 1925 to 1929.[69] Valued at 1.4 million Reichsmark and maintaining a workforce of about 100,000 people in 1926, the chemical group was the largest private enterprise in all of Europe and the fourth largest in the world. The IG factories returned to their prewar focus on dye production, but they retained the ability to easily manufacture large amounts of chemical weapons from their raw materials and byproducts.

The IG companies also forged connections to the pesticide industry, buying 42.5 percent of Degesch in 1930 and contributing a stabilizing agent for the production of Zyklon B.[70] This allowed Degesch to sell Zyklon B throughout the 1930s as an industrial strength delousing product that could be released in a patented sealed chamber. None of this activity, however, raised significant international suspicion that the IG conglomerate might renew production of chemical weapons. Indeed, in a 1925 discussion with a German journalist, a French chemical weapons inspector proclaimed that "Everything is all right with the German chemical industry ... There can be no question of it manufacturing poison gas or being in any way not completely back on a peacetime footing."[71]

During this same mid-decade period, the German gas mask companies also sought to stabilize their business. With the reduced demand for gas

[68] Agfa represented 9 percent, Chemische Fabrik Griesheim-Elektron 6.9 percent, and Chemische Fabrik vorm. Weiler Ter Meer 1.9 percent. Helmuth Tammen, *Die I.G. Farbenindustrie Aktiengesellschaft (1925–1933): Ein Chemiekonzern in der Weimarer Republik* (Berlin: H. Tammen, 1978), 195.

[69] Peter Hayes, *Industry and Ideology: IG Farben in the Nazi Era* (Cambridge: Cambridge University Press, 2001), 17.

[70] Hayes, *From Cooperation to Complicity*, 275.

[71] Harry Kessler, *Berlin in Lights: The Diaries of Count Harry Kessler (1918–1937)*, trans. Charles Kessler (New York: Grove Press, 1999), 249.

masks, the Auergesellschaft focused on the production of lightbulbs through the cooperative OSRAM company, which started coordinating the production of Auer, Siemens & Halske, and AEG in 1923. At the same time, Auer continued a reduced level of gas mask production while also repurposing World War I gas masks for industrial purposes as part of the Industrial Mask Sales Cooperative with Hagenuk and the Drägerwerks. In 1922, Karl Quasebart joined Auer's managing board and succeeded in increasing industrial demand for Auer gas masks, especially in the mining industry.[72]

The Drägerwerks, on the other hand, was forced to lay off most of its employees and end its gas mask production at the end of World War I. The company lost its international patents and inflation continued to hurt its profits through 1923. In order to remain solvent, the company produced linens, clothing, and curtains until its revenue began to stabilize. In 1924, it was able to return to respirator production with 200 employees.[73] Dräger then began to market its products more exclusively to the mining industry, thus coming into direct competition with the Auergesellschaft.[74] Throughout the remainder of the 1920s, the two companies entered into numerous cartel agreements, patent disputes, joint contracts, and legal battles in the process of negotiating the reduced national gas mask market.[75] But beyond the borders of Germany, Dräger developed a greater focus on rebuilding its international markets, and by 1928, Heinrich Dräger had reopened offices in the United States and the Soviet Union.[76] Alongside these legitimate enterprises, both Dräger and Auer received several small gas mask contracts from the Reichswehr that violated the terms of the Treaty of Versailles.[77] Taking these contracts without much apparent hesitation, Auer developed a new military gas mask called the Gas Mask 24 (GM 24) that allowed for greater movement, featured a longer-lasting filter, and reportedly reduced the interior temperature of the mask.[78] Further still, Auer created gas training grounds at their factory in Oranienburg and provided ancillary products for the production of mustard gas.[79]

[72] Formerly an engineer in a mirror factory and a member of the commission that organized the surrender of German heavy machinery after World War I, Quasebart received his doctorate in chemistry from the Technical University in Aachen. He then became a professor of metallurgy at Aachen and helped to found the German Glass Technology Society before joining Auer. Hayes, *From Cooperation to Complicity*, 127.

[73] Michael Knoll, *Die Drägerwerke im Dritten Reich* (Norderstedt: GRIN Verlag, 2004), 6.

[74] Haase-Lampe, *Sauerstoffrettungswesen und Gasschutz*, 13.

[75] Lorentz, *Industrieeltie und Wirtschaftspolitik 1928–1950*, 151. [76] Ibid., 39.

[77] Chemische Abteilung II, 1927 (BA-MA) RH/12/4. [78] RH/12/9/20, 1930 (BA-MA).

[79] Bestimmungen Ausstattung des Heeres mit Gasschutzgerat- Beschaffung, Versendung, Verwaltung und Unterbringung, 1920 (BA-MA) RH/12/4.

Anticipating future competition over these military contracts, Auer and Dräger negotiated the division of the gas mask market in 1929. In all forthcoming government contracts, Auer would produce 65 percent and Dräger 35 percent of gas masks ordered by the Reichswehr. The two companies would split all contracts for other gas protection devices equally and Dräger would produce 65 percent of all other oxygen-providing equipment.[80] This expectation of future military contracts would prove decisive in keeping the two companies afloat during the Great Depression.[81]

Since 1916, the German government had discussed the protection of civilians in the event of a large-scale poison gas attack. But even in the 1920s, it remained unclear whether such large-scale attacks were possible or if the gas mask served as the best form of civilian defense. A military report from 1925 read:

> Gas protection for the population is urgently needed wherever enemy air raids are most likely to take place. The gas protection device for the civilian population must be easy to handle and durable. The gas mask does not meet these conditions: the mass provision of the various sizes, the ease of losing them, the need to check them in gas chambers for a good fit, the periodic tests for gas impermeability, and their ineffectiveness with the slightest damage to the mask fabric make them unsuitable as protection for the masses.[82]

Regardless of the mask's imperfections, it had certainly proved valuable for emergency services and the military. As such, in 1928, the Reichswehr Quartermaster Wilhelm Groener renewed large military gas mask contracts with Dräger and Auer. The first contract would pay 1.45 million Reichsmark for the fabrication of 170,000 new Gas Mask 24s.[83] The reintroduction of large-scale standardized gas mask production also encouraged a greater number of gas training courses. Both Auer and Dräger began to regularly provide such courses at their factories and through traveling seminars.[84] These courses and gas protection products were advertised in the companies' respective trade journals. The Dräger company produced the *Drägerheft*, published triennially under the director of the literature department, Wilhelm Haase-Lampe. And the Auergesellschaft produced a monthly sales publication called *Die*

[80] Hayes, *From Cooperation to Complicity*, 127.
[81] Auergesellschaft Oranienburg, 1929 (BL) Rep 43 Eberswalde Nr 129.
[82] Bestimmungen Ausstatlung des Heeres mit Gasschutzgerat (BA-MA).
[83] Berichte Korrespondenz von Bauer an Letcher und Strauss (Röhm Bestand) 1928–1933 (EA) RHT.2./6.
[84] Ausbildung der Johanniterschwestern im Gasschutz, 1932 (LBWH) P 7/2 Bü 466.

Gasmaske, which was transformed into a full trade journal in 1929, complete with long-form articles on firefighting, industrial chemistry, and the mining industry.[85] While these realms and the military remained the largest markets for gas protection technologies, both the *Drägerheft* and *Die Gasmaske* began to dedicate the majority of their pages to general civilian protection. And in doing so, they helped to both turn aerochemical attack into a national concern and to reignite debates over civilian gas mask distribution.

Restriction or Resistance: Waldus Nestler, Gertrud Woker, and the Pacifist Response

After weathering both state surveillance and general public distrust during World War I, pacifist political groups in Germany began to see an uptick in interest and support immediately following the war.[86] In 1920, pacifists began to stage so-called No-More-War Demonstrations, attracting hundreds of thousands of supporters across the nation.[87] Beyond making retrospective claims about the horrors and immorality of World War I, these rallies began the process of galvanizing transnational pacifist networks and linking individual pacifists to sympathetic organizations of feminists and Social Democrats across Germany. One such tangential group was the International Committee of Women for Permanent Peace, renamed the Women's International League for Peace and Freedom in 1919, which could count well-known peace activists such as Jane Addams and Helene Stöcker as members. Through publications such as the Women's International League's *Pax et Libertas*, the German Peace Society's *Völker-Friede*, and the Social Democrats' *Vorwärts*, pacifists throughout Germany were able to advocate for total disarmament, the end of conscription, and the banning of the right to declare war.[88] However, the domestic instability caused by the 1923 French occupation of the Ruhr region led to the prohibition of all public demonstrations in Germany, thus turning No-More-War rallies into much smaller private gatherings. By 1925, the No-More-War movement had lost its public recognition and the focus of the Social

[85] Welf Böttcher and Martin Thoemmes, *Heinrich Dräger: Eine Biographie* (Wachholtz Verlag, 2011), 59–60.
[86] Regina Braker, "Helene Stocker's Pacifism in the Weimar Republic: Between Ideal and Reality," *Journal of Women's History* 13, no. 3 (2001): 72.
[87] Shelley E. Rose, "The Penumbra of Weimar Political Culture: Pacifism, Feminism, and Social Democracy," *Peace & Change* 36, no. 3 (2011): 317.
[88] Benjamin Ziemann, *Contested Commemorations: Republican War Veterans and Weimar Political Culture* (Cambridge: Cambridge University Press, 2013), 241.

Democratic Party turned away from independent pacifist activities toward Marxist-inflected class struggle.

As such, Germans with pacifist convictions remained politically divided throughout the 1920s. The major point of contention was whether the ultimate arrival of socialism or communism justified the use of violence. Clearly the communists believed so, as did many of the more radical socialists; but those committed primarily to pacifism disagreed, often remaining completely unconvinced or unconcerned by Marxist theory. Not surprisingly, this ideological divide frequently fell along class lines, with more middle-class activists gravitating toward pacifist-first and moderate Social Democratic policies and more working-class activists drawn toward the Communist Party. This roughly defined distinction also had major implications for the rhetorical arguments surrounding disarmament throughout the 1920s.

While public concerns over disarmament lulled briefly in 1923, the momentous 1924 publication of Ernst Friedrich's *Krieg dem Kriege!* visually reminded Germans of the costs of industrialized war. Friedrich's photobook presented images of dead, dying, and mutilated World War I soldiers in order to refute the common postwar attempts to make sense of the fighting through stories of soldier honor and bravery.[89] The gruesome images of death and destruction helped to make Friedrich's book a national sensation and to designate 1924, the ten-year anniversary of mobilization, as the "antiwar year."[90] The first 70,000 copies of Friedrich's book sold out in the first few minutes of availability and the leftist journalist Kurt Tucholsky wrote of the book, "no written work can come near the power of these images … Whoever sees these and does not shudder is not a human being."[91]

Friedrich claimed that his photobook could reveal the true horrors of war and inspire Europeans to shun future warfare. Photography, he argued, maintained the ability to evoke greater emotion and was thus a more honest medium than text. Perhaps equally important to the book's success, Friedrich's self-proclaimed anarcho-pacifism was not strictly Marxist in nature and his work spoke directly to pacifists without necessarily alienating others on the political left.[92]

While the publicity surrounding Friedrich's book began to die down in 1925 and nationalist authors presented heroic counter-narratives of the

[89] Ernst Friedrich, *Krieg dem Kriege!* (Berlin: Freie Jugend, 1924), 130.

[90] Dora Apel, "Cultural Battlegrounds: Weimar Photographic Narratives of War," *New German Critique* 76 (1999): 49.

[91] Kurt Tucholsky, "Waffe gegen den Krieg," *Die Weltbühne* (February 23, 1926).

[92] Apel, "Cultural Battlegrounds," 54.

war that enjoyed great public appeal, *Krieg dem Kriege!* focused pacifist attention more acutely on the deadly methods of modern warfare. A significant part of Friedrich's critique was levied at poison gas, which he presented as one of the most feared and mythologized weapons of World War I. Interestingly, not all pacifists agreed with Friedrich's condemnation of a weapon that might be used as a form of deterrence against future wars, but nevertheless, by 1924, a small but dedicated antigas movement had developed out of the long-standing pacifist calls for national disarmament. Such reformers rightly claimed that the fighting on the Western Front had demonstrated exactly how these weapons could either asphyxiate humans immediately or slowly destroy the body from the inside out. They further insisted that European militaries still maintained massive chemical arsenals left over from the war.

The development of a semi-formalized German antigas movement led to a slew of fiction and nonfiction publications that attempted to awaken the public to the dangers of chemical weapons.[93] Pacifists who wished to appeal to the public's sense of justice called for international legislation that could ban future chemical warfare. This reflected a general commitment to the international and transnational networks that had helped to build the formal pacifist groups before and after World War I. However, many Germans were skeptical of the effectiveness and fairness of international cooperation. Certainly, to their minds, the 1899 and 1907 Hague Conventions' attempts to preemptively ban chemical weapons did little to prevent the outbreak of chemical warfare in World War I. Even the pacifist Karl Helmut Pfuhl admitted, "The danger of the violation of such a [gas warfare] prohibition would not be great; every city, due to its preparation for civilian gassing, would be able to respond to an attack on its civilian population without having to make any special arrangements."[94]

In the quest to provide popular support for international arms limitations, antigas pacifists regularly criticized the monetary cost and difficulties associated with gas protection. Relying on the public's broad postwar distrust of war profiteering, the pacifist Franz Carl Endres claimed that the chemical industry stood to make immense sums of money by outfitting Germans with gas protection devices such as the gas mask. According to Endres, gas protection was simply a scam that

[93] Andreas Pehnke, *Der Hamburger Schulreformer Wilhelm Lamszus (1881–1965) und seine Antikriegsschrift "Giftgas über uns": Erstveröffentlichen des verschollen geglaubten Manuskripts von 1932* (Beucha: Sax Verlag, 2006), 52.

[94] Karl Helmut Pfuhl, *Gaskrieg Und Völkerrecht* (Würzburg: Handelsdruckerei Würzburg, 1930), 57.

could never protect civilians in the event of a gas bombing. He wrote, "Industry dictates, and scholars and militants are not astute enough to realize that they are just puppets of big business."[95] In the end, Endres claimed, only broad international cooperation could truly protect civilians from chemical attacks.

A second major rhetorical strategy for German disarmament relied on scare tactics and dubious assertions. Reflecting on the technological advancement of airplanes during and after the war, several antigas reformers argued that gas could now be aerially dropped into any major civilian population center, effectively wiping out large numbers of innocent people. For instance, the pacifist Kurt Hiller wrote:

> The poison gas war of the future will overshadow the uniform and the fronts. Air squadrons will thrust to the center of a country and gain access to big cities. Nobody will be immune to the annihilation anymore – not women, not the elderly, and not children.[96]

While European militaries certainly sought the ability to conduct devastating aero-chemical raids, scientific studies from the 1920s revealed that gas bombs struggled to create dense enough clouds to affect people on the ground. As military engineers had learned in World War I, gas clouds had to be extremely thick in order to hang together and to have any serious effect.[97] Furthermore, weather and topography still had major impacts on the success of gas attacks.[98] One German chemical weapons report from 1923 read:

> Experiments in the test area, however, had already shown that with the materials so far discovered it would be almost impossible to spray a whole town with the chemical in such a way that a sufficient number of mouths would be reached simultaneously causing a panic-like terror.[99]

But due to either a lack of knowledge or a willingness to play fast and loose with the technical facts, pacifists continued to rely on apocalyptic visions such as Hiller's to express the dire consequences of chemical warfare. In another instance, Hans Harder-Hamburg wrote:

[95] Franz Carl Endres, *Giftgaskrieg: die grosse Gefahr* (Zürich: Rascher & Cie, 1928), 39.

[96] Kurt Hiller, *Der Sprung ins Helle* (Leipzig: Wolfgang Richard Lindner Verlag, 1932), 95.

[97] *Dräger Gasschutz im Luftschutz: Individual-Gasschutz Kollektiv-Gasschutz; Charakter des Chemischen Krieges Chemische Kampfstoffe; Organisation des Luftschutzes Städtebau und Luftschutz*, vol 2 (Lübeck: Kommissionsverlag H. G. Rahtgens, 1933), 50–51.

[98] Hugo Stoltzenberg Fabrik, "Gibt es ein Gaskampf?" n.d. Leibniz-Informationszentrum Wirtschaft, http://zbw.eu/beta/p20/company/42006/00001/about.en.html.

[99] Alexander Kluge, *The Air Raid on Halberstadt on 8 April 1945*, trans. Martin Chalmers (London: Seagull, 2014), 121.

Only a few today doubt, despite some melodious declarations, that a future war will feature the use of poison gas. Nevertheless, some still argue that war is uglier and more brutal for all of civilization when poison gas is not used. In our opinion, there can be no doubt that the advances in toxic gas chemistry and aircraft technology since 1918 are so great that the brutal effects of a poison gas can hardly be understood.[100]

Two of the most strident antigas reformers, Waldus Nestler and Gertrud Woker, broadly subscribed to Hiller's alarmist approach. Nestler and Woker's uniqueness, however, lay in their realization that the field of gas protection posed an equally dangerous threat to international peace. With their assurances of national security through seemingly no-loss technological solutions, gas protection scientists and engineers offered Germans a tenuous peace that would inevitably rely on a defensive arms race. Only the nation that could produce the most gas masks, air raid shelters, antiaircraft weapons, etc., would be able to survive what seemed like an unavoidable chemical war. Nestler wrote in his widely read pacifist tract, *Giftgas über Deutschland (Poison Gas Above Germany)*, that "the air raid people are the biggest enemies of our security today, because they calm us down with silly gimmicks that prevent humanity from becoming aware of the tremendous danger and to take corresponding measures against it."[101] Nestler further pointed to the absurdity of training civilians to use gas masks:

Imagine children turning to the psychic discipline that the battle-hardened men sometimes lacked – the children crying, moaning, wheezing. I would like to see the mother staying calm and not losing her head! They will pull the mask from their faces to comfort the little ones, and death will seize the infants! For the babies, you need to construct a gas-safe box with filter and ventilation in which you can put their cribs. And the poor mother then stands by for hours and pumps air in and out, unaffected by the screaming and whimpering. No, the emotional situation will be so terrible that few people will cope with it.[102]

Like in Nestler's passage, it was not unusual for pacifist appeals to rely on gender stereotypes in order to make their case for international peace. By describing women and children as completely helpless, Nestler could point to the impossibility of complete gas protection and call on German men to defend a patriarchal hearth and home through the only means

[100] Hans Harder-Hamburg, "Der Gaskrieg im Völkerrecht," *Die Friedens-Warte* 27, no. 5 (1927): 137.

[101] Waldus Nestler was born in 1887 in Meissen. He studied theology at several universities before serving as a gas protection officer in World War I. After the war, he taught Latin, German, and Practical Philosophy in Leipzig while increasing his participation in religiously minded pacifist organizations. Pehnke, *Der Hamburger Schulreformer Wilhelm Lamszus*, 57.

[102] Nestler, *Giftgas über Deutschland*, 17.

available, namely international legislation. Such arguments elided the experiences of the thousands of female nurses who regularly and effectively dealt with gas in the field hospitals of World War I. These women were fully aware of the misery of gas death and had used the gas mask with sufficient skill to avoid a similar fate.

The image of helpless women and children was quite popular and not all antigas activists who appealed to patriarchal and familial protection were men. Harkening back to the World War I ideal of the "motherly heroine" in her antigas writing, Gertrud Woker tended to depict German women as mothers, only valued insofar as they served a nurturing role during the moment of attack. Woker, a professor of chemistry at the University of Bern, cofounded the Women's International League for Peace and Freedom (WILPF) in 1915.[103] She then served as the coleader of the organization's Swiss branch with Clara Ragaz.[104] Given her credentials in academic chemistry, Woker was particularly qualified to discuss the nature of poison gases, and she became especially dedicated to the eradication of chemical weapons. After visiting the American Gas Armament Center only months before a 1924 WILPF Conference in Washington, DC, she began to fully realize the dire threat of national chemical arscnals and agreed to serve on the WILPF's Committee on Chemical Warfare.[105] This committee sent memorandums to delegates at international disarmament conferences in an attempt to create legislation that might limit future chemical weapons use. While the WILPF normally advocated for the end to all forms of violence, Woker convinced the organization to support international legislation that merely sought limitations on poison gas because, as she saw it, chemical weapons produced such a uniquely horrifying form of warfare.[106]

Reflecting on her scientific training, Woker's writings and public presentations regularly spoke directly to the gas specialists who claimed

[103] Gertrud Woker was born in 1903 in Bern, Switzerland. The daughter of a theology and history professor, Woker received an exemplary education and was able to study organic chemistry at the University of Bern. She was the first Swiss woman to earn a doctoral degree in chemistry and then one of the first women to obtain a professorship in a German-speaking country. While she pioneered the study of biochemistry, Woker remained deeply committed to both women's suffrage and pacifism.

[104] For more on Gertrud Woker's life and work, see Gerit von Leitner, *Wollen wir unsere Hände in Unschuld waschen? Gertrud Woker (1878–1968) Chemikerin & Internationale Frauenliga 1915–1968* (Berlin: Weidler Buchverlag, 1998).

[105] In 1924, Woker also produced a widely distributed pamphlet entitled *A Hell of Fire and Gas* that warned of a future gas war alongside illustrations from Käthe Kollwitz.

[106] Jo Vellacott, "Feminism as If All People Mattered: Working to Remove the Causes of War, 1919–1929," *Contemporary European History* 10, no. 3 (2001): 390.

expert knowledge on poison gas. She accused these men of downplaying the dangers of chemical weapons and providing false promises of gas protection. In fact, in 1925, Woker declared that the structures of the increasingly specialized German scientific community were inherently tied to militarism, nationalism, and war.[107] She further pointed out that gas protection for an entire nation was still impossible, writing, "Anyone who is able to grasp the great importance of this question for the future of the human race will be convinced that we must try to explain to all people the fearful dangers of the future war."[108] Ultimately, Woker claimed that offers of gas mask distribution and other forms of gas protection only served to normalize and justify a future aero-chemical war.

Woker thus chastised any scientist who supported Fritz Haber's claim that chemical weapons were more humane than traditional explosives. After her trip to the American Gas Armament Center, Woker was convinced that effective aero-chemical attack was technologically possible, arguing in scientific literature reviews that research from World War I supported this claim.[109] Furthermore, Woker correctly stated that gas casualties from the war were underreported both because many gassed men were sent back to the front with mild symptoms and because many of the asphyxiation casualties were recorded as shell deaths. In her most comprehensive text on chemical warfare, *Der kommende Giftgaskrieg* (*The Future Gas War*, 1925), Woker claimed that scientists who downplayed the pain and suffering of gassing did so out of a repressed sense of shame. According to Woker, these gas specialists had personal financial interests in the continued development of poison gas and they chose to soothe the public's conscience in order to resume their lucrative work.[110]

At a 1926 WILPF conference in Frankfurt, Woker succeeded in passing an organizational resolution on chemical warfare that unequivocally stated the following: there was no effective gas protection for civilian populations; international agreements were not guarantees against chemical warfare because nations continued to produce chemical weapons; and a future gas war would surely destroy all civilization. In coordination with this resolution, the WILPF proclaimed its intention to educate civilians about the dangers of gas, express the impossibility of civilian gas

[107] Gertrud Woker, *Wissenschaft und wissenschaftlicher Krieg* (Zürich: Zentralstelle für Friedensarbeit, 1925), 3.

[108] Gertrud Woker, "Die Gaskampfinteressenten," *Die Friedens-Warte* 27, no. 2 (February 1927): 42.

[109] Gertrud Woker, "Die Wahrheit über den Gaskrieg," *Die Friedens-Warte* 27, no. 1 (1927): 6.

[110] Gertrud Woker, *Der kommende Giftgaskrieg* (Leipzig: Ernst Oldenburg Verlag, 1932), 243. This was Woker's most widely read book, going through nine editions by 1932.

protection, and wake people to the desperate need for lasting peace.[111] This ambitious agenda appears to have influenced the discussion of chemical weapons bans in international peace talks during the 1920s. For instance, after receiving a memorandum from the WILPF, the American delegate to the League of Nations Conference on Control of Traffic in Arms initiated an unscheduled discussion on chemical weapons that ultimately led to the 1925 Geneva Protocol's prohibition of the future use of chemical weapons.[112]

Like Nestler, Woker did discuss the need to protect helpless women in her calls to action. But gender was not the only category of identity that Woker employed to stoke public fear of poison gas. Rallying Germans around an essentialized ethno-racial concept of the nation, Woker warned of future global gas warfare. In a 1923 article, she wrote, "The agents that will play the most central role in the self-destruction of the white race are poison gas and modern aviation technology."[113] Further playing on long-standing German fears of Social Darwinian competition, Woker claimed that it was poison gas that would lead to the political rise of the "black race." This tapped into German doubts about the ability to completely control poison gas, whether it be in a future national, ethnic, and/or racial conflict. She made this explicit in a subsequent article from 1925, writing:

> From his first-hand knowledge of colonial troops [in World War I] [J.B.S.] Haldane stated that Negroes are relatively immune [to mustard gas]. It may be comforting for the colored population to know that once the European and American cities are transformed into Negro villages, the citizens no longer need to worry too much about mustard gas. It is only a matter of time before the Negro and other colored people realize the import of their immunity while the white race destroys itself. Poor white folks! It is not the height of your culture that determines your fate, but the mustard immunity of your black slaves. You have introduced the blacks to your murderous tools; when will these murderous tools be directed against you?[114]

Ultimately, antigas activists such as Gertrud Woker and Waldus Nestler correctly realized that calls for national gas defense only justified

[111] Gertrud Woker, "Im Zeichen der Wissenschaft dem Abgrund entgegen. (Betrachtungen zum chemischen Krieg)," *Die Friedens-Warte* 25, no. 1 (January 1925): 20.

[112] Jo Vellacott, "Feminism as If All People Mattered," 390.

[113] Gertrud Woker, "Ueber Giftgaskrieg," *Die Friedens-Warte* 23, no. 11/12 (November 1923): 393.

[114] Getrud Woker, "Erwiderung," *Die Friedens-Warte* 25, no. 9 (September 1925): 268. The idea that black skin was immune to the effects of poison gas has a long and complex history. See Susan L. Smith, *Toxic Exposures: Mustard Gas and the Health Consequences of World War II in the United States* (New Brunswick, NJ: Rutgers University Press, 2017), 10.

further chemical weapons production. However, with their use of both gendered and racialized imagery as well as overwhelmingly apocalyptic prognostications, their visions simultaneously left little room for political agency and fed into certain societal divisions based on essentialized categories. Even if German civilians could resist the offers of military protection in these future visions, their lives would theoretically hang in the balance, requiring a similar level of disarmament in every other industrialized world power.

As such, Woker and Nestler apparently provided very little hope for peace in an age in which aerial bombing could be initiated by the few who embraced it. In the end, these antigas activists were caught between the political pincers of German nationalists and militarists. If they denied citizens the right to any form of poison gas protection, they tended to appear as traitors to both the German state and the German people. On the other hand, they could accept a certain level of ethno-nationalist rhetoric in order to assert the collective need to resist any form of rearmament, but in doing so, they hamstrung the level of commitment to broad international cooperation.

Conversely, radical socialists and communists argued more stridently for armed resistance, calling on the proletariat to rise up and destroy the German chemical industry. For instance, the communist Günther Reimann wrote, "Chemical weapons are becoming the most important weapons of war for the future. Already, the horrors of the past chemical war imperiously demand that the leaders of the working class … stop the new war preparations of the imperialists."[115] While such calls to action were clearly distinct from those of the pacifists, Marxist-inspired antiwar supporters similarly tended toward dramatic and nightmarish depictions of future chemical warfare. The novelist Johannes Becher was particularly explicit:

> Thousands of gas-poisoned people are now slowly burning to death in the heat. Many corpses began to seethe. Flies buzzed, beetles and lizards crawled over them. There lies such a fleshy heap: the intestines from the belly poured down next to him: a strange life begins to stir in the carcass.[116]

Of course, such horrible descriptions ran the risk of destroying any sense of human agency in the fight to end chemical warfare, but communist writers such as Becher juxtaposed such scenes against powerful depictions of the collective proletariat, often described as a propulsive

[115] Günther Reimann, *Giftgas in Deutschland: Die Machtstellung der IG Farbenindustrie A.G.* (Berlin: Vereinigung Internationaler Verlagsanstalten, 1927), 5.
[116] Becher *(CHCl=CH)3As (Levisite)*, 164–165.

machine of justice. Appeals to proletarian resistance ultimately led to leftist surveillance of both German chemical companies and gas mask manufacturers. In 1929, this surveillance correctly reported that the Dräger company was illegally producing gas masks for the Reichswehr. Both Dräger and the German government denied the accusation and subsequently pressed charges against the newspapers that printed the claims.[117]

Ultimately, communists who sought to prevent future chemical warfare relied on the redemptive value of collective violence to chart their course of action. While the revolution of the proletariat would require an initial embrace of brutality, it would eventually lead to a lasting peace. This call certainly appealed to those who desired immediate action or already harbored communist sympathies, but it also created deep political divisions among antiwar supporters by alienating strict pacifists who abhorred any form of violence and by sowing doubt over the ultimate motivations for international peace conferences, which many communists viewed as bourgeois platitudes.

A Suspicion of Cooperation: International Disarmament Treaties in an Age of Air-Minded Geopolitics

While the Treaty of Versailles was the first postwar international agreement that explicitly treated chemical weapons, it only banned German chemical weapon production and air protection, leaving the Allies free to continue their chemical research and manufacturing.[118] Theoretically, the Hague Conventions still prohibited the use of chemical weapons, but the recent experience of World War I seemingly nullified those prewar agreements.[119] Regardless, the German government was not especially concerned with aero-chemical protection in the tumultuous years immediately following the war. Germany was far more preoccupied with the more tangible, often financial, difficulties that the war precipitated and it was not until 1921 that the *Verein ehemaliger Angehöriger der Flugabwehr*

[117] Justizministerium Untersuchungsverfahren, 1929 (GStA) I. HA Rep 84 a Justizministerium Nr 52373.

[118] General J. F. C. Fuller and the military theorist Basil Liddell Hart were the chief proponents of chemical weapons in Great Britain. Throughout the interwar period, the British continued to stockpile arsenic agents and mustard gas. In 1919, the British opened the Sutton Oak Chemical Defense Research Establishment. Peter J. Hugill, *Global Communications since 1844: Geopolitics and Technology* (Baltimore: Johns Hopkins University Press, 1999), 168.

[119] In 1920, The International Committee of the Red Cross again failed to ban all uses of poison gas through the League of Nations.

e. V. (The Association of Former Members of the Air Defense) replaced the Reichswehr as the central authority for educating the German public on air defense.[120] Over the course of the 1920s, this private interest group held lectures, wrote essays for various newspapers, and published their own journal, the *Luftschutz-Nachrichtenblatt* (Air Protection Newsletter), to impart the importance of air and gas protection to civilians.[121] In 1927, leadership of the group was taken up by the former general of flak artillery, Hugo Grimme.

Germany was not the only nation to express concerns over both air and chemical protection in 1921. An American subcommittee on suffocating gases initiated discussions surrounding the impact of chemical weapons in World War I with the intention of drafting international restrictions on such weapons. This culminated in President Warren G. Harding's 1922 Washington Arms Conference, which served as a response to the American public's broad appeal for a lasting peace.[122] The conference, attended by delegates from the United States, Britain, France, Japan, and Italy, attempted to craft a disarmament treaty that would prohibit the wartime use of asphyxiating, poisonous, or any other form of gas. While the French refused to ratify the treaty, viewing its clauses related to submarine warfare as far too favorable to the English and Americans, concern over chemical weapons and aero-chemical attacks now became a staple part of interwar international diplomacy.[123]

The year 1922 also saw the publication of Italian General Giulio Douhet's treatise on air power, *The Command of the Air*. Douhet's influential book reinforced the belief that both airplanes and chemical weapons would play a central role in any future war by pointing out that air power opened a third dimension on the battlefield, thus allowing attacking forces to entirely skirt land defenses.[124] Because air space is so vast, Douhet claimed that truly effective defensive measures were practically impossible. Rather, nations would need to establish command

[120] Bernd Lemke, *Luftschutz in Grossbritannien und Deutschland 1923 bis 1939* (München: Oldenbourg Verlag, 2005), 125–126.

[121] Hampe, *Der Zivile Luftschutz im Zweiten Weltkrieg*, 9.

[122] According to a 1922 survey, 99.9 percent of Americans favored the abolition of gas warfare. Joris Mercelis, *Beyond Bakelite: Leo Baekeland and the Business of Science and Invention* (Cambridge, MA: MIT Press, 2020), 119.

[123] Donald S. Birn, "Open Diplomacy at the Washington Conference of 1921–2: The British and French Experience," *Comparative Studies in Society and History* 12, no. 3 (1970): 319.

[124] Douhet's prognostications were largely supported by other contemporary air-war theorists. Most notably, Basil Liddell Hart popularized his writings on aerial warfare in several widely read British newspapers throughout the 1920s. Hugill, *Global Communications since 1844*, 168.

of the air by preemptively destroying enemy air forces.[125] Once they controlled a contested air space, they could then methodically bomb the military, industrial, and civilian centers of an enemy nation until it was forced into submission.

In order to give his claims a sense of reality, Douhet ended his book with a fictional scenario in which the French air force bombed the German cities of Cologne, Mainz, Koblenz, and Frankfurt until they were reduced to rubble.[126] Douhet further argued that air forces would undoubtedly drop gas bombs in this envisioned scenario. He wrote, "First would come explosions, then fires, then deadly gases floating on the surface and preventing any approach to the stricken area. As the hours passed and night advanced, the fires would spread while the poison gas paralyzed all life."[127] Douhet dubiously claimed that any major city could be completely enveloped by 80–100 tons of poison gas, effectively shutting down all urban services and slowly choking out all life. Playing on the fundamental fears of gas, Douhet further wrote:

> the poison gas is silent and often invisible but poison gas penetrates, expands, enters any crack or crevice, permeates the element man cannot do without for a moment, and can therefore kill simultaneously masses of men over wide stretches of ground. When we consider that everything in this world undergoes improvement, it is clear that the atrocious gas attack of April 25, 1915, will be child's play to soldiers and civilians of tomorrow.[128]

In 1923, the German Ministry of Defense responded to the general sense of anxiety inspired by Douhet's apocalyptic vision by writing the first comprehensive instructions for civilian air protection. The Weimar government envisioned a central ministry for air protection that would serve as a nationwide warning service in case of aerial attack. Furthermore, the ministry would distribute protective devices and educate the public on correct defensive actions.[129] While Germany's possible actions were limited by the disarmament stipulations in the Treaty of Versailles, it could organize evacuation routines and create bomb shelters.

Alongside such national planning, disarmament talks intensified across the international arena. The United States, France, and Italy discussed restricting chemical weapons at a 1924 conference in Paris and then again in Geneva in 1925. The second meeting produced the Geneva Protocol, which again formally prohibited the use of chemical and biological

[125] Ernst Jünger, *Luftfahrt ist Not!* (Leipzig: Wilhelm Andermann Verlag, 1930).
[126] Giulio Douhet, *The Command of the Air*, trans. Dino Ferrarri (Washington, DC: Air Force History and Museums Program, 1998), 390.
[127] Ibid., 58. [128] Douhet, *The Command of the Air*, 181.
[129] Hampe, *Der Zivile Luftschutz im Zweiten Weltkrieg*, 10.

weapons in all future international armed conflicts.[130] Thirty-eight nations, including Germany, signed the Protocol on June 17, 1925, but not all signatories ultimately ratified the agreement.[131] The Geneva Protocol was followed by three more conferences convened by the International Committee of the Red Cross in 1928 (the Hague), 1929 (Rome), and 1930 (Brussels). These meetings aimed at clarifying the language of the original Protocol and creating more protections for civilians. The 1928 conference at the Hague particularly focused on the measures that states could take to protect civilians against chemical weapons. Delegates created suggestions for participating nations to build gas protection shelters, train sanitary troops, provide gas protection technologies, and educate people about poison gas.[132] They even commissioned scientific studies of poison gas to determine the danger involved in an aerial bombing attack.[133] All of these chemical weapons talks were part of the larger movement for diplomatic peace that fed into the 1928 signing of the Kellogg-Briand Pact, which attempted to renunciate the use of warfare as a means for solving international disputes.

In light of public fears over the future of warfare, legislation to limit chemical weapons clearly maintained a prominent place in 1920s international diplomacy. Pacifists and the governments of the major world powers seemed to agree that international legislation and arbitration was the best way to avoid a future chemical war. However, there were considerable public doubts about the efficacy of international law, substantiated by the earlier failure of both the Hague Conventions and the postwar weakness of the League of Nations. These doubts came not only from communists and socialists, who questioned the justice of the established world order, but also from the interwar political Right.[134] Tapping into the language of national geopolitical struggle that permeated right-

[130] The International Committee of the Red Cross contributed directly to the drafting of the Geneva Protocol, which included the banning of tear gas in military combat. The protocol said nothing about stockpiling poison gas or using chemical weapons within a state's own borders. Further weakening the protocol's power, several signatories only viewed the restrictions on chemical warfare as applying to treaty members. This encouraged the continued use of chemical weapons against both peoples outside of the Western political order and against colonial populations. Rudolf Hanslian, *Die internationale Gasschutzkonferenz in Brüssel* (München: Verlag für das gesamte Schiess- und Sprengstoffwesen, 1928), 30.

[131] The Germans did not ratify the agreement until April 25, 1929. The United States signed the Geneva Protocol, but the Senate refused to ratify it until 1975.

[132] Leo van Bergen, "The Poison Gas Debate in the Inter-War Years," 181.

[133] The gas specialist Rudolf Hanslian served as one of the German representatives to the Brussels conference. Hanslian, *Die internationale Gasschutzkonferenz in Brüssel*, 30.

[134] David Thomas Murphy, *The Heroic Earth: The Flowering of Geopolitical Thought in Weimar Germany, 1924–1933* (Kent: Kent State University Press, 1990), 19.

wing rhetoric, a 1920s military report made it clear that German civilians could not rely on international treaties to protect them from new weapons. The report read:

> It is unlikely that militaries would stop shooting gas due to feelings of humanitarianism. If a chemist succeeds in inventing a gas which not only causes terrible injuries, such as mustard gas, but also death, then history shows that humans will use it. If human rights legislation did not protect us from dum-dum bullets, then it will not protect us from poison gases in the future, and we will only be disadvantaged to rely on an illusory international law.[135]

If, as many ardent German nationalists asserted, international treaties were not to be trusted, then the state had the duty to prepare itself for a future war. In fact, as right-wing interwar political theorists argued, the state was defined by its ability to defend its borders. Douhet's *The Command of the Air* made it quite clear, however, that border defense would now experience a new type of warfare, one fought with airplanes and gas bombs.[136] Pointing to the strategic problems that this new form of fighting created, General Hermann Geyer wrote that "war is becoming more complicated due to the use of gas. We must accept this. This is a logical consequence of scientific and technical developments. The old idea that simplicity is best in warfare is now outdated."[137]

Not surprisingly, even after the signing of the Geneva Protocol in 1925, German military theorists and fervent nationalists remained concerned by the continued development of chemical weapons abroad.[138] The minutes from an interwar military discussion on chemical disarmament claimed that enemy nations were still training their troops with poison gas weapons and therefore, "international regulation [was] impossible."[139] Such assertions further encouraged General Seeckt to continue both the secret development of German poison gas and military gas protection training.[140] For both military men such as Seeckt and the wider German public, Germany's geographic vulnerability at the center of continental Europe created the sense that the nation was surrounded by potential enemies. If war was to break out again, then enemy aircraft could presumably fly unimpeded over German cities and bomb them at will.[141] Stoking

[135] Erfahrungen im Gaskampf, 1920 (BA-MA) PH 14/229.

[136] Stig Förster, ed., *An der Schwelle zum Totalen Krieg: Die militärische Debatte über den Krieg der Zukunft 1919–1939* (Paderborn: Ferdinand Schöningh, 2002), 342.

[137] Geyer, *Der Weltkampf um Ehre und Recht*, 528.

[138] RH/12/9 Sig 21, 1931–1933 (BA-MA).

[139] Gas und Nebel im Volkerrecht, 1933–1936 (BA-MA) RH/12/4.

[140] RH/12/9 Sig 33 (BA-MA).

[141] Friedrich von Tochenhausen, ed., *Wehrgedanken: Eine Sammlung wehrpolitischer Aufsätze* (Hamburg: Hanseatische Verlagsanstalt, 1933), 59.

these fears throughout the decade, the aviation magazine *Die Luftwacht* (*The Air Guard*) ran stories about aerial bombings in World War I, listing the number of dead Germans in each year of the war.[142] Another article claimed that 8,000 war planes were surrounding Germany, ready to reach any location in under two hours.[143] Explaining the danger of this geopolitical situation, a German gas defense memo read:

> the gigantic French fortifications on our Western frontier make attack by infantry seem quite (and artillery attack almost) hopeless. Consequently there remains only the most intensive development and extension of the air weapon, in order that air warfare may be waged effectively and ruthlessly against important military and industrial centers and, above all, also against the civilian population of the large cities.[144]

It is worth noting that those who supported aero-chemical protection measures justified the rearmament of Germany with the same arguments that Fritz Haber had first used in World War I. While proponents of gas protection asserted that poison gas was an existential threat to the national population, they simultaneously argued that it was far less dangerous than most pacifists claimed. For instance, the gas scientist Carl Besse wrote, "Without underestimating the dangers of an aerial gas attack, one cannot warn the public enough about exaggerated pacifist claims. Aided by sensational magazines, the pacifists create turmoil among the population, leading to irresponsible action."[145] The former artilleryman Albert Benary agreed, claiming that pacifists relied too much on apocalyptic visions and not enough on the practical knowledge gained in World War I. To illustrate this point, Benary argued that a successful gas attack on Berlin would require about 5,400 aircraft and an unfathomable number of gas bombs.[146]

Fritz Haber also continued to maintain a similar argument into the interwar years, writing in 1924, "Gas as a weapon is no crueler than flying pieces of metal. To the contrary, the percentage of fatal gassings is comparably small and there are no mutilations. Based on these objective arguments, one would not easily ban gas warfare."[147] Still other commenters

[142] *Die Luftwacht* was the official magazine of the Flakverein. *Die Luftwacht: Zeitschrift für das Weltflugwesen*, Heft 10 (October 1929), 499.
[143] K. Gemeinhardt, "Gasschutz und Luftschutz der Zivilbevölkerung," *Archiv Der Pharmazie* 270, no. 4 (1932): 232.
[144] Jeanne Guillemin, *Biological Weapons: From the Invention of State-Sponsored Programs to Contemporary Bioterrorism* (New York: Columbia University Press, 2006), 41.
[145] Carl Besse, *Gaskampf und Gasschutz* (Berlin: Offene Worte, 1932), 22.
[146] Albert Benary, *Luftschutz* (Leipzig: Verlag von Philipp Reclam, 1933), 22, 24.
[147] Fritz Haber, *Fünf Vorträge aus den Jahren 1920–1923* (Berlin: Verlag von Julius Springer, 1924), 37.

maintained that weaponized gas was a distinctly German invention and that the Germans had the right to use it for their own protection.[148] Such assertions were certainly attempts to assuage feelings of doubt raised by antigas activists such as Gertrud Woker over the German ability to control chemical weapons. By assuring Germans that they were indeed the masters of the chemical battlefield, gas protection advocates attempted to spur national pride and confidence in the chemical armaments industry.

As part of this growing interest in a future war from the air, the German Foreign Office collected substantial files on chemical weapons development in foreign countries.[149] Comprised primarily of foreign press clippings and diplomatic reports, these files described both the chemical weapons capabilities and the level of civilian protection in countries such as Great Britain, France, the Netherlands, Italy, and Poland.[150] The largest files, however, covered chemical weapons in the United States and the Soviet Union.

Since the final year of World War I, the Americans had substantially scaled up their chemical weapons research at the Edgewood Arsenal in Maryland.[151] By 1918, American scientists were isolating and producing 20,000 tons of the new poison gas called Lewisite. A stronger version of mustard gas, Lewisite inspired great anxiety among both German troops and military theorists. While the American government expressed an ostensible desire to avoid the future use of chemical weapons through international treaties, it continued to increase its budget for chemical weapons research and production in the 1920s.[152] German reports claimed that the American military even conducted war games in which airplanes dropped mustard gas on participating soldiers.[153] Simultaneously, American military theorists continued to defend the use of chemical weapons, relying on many of the same rhetorical arguments that the Germans had first employed. For instance, the Chief of the American Chemical War Service, General Amos A. Fries, claimed that gas was far more humane than explosives

[148] Nederland I, 1924–1929 (BA-MA) RH/12/4.

[149] Sheldon Garon, "Defending Civilians against Aerial Bombardment: A Comparative/Transnational History of Japanese, German, and British Home Fronts, 1918–1945," *The Asia-Pacific Journal* 14, 23 (2016).

[150] See, for example, Polen IV, 1929–1931 (BA-MA) RH/12/4; Nederland I, 1924–1929 (BA-MA) RH/12/4; Berichte, Korrespondenz Berufsberichte Dr. Bauer (Röhm Bestand), 1932 (EA) RHT.2./6.

[151] This was part of a larger American desire to build their own version of Haber's KWI and to achieve market independence in synthetic organic chemicals. Kathyrn Steen, *The American Synthetic Organic Chemicals Industry: War and Politics, 1910–1930* (Chapel Hill: University of North Carolina Press, 2014), 119.

[152] RH/12/9 Sig 66, 1927–1928 (BA-MA).

[153] RH/12/9 Sig 69, 1931–1932 (BA-MA).

and that calls to ban chemical warfare were part of a communist plot to weaken national governments.[154]

To many German military strategists, only the Russians surpassed the Americans as the greatest international chemical threat. Even though the Reichswehr was conducting cooperative weapons production and training with the Russians, German fears over future Soviet aggression were never fully quieted.[155] Soviet international policy appeared inherently hostile to western Europe and their lack of involvement in any international treaties meant that they had never agreed to ban the use of chemical weapons. During this period, Germans were liable to invoke old ethno-racial stereotypes in expressing their fear of the Soviets, claiming that their inherent Slavic barbarism would encourage them to indiscriminately use chemical weapons.[156]

Such fears ignored the fact that the Russians had been late to develop their chemical warfare program and that they lacked a strong chemical industry in the early 1920s. Nevertheless, German reports from 1927 stressed that the Russians had indeed begun to test aero-chemical bombs.[157] In that same year, Russia's Society for Chemical and Air Defense, the Ossoawiachim, made it mandatory for residents in large cities such as Leningrad to buy a gas mask and attend gas training.[158] By 1931, the Ossoawiachim maintained a membership of 12 million Russians, justifying its mobilization of Russian society by claiming that "nobody today can say if the war will break out in one, three, or even six years, but it is clear that one day the bourgeoisie of the world will have to face the proletariat of the Soviet Union in a final battle" (Figure 5.1).[159]

In the late 1920s, such fears of the American and Russian chemical weapons programs reinforced the belief that German civilians required gas protection. Often overstating the reality of the foreign threat, militarists and proponents of chemical warfare could advocate for the

[154] Thomas Faith, "'As Is Proper in Republican Form of Government': Selling Chemical Warfare to Americans in the 1920s," *Federal History* 2 (2010): 31–32.

[155] If anything, the joint German-Russian chemical weapons production program only proved to the German military that the Russians now had the capability to industrially produce modern chemical weapons.

[156] Annemarie H. Sammartino, *The Impossible Border: Germany and the East 1914–1922* (Ithaca: Cornell University Press, 2010), 148.

[157] In fact, in 1926, the Soviets had begun to test aero-chemical attacks in far-flung regions of the Eurasian steppes and to increase public awareness of future aero-chemical warfare. Scott Palmer, *Dictatorship of the Air: Aviation Culture and the Fate of Modern Russia* (Cambridge: Cambridge University Press, 2006), 160.

[158] Russland III, 1927–1939 (BA-MA) RH/12/4.

[159] Russland IV, 1929–1932 (BA-MA) RH/12/4.

Figure 5.1 The German imagination of Soviet gas. The "Ghost from Moscow" says: "I will pretend to be upset about the production of poison gas." This 1928 political cartoon implies Russian connection to the gas accident in Hamburg.
Source. Giftgas Unglück auf der Veddel, 1928 (SH) 135-1 I-IV 4069

continuation of a German chemical armaments program by stressing its essential role in national protection. When laying out the requirements for such protection, one Major Zanetti reminded Germans that it was:

questionable whether one can expect countries to freely renounce a modern means of warfare. The strength of the home front, which was decisive in the First World War, will play a major role in the war of the future. The better a state's air supremacy, the greater chance it has to protect itself in peace by organizing the air defense and training the masses in gas protection.[160]

By claiming that air and gas protection was a matter of national existence, advocates argued that this was not a matter for political debate. Rather,

[160] Vortäge der Gasschuztlehrgänge Teil II, 1930 (BA-MA) RH/12/9.

every citizen had an irrefutable right to protection.[161] The vice chief of the state emergency services, Erich Hampe, wrote:

> But it would also be just as wrong to see the civilian air defense as a matter that could or should be measured on a political scale. The plight of man and his rescue from distress has never been a matter of politics. Luckily, all the slogans of party politics have been overturned, and civilian air defense is beginning to take effect because politicians have ceased to be effective. Civil air defense is not a military or political matter, but a self-evident act of affirming and protecting life. All people are threatened by the danger of aerial attack.[162]

Such arguments for air and gas protection were not limited to military reports and specialized journals. Throughout the late 1920s, opinion pieces asserting the necessity of national protection flooded local and national newspapers including, but not limited to, leading publications such as the *Deutsche Allgemeine Zeitung*, the *Deutsche Tagezeitung*, the *Berlin Lokal-Anzeiger*, the *Vossische Zeitung*, and the *Berliner Illustrierte Zeitung*.[163] Expressing the omnipresence of such op-eds, a Stoltzenberg chemical company pamphlet read, "Today, every newspaper writes almost daily of poison gas."[164] Turning air and gas protection into a civic concern, the publication of these various articles forced the Weimar government to more actively address possible protective measures and limited rearmament.

In 1926, Germany signed the Paris Air Agreement, which allowed the Germans to resume commercial air travel and to control their own air space. The Allies agreed to this so that they could use German airports for fuel and maintenance, a privilege that the Germans had been strategically withholding throughout the early 1920s. This agreement further allowed the German Ministry of Defense to set up a military air defense service in 1927 under the direction of members from the *Flakverein* (Anti-Aircraft Association), a private organization that lobbied for greater national air protection. For civilian measures, the Ministry of the Interior became the overarching organization for air protection, often supervising the private air protection association *Deutscher Luftschutz e. V.* (The Association for German Air Defense) in the implementation of nationwide plans. *Deutscher Luftschutz*, under the direction of Rudolf

[161] Dietmar Süss, *Death from the Skies: How the British and Germans Survived the Bombing in World War II*, trans. Lesley Sharpe and Jeremy Noakes (Oxford: Oxford University Press, 2011), 35.

[162] *Luftschutz Nachrichtenblatt* Heft 1 (January 1931), 15–16. The *Luftschutz Nachrichtenblatt* was the successor of the Flakverein's *Die Luftwacht.*

[163] Fritzsche, "The Economy of Experience in Weimar Germany," 375–376.

[164] Chemische Fabrik Stoltzenberg, "Instruktionskasten fur chemische Kampfstoffe und ihren Sanitatskoffer fur Kampfgaserkrankungen."

Kröhne, began to conduct studies on the most effective forms of civilian protection and to present its findings through a traveling air protection exhibition. During this same period, several other private air defense organizations, including the *Deutscher Luftschutzliga* (German Air Defense League), the *Flakverein*, and the Ring of Fliers, produced their own sensationalized exhibitions and literature.[165] To them, air raid exhibitions and public lectures were opportunities to imbue German civilians with the martial spirit necessary to resist both aerial bombings and an insidious weapon like gas.

However, due to a lack of funds in 1928, the Ministry of the Interior was unable to create a government service that solely dealt with air protection. Instead, national air protection guidelines put state-approved air defense in the hands of the preexisting police, firefighter, and ambulance services.[166] These largely disconnected services set to work, creating both air raid shelters in major German cities and industrial areas as well as educational materials for the press and schools. At the national level, the Ministry of Foreign Affairs followed developments in foreign air protection while the Ministry of Economic Affairs assessed the possibility of building large collective air raid shelters. The Ministry of Labor was given the task of creating civil air protection jobs, and the Reichswehr was charged with testing air raid protection devices such as gas masks. Finally, the Ministry of Transportation worked to protect train lines, water services, and energy delivery in the event of an attack.[167] Thus, while air protection gained a significant public presence in Germany after 1926, the implementation of aero-chemical protections was apportioned to various emergency services, private interest groups, and levels of government.[168]

Normalizing Contingency: The Rise of the Gas Specialists

Throughout the 1920s, the men who had studied or worked with chemical weapons in World War I created increasingly tighter-knit circles that could speak to issues of national gas protection. Many of these same men

[165] The *Verein Deutscher Chemiker e. V.* (The German Association of Chemists) also made suggestions for correct gas protection. In September 1932, they formed the Professional Group for Air Protection to continue an advisory role in the Weimar government. Vergiftungen durch Gase, 1932 (BArch-B) R/86/.

[166] The *Technische Nothilfe* (federal emergency services), German Red Cross, and the *Arbeiter-Samariterbund* (worker aid agency) also served in this endeavor. Gasschutzmassnahmen, 1931–1932 (LBWK) 233, 12570.

[167] Literatur, Kriegsfuhrungen, Auslandsnachrichten, 1931–1933 (BA-MA) RH/12/4.

[168] Hampe, *Der Zivile Luftschutz im Zweiten Weltkrieg*, 14.

joined the private organizations that promoted air protection such as the *Verein ehemalige Angehörer der Flugabwehr* (The Association of Former Members of the Air Defense) or the later-formed *Deutscher Luftschutzliga* (The German Air Defense League).[169] Those who remained in the field of applied chemistry primarily built and maintained relations through the *Verein Deutscher Chemiker* (The German Chemists Association).[170] Former gas pioneers, medical doctors specializing in gas poisoning, and chemists from Fritz Haber's Kaiser Wilhelm Institute began to write for the journals produced by both these private associations and the major gas mask manufacturers. While there were hundreds of contributors to these journals over the course of their existence, a cohort of about fourteen men began to formalize the growing field of gas protection through regular publication. This group included Hugo Stoltzenberg, the chemists Fritz Wirth, Albrecht Hase, and Heinrich Remy, the gas-school instructor Julius Meyer, the toxicologist Ferdinand Flury, Wilhelm Haase-Lampe of the Drägerwerks, Karl Quasebart of the Auergesellschaft, the former policeman Heinrich Paetsch, General Hans-Georg von Tempelhoff, the politician Rudolf Kröhne, the vice chief of the *Technische Nothilfe* Erich Hampe, and the former military doctors Otto Muntsch and Hermann Büscher. But of these gas specialists, the most publicly decorated was Rudolf Hanslian.[171]

Born in 1883 to a brewer in Moravia, Rudolf Hanslian earned his doctorate in chemistry from the University of Leipzig and became an *Assistent* at the University of Halle. At the outbreak of World War I, he served as both a troop pharmacist and a gas officer, eventually securing a postwar position as a military pharmacist in Berlin. While serving in this position, Hanslian was selected as a clerk for the International Commission of Red Cross Experts on Civil Gas Protection. This allowed him to represent Germany at the meetings of the International Red Cross throughout the 1920s. By 1926, he was known as one of the "most

[169] The Association of the Former Members of the Air Defense was created in 1920 under the former artillery general Hugo Grimme. The Association held lectures and published essays on air protection, but their actions were limited in scope and impact. Hampe, *Der Zivile Luftschutz im Zweiten Weltkrieg*, 9. The German Air Defense League united a number of smaller air defense organizations with tens of thousands of members in 1931. Edward Westermann, *Flak: German Anti-aircraft Defenses, 1914–1945* (Lawrence: University of Kansas Press, 2001), 36.

[170] The VDCh was founded in 1887 and was mostly made up of chemists working in private industry.

[171] While the gas specialists tended to come out of Fritz Haber's KWI, their diverse mid-career working fields allowed them to champion the importance of aero-chemical protection in the military, industry, health sciences, and professional chemistry. For a comprehensive list of chemists who worked on gas protection, see Maier, *Chemiker im "Dritten Reich"*, 263.

prominent figures in public gas protection," playing a key role in bringing various gas specialists together into a formalized professional group.[172] This was primarily done through the journal *Gasschutz und Luftschutz*, which Hanslian coedited with Heinrich Paetsch (Figure 5.2).[173]

In the mid-1920s, Hanslian and his fellow gas specialists largely advocated for increased public knowledge of poison gas protection. Simultaneously, they responded to the criticism of pacifists such as Gertrud Woker, who claimed that they were simply driving German rearmament through indirect means. As part of their retort, the gas specialists were quick to use ad hominem attacks, claiming that pacifists were covert communists or anarchists who desired to weaken Germany by fanning public fears of gas attacks.[174] According to Hanslian and his colleagues, the pacifist visions of apocalyptic gas attacks were largely fiction.[175] Relying on their wartime experience, the gas specialists repeatedly pointed out that effective gas attacks required thousands of gas shells to produce a sufficiently dense cloud over a small section of battlefield. Given this fact, they asked how could such an effect possibly be produced by aerial bombing. They further attempted to place poison gas within the larger context of industrialized violence, arguing once again that gas was no more morally repellant than other means of total war. For instance, the former artilleryman Wilhelm Hartung wrote:

> Today, the pacifists use chemical weapons to arouse disgust for war. Such a point of view is incomprehensible to me as a former front-line soldier … is [gas] not a thousand times more humane than an enemy hunger blockade, starving a whole nation of women and children?[176]

Given that the gas specialists, on the whole, did not believe that international peace treaties would necessarily prevent a gas war, they claimed that public knowledge was the best form of future protection.[177] Through explanations of the physical nature of various poison gases and

[172] Hanslian received popular acclaim after the 1925 publication of his first book, *Der chemische Krieg*. In the Weimar and Nazi periods, the book was the foremost historical and scientific account of German gas warfare in World War I. In it, Hanslian normalized many of Fritz Haber's rhetorical justifications for the German deployment of chemical weapons. Wilhelm Haase-Lampe, *Handbuch für das Grubenrettungsweser Sauerstoffretttungswesen und Gasschutz, Gerätebau und Organisation seit 1924* (Lübeck: H. G. Rahtgens, 1929), 694.

[173] Paetsch would later become a government liaison for the Auergesellschaft in the 1930s.

[174] Haeuber and Gassert, *Der Kampf um den Luftschutz*, 9.

[175] Fessler, Gebele, Prandtl, *Gaskampstoffe und Gasvergiftungen: Wie schützen wir uns?* (München: Verlag der Ärtzlichen Rundschau, 1931), 59.

[176] Hartung, *Großkampf Männer und Granaten*!, 448.

[177] Margarete Stoltzenberg-Bergius, *Was jeder vom Gaskampf und den chemischen Kampfstoffen wissen sollte* (Hamburg: Chemische Fabrik Stoltzenberg, 1930), 3.

Figure 5.2 The gas specialist Rudolf Hanslian.
Source. Wilhelm Haase-Lampe, *Handbuch für das Grubenrettungsweser Sauerstoffretttungswesen und Gasschutz, Gerätebau und Organisation seit 1924* (Lübeck: H. G. Rahtgens, 1929)

presentations on the best technical solutions for gas protection, the gas specialists cast themselves as humanitarians and pacifists of a different stripe. With a more skeptical approach to the limited capabilities of both poison gases and international treaties, they claimed that they could present the German public with their best options for salvation in a future war. For instance, one gas protection article read, "Like the pacifists, we hope that gas warfare will never come back, but we do not want to be defenseless if fate, which does not care about hopes or desires, should plague our people again. Preventative protection is our elementary right."[178] By casting gas protection as a new human right, the gas specialists followed the German military in insisting that their calls for protection were ethical, rather than political in nature. In fact, in one representative press snippet, the Dräger company wrote:

[178] *Die Gasmaske* Heft 3 (Juli 1929), 59.

It is not the task of the gas protection technician to make major political considerations. He is a pacifist by nature; his work is directed against the horrors of war, against the annihilation of human and animal life by aero-chemical war. It is quite secondary to his work if gas protection equipment is also a weapon of aggression. One way or another, the purpose remains human protection. However, today's gas protection engineer is forced to comment on the question: "Will a future war also be a chemical war?" He cannot deny this question, and he is filled with skepticism about efforts to make war politically impossible ... Only gas protection can resist the chemical war.[179]

Between 1925 and 1927, antigas pacifists and gas specialists continued to aggressively vie for trust and support from the German public. But on May 21, 1928, a gas leak at Hugo Stoltzenberg's Hamburg chemical factory proved decisive for shifting public opinion and mobilizing political influence. That morning, three large tanks of phosgene leaked onto the industrial docks on the south side of Hamburg's harbor.[180] A fortunate breeze blew the eleven-cubic meter gas cloud westward away from the city center. Nevertheless, about 200 people ended up at local hospitals, some of them seriously poisoned. Authorities evacuated the west-lying town of Wilhelmsburg, where the gas killed off the low-lying vegetation and animal life.[181] Ultimately, eleven civilians, mostly industrial workers and rescue team members, died. The *Hamburger Stimmen* reported that news of the gas leak created a "terrible panic," and more residents reported gas sickness over the following days.[182] *The London Times* wrote that the gas leak has "given the appearance of a plague-stricken area, where men and women, haggard from sleeplessness, walk about haunted by the fear that they will yet be the victims of the invisible peril."[183]

In the wake of the gas leak, Hugo Stoltzenberg did not have the money to compensate the victims or pay for the environmental damage.[184] He had lost most of his wealth after being removed from the Bersol project in Russia. Investigations and a trial followed, but Stoltzenberg emerged relatively unscathed. The major fallout from the event, however, was that it revealed to the public that chemical weapon stockpiles remained in

[179] Draeger, *Gasschutz im Luftschutz*, 27.
[180] Konstruktion, Erfindungen, Erfahrungen uber Gasschutzgerat, 1927–1929 (BA-MA) RH/12/4.
[181] Klaus Meise, "Der Giftgas-Unfall am 20. Mai 1928: Phosgen-Explosion in der Chemischen Fabrik Stoltzenburg am Hovenweg," *Die Insel: Zeitschrift Des Vereins Museum Elbinsel Wilhelmsburg* 5314 (2012): 63.
[182] *Hammburger Stimmen* 119 (May 13, 1928).
[183] *The London Times* (May 23, 1928) in Giftgas Unglück auf der Veddel, 1928 (SH) 135-1 I-IV 4069.
[184] Phosgen Ermittlung, 1928 (SH) 311-2 IV Vuo II C 5 a II A 34.

Germany. The pacifist outrage over this scandal led to the German press' eventual discovery of Stoltzenberg's connections to the Rif War in Spain and chemical production in Russia.[185] Many on the more radical Left blamed the eleven deaths on broader societal failure to destroy German militarism and its related industries.[186] For instance, pacifist Carl von Ossietzky wrote, "[T]his poison gas attack on the great city of Hamburg, brought on by the irresponsible stupidity of the public authorities and the criminal profiteering of business-savvy former military brass is an obvious education on the methods of the next war."[187] On the other hand, the more moderate Social Democrats in office stressed Stoltzenberg's links to Russia, insinuating that the leak was an act of communist terrorism.

Renewed fears over poison gas drove increased reporting on carbon dioxide and heating gas poisonings through 1929.[188] Newspaper articles blamed local authorities for their oversights in industrial inspection and their slow reactions during emergencies.[189] This created a palpable politics of risk that made many Germans seriously question whether they were sufficiently protected from poison gas. The *Hamburgischer Correspondent* subsequently wrote of the Hamburg gas leak that:

> This incident gives cause to discuss the need of a civilian gas mask. As long as the country produces poison gas for military or industrial reasons at all, as the Hamburg catastrophe shows, the danger of poison gas still exists in the midst of peace for the civilian population. For such moments, the possibility of a defense must exist.[190]

The Hamburg gas accident moved the simmering poison gas debates into the national spotlight and essentially forced the Weimar government into action.[191] As historian Michael Geyer has pointed out, fear-inducing events in the late 1920s (including the Hamburg gas leak) turned earlier calls for German rearmament into a larger social issue in which a middle-

[185] Kunz and Müller, *Giftgas Gegen Abd El Krim*, 72.

[186] Johnson, "A Suffocating Nature," 195.

[187] Carl von Ossietzky, "Gasangriff auf Hamburg," Die Weltbühne, No. 22 (May 29, 1928): 814.

[188] Vergiftungen durch Gasse, 1928 (Barch-B) R/86/; Christian Goeschel, *Suicide in Nazi Germany* (Oxford: Oxford University Press, 2009), 25.

[189] Similar blame was spread after a 1930 mining explosion in Silesia, in which poison gas broke through the walls of the mine and asphyxiated hundreds of miners. Otis C. Mitchell, *Hitler's Stormtroopers and the Attack on the German Republic, 1919–1933* (Jefferson, NC: McFarland & Company, 2013), 105.

[190] *Hamburgischer Correspondent* 238 (May 23, 1928).

[191] In 1931 alone, lengthy opinion pieces on the necessity of increased gas protection appeared in the *Deutsche Allgemeine Zeitung* 571, *Berlin Lokal-Anzeiger* 584, *Deutsche Tagezeitung* 536, *Deutsche Zeitung* 296, *Der Tag* 296, *Der Reichsbote* 296, *Germania* 525, and *Vossische Zeitung* 583. For these references, see *Luftschutz Nachrichtenblatt* Heft 1 (1931).

class consensus began to push for greater national security.[192] The gas specialists stepped into these debates as the highest authorities on the best practices for national gas protection. Public fears over poison gas tended to now outright trump pacifist critiques of defensive rearmament, thus making international treaties a secondary concern. This climate of fear was best demonstrated across Europe by the 1932 Geneva Disarmament Conference, which initially intended to broker a major inter-European disarmament deal but ended with merely a vague proclamation of diplomatic peace.[193]

In Germany, the gas specialists were not regularly lumped in with chest-beating militarists who made various arguments for renewed poison gas production. In what was essentially a 1932 proclamation of their political victory over the pacifists, Rudolf Hanslian wrote, "From here on out I note the following: the gas protection industry is first and foremost a peace industry."[194] But as early as 1928, many Germans began to willingly accept such a claim, increasingly viewing the gas specialists as the only possible saviors of the nation. This ignored the fact that from 1925 onward, the Reichswehr had set up decentralized scientific networks that could pull on the expertise of gas specialists such as Ferdinand Flury for developing offensive chemical weapons capabilities.[195] Respected gas scientists often maintained similar direct or indirect connections to poison gas production and many of these men clearly believed in a form of zero-sum geopolitics that quietly merged national protection with aggressive militarism.

The political consolidation of the gas specialists truly began in Hamburg, where Heinrich Remy gathered forty-three members of the local chapter of the German Chemists Association after the Stoltzenberg phosgene leak. Eight members of this local professional chemists' association were already sitting on the Air Protection Advisory Board of the city of Hamburg, but now Remy pushed for a permanent "specialist group for air protection" comprised exclusively of association chemists.[196] The city created this "specialist group" in 1932 without going through the

[192] Michael Geyer, "The Dynamic of Military Revisionism in the Interwar Years: Military Politics between Rearmament and Diplomacy," in *The German Military in the Age of Total War* (Warwickshire: Berg, 1985), 115.

[193] Geyer, "The Dynamic of Military Revisionism in the Interwar Years," 114.

[194] *Gasschutz und Luftschutz* Heft 4 (1932).

[195] Maier, *Chemiker im "Dritten Reich"*, 265.

[196] This should remind us that profession and occupation were not always the same for German chemists in the early twentieth century. On the whole, the chemists of the German Chemists Association were happy to act in advisory roles on chemical defense given the potential for a certain level of political influence.

normal channels of approval, citing the "urgency of the matter."[197] Writing in support of the decision, Remy claimed that:

While in the older form of warfare the chemist as such was only indirectly involved. In a war waged using combat poisons – the so-called "gas wars" – the chemist will have to be directly involved. His knowledge and experience are just as indispensable for the measures to prevent or protect against fighting poisons as for the correct assessment of the special attack and defense possibilities given by using weaponized poisons.[198]

Chemistry advisory groups, such as the one set up in Hamburg, were also assigned to various chapters of the German Air Defense League and the "Air Advisory Boards" of other major cities such as Berlin. Through national newspapers, traveling air and gas protection exhibits, and their own gas protection journals, the gas specialists now began to put greater political pressure on the Weimar government's federal air protection services.[199] They publicly claimed that the German authorities were not doing enough to protect their citizens from a future aero-chemical war. For instance, a leader in the *Deutscher Luftchutz*, Rudolf Kröhne, wrote that "it has not yet been possible to bring the negotiations between the government and the *Deutscher Luftschutz* to a satisfactory conclusion … they are burdening us and compelling us to make a clear statement, to clearly state 'air protection is necessary!'"[200]

Leaders in the gas protection movement such as Heinrich Remy, Ferdinand Flury, Rudolf Hanslian, Albrecht Hase, Otto Muntsch, Karl Quasebart, and Fritz Wirth all seemingly agreed that German civilians needed a better understanding of poison gases and the possible forms of protection against them. Consequently, they made it their goal to produce standardized educational literature and to open civilian gas protection schools. Funded by the private air protection organizations, they also sought to refit and test old military gas masks for civilian use.[201] Recognizing both the value and danger of the gas specialists' activities, the Reich Ministry of the Interior continued to try to fold the various air protection groups into a complex and often localized network of air protection.[202] However, the gas specialists felt that greater public funding, the inclusion of research chemists and engineers, and a more hierarchical organizational structure were needed. Many stressed the

[197] Maier, *Chemiker im "Dritten Reich"*, 268.
[198] Heinrich Remy, *Denkschrift, Aufgaben des Chemikers im Luftschutz*, 16.5.1933.
[199] *Die Gasmaske*, Heft 1 (1929); *Gasschutz und Luftschutz* Heft 7 (1932)
[200] Rudolf Krohne, *Luftgefahr und Luftschutzmöglichkeiten in Deutschland* (Berlin: Verlag Deutscher Luftschutz, 1928), 81.
[201] Bestimmungen Ausstatlung des Heeres mit Gasschutzgerat (BA-MA).
[202] *Gasschutz und Luftschutz* Heft 3 (1932).

difference between individual and collective protection, arguing that while the government attempted to create collective solutions such as gas-resistant shelters, it was also necessary for individuals to learn to protect themselves through chemical knowledge and gas discipline.[203] It was individual awareness and resistance, according to this argument, that could ultimately add up to large-scale national protection. Or, as the *Deutscher Luftschutz* put it, "The possibility of aerial attack threatens each individual. Therefore, the entire German people must join together in an air raid community."[204]

At the procedural level, the gas specialists' greatest fear was civilian panic, which they believed would entirely incapacitate German urban spaces in the event of an aero-chemical attack. However, some believed that civilians could internalize the same form of gas discipline that gas officers had demanded from World War I troops. In a time when the home front was now seen as a potential battleground, certain gas specialists argued that civilians would need to control themselves if they were to survive.[205] A representative article in the *Luftschutz-Rundschau* read:

> Because discipline is the first civil duty in an air raid, the population is educated to keep calm, to avoid panic and not to lose nerve. Nowhere is peace as necessary as in a gas attack. Although many people became sick in the phosgene poisoning in Hamburg a few years ago, only a few deaths occurred. And these deaths were people who wanted to quickly escape the gas cloud. Their lungs were exhausted and they inhaled large quantities of poison gas and died. Had these people kept calm, walked slowly, or lay down quietly, they would still be alive today![206]

While most gas specialists accepted the broad importance of gas discipline, it was the gas mask that served as the main source of internal strife in debates over civilian gas protection measures. Air protection journals had previously announced that various segments of the British, French, Russian, and Polish citizenry had received civilian gas masks between 1925 and 1927.[207] However, many gas specialists insisted that only members of the German air protection services should be outfitted with masks. The provision of gas masks to the entire populace, some argued, could induce nationwide panic. Furthermore, the cost of such a plan would be astronomical, thus counteracting the seemingly no-loss political solutions of gas protection training.

[203] Einzel und Sammelschutz, 1931–1932 (BA-MA) RH/12/4.
[204] Vergiftungen durch Gase, 1932 (BArch-B) R/86/.
[205] Maier, *Chemiker im "Dritten Reich"*, 269.
[206] *Luftschutz Rundschau* Heft 4 (Dezember 1932) "Psychologische Wirkungen von Luftangriffen" von Max Peachmann.
[207] Die Luftwacht: Zeitschrift für das Weltflugwesen Januar Heft (1927).

Even if only public officials were to be given masks, high prices and production capacity remained a challenge for the Ministry of the Interior.[208] The Weimar government had been attempting to solve this problem since 1925 when they paid the gas specialist Fritz Wirth 45,000 Reichsmark to refit and test 25,000 World War I gas masks at the Technical University of Berlin.[209] In 1929, the government ordered over 30,000 new GM 24 masks from the Drägerwerks to try to meet increasing demand from both the military and the Ministry of the Interior.[210] While General von Seeckt was among those who wanted to provide each German civilian with a gas mask in the mid-1920s, this remained a sheer impossibility due to cost.[211] Guidelines for the aerial defense of Lübeck claimed that:

> Gas masks would be the best protection for individuals, of course, but the cost of supplying the population with gas masks was far too high, since the gas mask, even when mass produced, was still at least 10 Reichsmark per piece. The gas mask has to fit exactly and be maintained, otherwise it will leak and spoil the filters, which is why it is out of the question for the majority of the population.[212]

Even Professor Karl Quasebart of the Auergesellschaft admitted in a 1931 *Gasschutz und Luftschutz* article that civilian masks were "not worth the expense."[213]

Beyond financial concerns, the gas specialists also tended to think that aerial gas attacks were not yet sophisticated enough to require individual respiratory protection. Gas-tight shelters in subway stations, sewers, and apartment building basements would provide sufficient collective protection while simultaneously making individual civilians responsible for the safety of their entire community.[214] This collective duty would then further remind Germans of the dire nature of the chemical threat.

Alongside these concerns over protection guidelines, not all of the gas specialists were convinced that the gas mask was a simple enough device for the average civilian to effectively operate. A 1932 article in *Gasschutz und Luftschutz* related the story of an eighteen-year-old boy who suffocated in a gas mask with a rusted filter.[215] A number of similar anecdotes further prevented the gas specialists from demanding gas masks for every

[208] Gasschutz-Volk, 1932–1933 (BA-MA) RH/12/4.
[209] Gasschutz fur die Zivilbewolkerung, 1925–1933 (BArch-B) R/43/I.
[210] Berichte Korrespondenz von Bauer an Letcher und Strauss (Röhm Bestand), 1928–1933 (EA) RHT.2./6.
[211] Gasschutz fur die Zivilbewolkerung, 1925–1933 (BArch-B) R/43/I.
[212] Luft und Gasschutzeinrichtungen bei Fliegergefahr, 1932 (SL) 18 4.6-6 Tiefbauamt 779.
[213] *Gasschutz und Luftschutz* Septemberheft (1931).
[214] Gasschutz fur die Zivilbewolkerung, 1925–1933 (BArch-B) R/43/I.
[215] *Gasschutz und Luftschutz* Heft 10 (1932).

German man, woman, and child. Nevertheless, Karl Quasebart called for an increased civilian familiarity with the mask, writing in the December 1932 *Gasschutz und Luftschutz* that "[j]ust as a child learns to clothe itself or the car owner learns to drive, every citizen should try on a gas mask in a gas shelter at least ... once a year, in order to see how it works ... to get rid of the unjustified aversion to this new piece of twentieth-century clothing."[216]

While technical questions regarding the nature of civilian air and gas protection remained open ended through the early 1930s, the gas specialists succeeded in garnering significant influence over these public issues. This, in turn, led to a certain level of power and expertise within the federal and state governments. Throughout the 1920s, gas specialists such as Rudolf Hanslian had fought to build their reputations as scientific experts and to repel accusations from pacifists such as Gertrud Woker. By 1928, the gas specialists were able to capitalize on both the Hamburg phosgene accident and general feelings of national vulnerability to make air and gas protection both a pressing issue of national importance as well as a "knowledge community" centering on certain technical questions that the gas specialists created and shaped according to their own social and political expectations.[217]

In this atmosphere of anxiety, the gas specialists offered their specialized knowledge as a form of salvation and catapulted themselves into the national spotlight.[218] Playing with the pervading sense of contingency and risk that visions of aero-chemical warfare infused into the Weimar era, the gas specialists proposed a mix of technology and trained vigilance for the survival of the German people. Based on the experience of the Great War, they claimed, the German people could indeed learn to discipline and prepare themselves for the war to come.[219]

216 *Gasschutz und Luftschutz* Heft 12 (1932).

217 Ann Johnson, *Hitting the Brakes: Engineering Design and the Production of Knowledge* (Durham: Duke University Press, 2010), ixx.

218 Other than in a few published books like Hanslian's *Der chemische Krieg,* most of the gas specialists work appeared in technical journals. For this reason, they rarely had direct influence on the German public. Their political sway was derived from controlling the contours of a scientific and technological concern that impacted the entire nation through governmental policy.

219 Peter Fritzsche, "Machine Dreams: Airmindedness and the Reinvention of Germany," *American Historical Review* 98, no. 3 (1993): 686–688.

6 Technologies of Fate: Cultural and Intellectual Prophesies of the Future Gas War

In the March 1930 edition of *Heerstechnik*, a Berlin-based journal for military technology, the gas specialist Hermann Büscher wrote about what he recognized as a renewed political interest in poison gas. In Büscher's opinion, the mounting gas debate was not merely a product of the 1928 Hamburg gas accident and the growth in power of private air protection groups but also derived from increased attention from German intellectuals and artists. Büscher suggested that this work bled into so much post-1928 cultural production that regular Germans could not help but haphazardly ruminate on the nature of poison gas, gas protection, and the future of warfare. And in lamenting the lack of specialized gas knowledge in this swirling speculation, he complained that "the whole field of battle gases has become a playground for the literati."[1]

As Büscher claimed, poison gas had indeed become a central theme in German cultural production and debate after 1928. Some of this attention was predicated on the larger boom in both the reading and writing of World War I literature that started around that same year. Historian Modris Eksteins has expressed the difficulty in pinpointing the cause of this influx in war literature ten years after the conflict. The most common explanation states that postwar European societies were simply too exhausted and/or traumatized to discuss the war in its immediate aftermath.[2] As evidence, Eksteins provides the observations of journalist Ilya Ehrenburg, who visited Berlin in 1921 and wrote that:

> catastrophe ... presented a well-ordered existence. The artificial limbs of war cripples did not creak, empty sleeves were pinned up with safety pins. Men whose faces had been scorched by flame-throwers wore large black spectacles. The lost war took care to camouflage itself as it roamed the streets.[3]

[1] Vergiftungen durch Gasse, 1928 (BArch-B) R/86/.
[2] Modris Eksteins, "All Quiet on the Western Front and the Fate of a War," *Journal of Contemporary History* 15, no. 2 (April 1980): 346.
[3] Ilya Ehrenburg, *Men, Years – Life, III* (London: MacGibbon & Kee, 1964), 11–12.

As Ehrenburg went on to suggest, this manifestation of war fatigue and war denial was most apparent in Germany.[4] Beyond the trauma of realizing the power of newly industrialized warfare, Germans further faced the difficulties associated with postwar inflation, international censure, and the instability of their newly democratic government. In a struggle for a sense of mere normalcy, many had little time or desire to reflect on the immediate past. Eksteins adds that most early reflections led to utter disbelief or confusion and many Germans, especially combat veterans, remained unsure of the true motivations for the war. Nationalist propaganda argued that Germany had responded to British and French aggression, fighting for German *Kultur* against the continued cultural and political pressure of a soulless Western materialism.[5] But in the postwar years, the envisaged spiritual destruction of Germany remained difficult to prove, and many began to wonder whether they had overestimated the threat. If the reasons for war were complete fabrications, then millions of men had rushed to their deaths either to line the pockets of European elites or for no purpose at all. The existential difficulties that such questions posed provided, for some, yet another reason to avoid discussion of the war.

The material and psychological stresses of the postwar years may help to explain why it took nearly a decade before German intellectuals began to grapple with the war, but they do not reveal why 1928 was the specific year in which this process began in earnest. Eksteins argues that the serialized appearance of Erich Maria Remarque's *All Quiet on the Western Front* set off a literary debate that inspired many former soldiers to write their own personal accounts. Certainly, the arrival of *All Quiet* in 1928 created a public stir that accelerated the publication of novels that attempted to explicitly narrate wartime experience, but Eksteins is forced to admit that "a number of [war-related] books appeared immediately before [Remarque's novel]."[6] It should also be noted that visual artists such as Otto Dix had already produced explicit war art well before 1928, although it certainly did not receive as much public attention as the later war novels. These earlier works force us to ask why some German soldiers chose to recount their wartime experiences before the publication of *All Quiet* and what inspired a larger reckoning with the war in 1928.

[4] Peter Jelavich, "German Culture in the Great War," in *European Culture in the Great War: The Arts, Entertainment, and Propaganda, 1914–1918*, eds. Aviel Roshwald and Richard Stites (Cambridge: Cambridge University Press, 1999), 36–37.

[5] Eksteins, "All Quiet on the Western Front and the Fate of a War," 347.

[6] Ibid., 346.

The previous chapter has pointed to 1928 as a turning point in discussions of German national security. In May of that year, the Hamburg gas accident brought these discussions to the forefront of public consciousness, inspiring a reorganization of the national aero-chemical protection services. This afforded both broad political power to private air defense organizations and more narrow professional influence for the gas specialists, who then further stoked public fears of a future aero-chemical war. These developments raised a slew of questions about the efficacy of international diplomacy, the inherent violence of humanity, and the role of technology in modern life. *Kriegserlebnis* (war experience) novels were one method of reflecting on such questions, often referring to the escalation of gas warfare in the final year of World War I as part of a prognosis for future conflict. As Eksteins points out, even Remarque's *All Quiet* "is more a comment on the postwar mind, on the postwar view of war, than an attempt to reconstruct the reality of the trench experience."[7] At the same time, however, *Kriegserlebnis* novels were not the only cultural medium used to tackle these topics. Visual art, film, and science fiction/dystopian novels from the Weimar period provided further avenues for exploring the sense of risk and contingency of any future aero-chemical war.

Of the numerous interwar depictions of both poison gas and the gas mask, a number maintain passages or images that appear strikingly similar to the first-hand descriptions of World War I gas warfare. Given that many interwar artists and intellectuals were themselves war veterans, this is not necessarily surprising. However, it was equally common for their vivid recreations of gas warfare to distort the realities of chemical warfare. It bears repeating that gas was not a very effective weapon until 1918, and most soldiers never suffered severe gas poisoning. But again, the postwar fixation on gas warfare points to the fact that artists and novelists were writing, painting, and drawing with distinctly interwar concerns in mind.[8] Reflecting on current anxieties over national vulnerability while also pulling on personal memories and shared war stories, many of these artists diachronically related the feeling of total victimization that gas produced among men in the trenches.

In this way, questions surrounding warfare technologies entered into larger contestations over the memory of war. The different ways in which interwar artists chose to represent chemical weapons often reflected both their personal reckoning with these technologies and their political

[7] Ibid., 351.

[8] Paul K. Saint-Armour, *Tense Future: Modernism, Total War, Encyclopedic Form* (Oxford: Oxford University Press, 2015), 305.

allegiances. While many former soldiers knew that human beings had released the gas, they often posited the weapon as a mindless and indiscriminate killer. It was a rolling fog, an eerie haze, a monster, or a creature that snuck into every crevice and choked out all life. Gas weaponized the atmosphere and created a sense of imminent risk that no other weapon could equal.[9] And as an extension of the gas, the gas mask could transform men into either the masters or monsters of this chemically impregnated world. Once pulled over the face, a soldier lost at least part of his humanity and was transformed into a bug, ghost, or alien that would either live or die in this toxic environment. For this reason, the human face served as both a literal and symbolic point of contention for interwar depictions of gas. Intellectuals debated whether a human mind and a human face lay behind both the poison gas and the gas mask. They asked themselves and their fellow Germans whether these technologies were the products of men who could be understood and/or controlled in any future conflict, or if poison gas and the masked creatures who lived among it were harbingers of a technologically determined power that would lead to an inescapable apocalypse.

Death or Rebirth: Poison Gas and Gas Masks in the *Kriegserlebnis* Novel

As Modris Eksteins had made clear, German war novels experienced a major publication boom between 1928 and 1930. Referred to as *Kriegserlebnis Romane* (war experience novels), these books "dominated" the lists of publishers in 1929.[10] When Erich Maria Remarque republished *All Quiet on the Western Front* in book form in January 1929, it quickly became Germany's, and subsequently the world's, best-selling novel. By April 1930, the book had sold 2.5 million copies, sometimes selling over 20,000 editions in a single day.[11] The massive popularity of the novel prompted one reviewer to write, "Remarkable! And a war book to boot, especially a war book! Who would have read war books a year ago."[12] Similar pronouncements proved that 1929 was a cultural "crossroads in the interwar era."[13] According to Eksteins, the years 1925 to 1929 saw an increase in both a form of optimistic humanism and a general sentiment of hope that largely stemmed from the successful

[9] Doris Kaufmann, "'Gas, Gas, Gaas!' The Poison Gas War in the Literature and Visual Arts of Interwar Europe," in *One Hundred Years of Chemical Warfare*, 185.
[10] Eksteins, "All Quiet on the Western Front and the Fate of a War," 345.
[11] Ibid., 353. [12] *Nouvelles Litteraires*, October 25, 1930.
[13] Eksteins, "All Quiet on the Western Front and the Fate of a War," 357.

brokerage of international peace treaties. However, the economic slump of 1929 and the ten-year anniversary of the Treaty of Versailles brought on a newly reflective mood that encouraged the writing and reading of war novels such as *All Quiet.* Thus, for Eksteins, *All Quite* was a "symptom rather than an explanation, of the confusion and disorientation of the postwar world."[14]

There is no denying the importance of Remarque's text for the popularization of the war novel genre in 1929. It is also fair to claim that external political, social, and economic problems helped to inspire greater interest in Remarque's book. German readers were not just attracted to depictions of soldiers' lives in the trenches, but they were also looking for answers to contemporary issues in the most immediately significant historical episode. However, it is worth considering external events other than the Great Depression in this grand societal reflection on World War I. This is especially the case because the stock market did not collapse until October 24, 1929, and the collapse's full effects were not fully felt in Germany until the 1930s. Remarque's novel, on the other hand, first appeared in serial form in late 1928, becoming massively popular by January of that year. Thus, while 1929 may have been a turning point for the increased sale of war novels, the inspiration for writing and publishing them came earlier. Remarque himself chose to write *All Quiet* in the summer of 1928, apparently stimulated by a feeling of exasperation over the battle to convey a retrospective yet authoritative meaning for the war.[15] For this reason, the novel provides a series of graphic, yet emotionally detached depictions of senseless violence, attempting to convey the veteran's sense of existential emptiness in the wake of his wartime experience.

In light of the importance of poison gas to both the soldier's experience of anonymous modern violence in the final year of the war and the 1928 German debates over civilian protection from aerial gassing, it is unsurprising that gas features prominently in Remarque's novel. Remarque presented poison gas as one of the modern technologies that made World War I truly horrifying for the average soldier.[16] He wrote, "Bombardment, barrage, curtain-fire, mines, gas, tanks, machine-guns, hand-grenades – words, words, but they hold the horror of the world."[17] Among these many weapons, however, gas created particularly horrible

[14] Ibid., 351. [15] Ibid., 349–350.

[16] Historian Eric J. Leed has claimed that the war encouraged many soldiers, like Remarque, to begin to view and write about technology as a major propulsive force in history. Leed, *No Man's Land*, 31–32.

[17] Erich Maria Remarque, *All Quiet on the Western Front* (Greenwich: Fawcett, 1967), 83.

moments of silent expectation followed by utter chaos. *All Quiet* maintains one such scene in which a gas attack leads to total battlefield confusion:

Gaaas – Gaaas – I call, I lean toward him, I swipe at him with the satchel, he doesn't see – once again, he merely ducks – it's a recruit – I look at Kat desperately, he has his mask ready – I pull out mine too, my helmet falls to one side, it slips over my face, I reach the man, his satchel is on the side nearest me, I seize the mask, pull it over his head, he understands, I let go and with a jump drop back into the shell-hole … Someone plumps down behind me, another.[18]

In this scene, Remarque does not describe the gas itself. In fact, the reader does not truly know if gas is actually present. We are rather presented with the soldiers' sheer anxiety and the din of human movement in expectation of gas. The young recruit who fails to ready himself is used to depict the attack as a deadly initiation ritual that can transform naïve boys into hardened veterans. The readying of defensive equipment is portrayed as both a hindrance and a possible solution to surviving this violent baptism. Helmets slip and gas masks subsequently become wet, forcing the men to consider the effectiveness of their tools. Remarque goes on:

I wipe the goggles of my mask clear of the moist breath … These first minutes with the mask decide between life and death: is it tightly woven? I remember the awful sights in the hospital: the gas patients who in day-long suffocation cough their burnt lungs up in clots. Cautiously, the mouth applied to the valve, I breathe.[19]

While this passage continues to present the gas as a disembodied and powerful enemy, the gas mask truly expresses the powerlessness of the soldiers. They grab it and exhort it to protect them, much like a magical talisman, but they ultimately have no assurance that it will work.[20]

Remarque's depiction aligns with most first-hand descriptions of gas attacks while simultaneously expressing postwar concern over chemical attacks and protection. In an imagined world in which everyone is a potential gas victim at any given moment, Remarque makes it clear that technological defenses such as the gas mask would not always suffice. Not only could the mask fail technologically, but its effectiveness relied

[18] Ibid., 44–45. [19] Ibid., 220.

[20] The creation of totems, talismans, and amulets was apparently common in the trenches of World War I. For instance, the Italian chemist Giuseppe Bellucci reported seeing amulets "formed of the copper rings from poison gas grenades." Paolo de Simonis and Fabio Dei, "Wartime Folklore: Italian Anthropology in the First World War," in *Doing Anthropology in Wartime and War Zones: World War I and the Cultural Sciences in Europe*, eds. Reinhard Johler, Christian Marchetti, and Monique Scheer (Bielefeld: Transcript Verlag, 2010), 92.

on quick-witted and collected operators. In both *All Quiet* and his second war novel, *The Road Back*, Remarque's gas attack passages reveal the difficulties inherent to remaining calm and composed.[21] Nevertheless, Remarque's central characters effectively employ the mask and survive the gas attacks; it is the background characters who remind us that not everyone is so lucky. In this way, Remarque attempted to create a repellent depiction of warfare by presenting scenarios in which soldiers appear as mere victims in an increasingly dangerous world. Technology such as the gas mask was not necessarily their salvation, but it was the best option in a new era defined by uncontrollable threats.

Interwar receptions of Remarque's novels most often classified them as pacifistic due to the lack of heroic passages and an apparent attempt to express the meaninglessness of war.[22] However, Remarque himself never fully claimed a political commitment to pacifism, and his writings never explicitly denounced violence.[23] In fact, historian Omer Bartov has argued that Remarque could not "avoid revealing [war's] fascination, its power to transport men to physical and mental states previously unknown to them. This was the intoxication of murder on a vast, unprecedented scale."[24] Regardless of this potential critique, Remarque certainly did not posit violence, gas-related or otherwise, as a wholly positive experience. Much like earlier pacifist works from activists such as Ernst Friedrich and Waldus Nestler, Remarque's novels relied on the graphic results of violence to convey a sense of untimely tragedy. At the same time, his texts provided few answers for avoiding future violence other than vague appeals to international cooperation.[25]

More explicitly pacifist novels such as Fritz von Unruh's *Opfergang*, originally released in 1919 and then republished in 1925, presented rather similar descriptions of poison gas. Unruh's novel depicts war gas as a metaphorical king that invites its subjects, the soldiers in the trenches, to suffocate. While the men don gas masks to protect themselves, the king "dances around the men like death."[26] In a separate but

[21] Erich Maria Remarque, *The Road Back* (New York: Grosset & Dunlap, 1931), 140.
[22] Hans Wagener, *Understanding Erich Maria Remarque* (Columbia: University of South Carolina Press, 1991), 33.
[23] Lawrence Rosenwald, "On Modern Western Antiwar Literature," *Raritan* 34, no. 1 (2014): 163.
[24] Omer Bartov, *Murder in Our Midst: The Holocaust, Industrial Killing, and Representation* (New York: Oxford University Press, 1996), 44.
[25] Hans-Harald Müller, "Politics and the War Novel," in *German Writers and Politics 1918–39*, eds. Richard Dove and Stephen Lamb (London: Macmillan Press, 1992), 113.
[26] Fritz von Unruh, *Opfergang* (Frankfurt am Main: Frankfurter Societäts Druckerei, 1966), 46.

related scene, gas is described as a "white, ghostly wall" that crept toward the soldiers. The men screamed, "Gas masks! And all covered their heads. [They became] eyes and nothing but eyes."[27] Again, Unruh portrays poison gas as an unstoppable weapon, relying on supernatural metaphors to express its uniquely destructive and even evil qualities. The gas mask is Unruh's last bulwark against such evil, allowing men to witness the end of the world through its celluloid eyepieces. Indeed, critical postwar authors such as Remarque and Unruh frequently depicted poison gas as a modern *Götterdämmerung*, employing biblical language to equate the destructive power of gas with the fires of hell or the vengeance of the four horsemen of the apocalypse. In his 1925 novel *Es lebe der Krieg!*, the avowedly pacifist writer Bruno Vogel used similar language to describe the World War I battlefield, writing, "The war, which, as a flood of steel and poisonous gases, will burn through the lands with fire, plague, and hunger, and drown them in blood and sullenness and misery, until all that man created is annihilated."[28]

Of the better-known antiwar writers, Edlef Köppen provided some of the most interesting depictions of the gas mask, putting his avowed hatred of the mask in tension with its potential to save life. Köppen pulled on his wartime experience as an artilleryman to write the 1930 novel *Heersbericht.*[29] The central character in Köppen's novel is gassed when, in the midst of an attack, a tree branch pulls off his gas mask. This injury reveals the fallibility of the mask, which cannot account for the accidents of war. Köppen further criticized the construction of the gas mask, writing:

> God, this gasmask! It is soaking wet, the water on the right and left collects from my breathing, and now it sloshes against my chin while running and sometimes gets in my mouth, ugh. And then the eye glasses fog over as if the whole landscape is covered with mist.[30]

But again, Köppen juxtaposed the failures of the mask against its absolute necessity, writing "but it's bad luck if you do not have the mask."[31] In referring to the concept of luck, Köppen again places technology beyond the total control of its user, giving the objects of gas warfare their own level of unpredictable agency. Like the interwar antigas activists, Köppen posited the gas mask as part of the militarist's arsenal. For the

[27] Ibid., 148. [28] Bruno Vogel, *Es lebe der Krieg!* (Leipzig: Verlag die Wölfe, 1925), 9.
[29] Brian Murdoch, *German Literature and the First World War: The Anti-War Tradition* (Farnham: Ashgate, 2015), 246.
[30] Köppen, *Heeresbericht*, 337–338.
[31] Ibid., 338. Köppen suffered several wounds during World War I, including mustard gas skin burns and gas inhalation.

common infantryman, the mask's offer of protection was never guaranteed; rather, it provided just enough defense to encourage the expansion of chemical warfare at the strategic level.[32]

Within the larger war literature boom, a second group of writers published accounts of their own war experience as a direct response to antinationalistic, pacifistic, and/or nihilistic commemorations. While Remarque's novels pointed to *Kameradschaft* (military comradery), as the only positive product of the war, distinctly nationalist and often reactionary writers insisted that this recently formed community of the trenches was part of a lager nationalist sentiment that could bind Germans together in the face of postwar political difficulties. Thus, while their *Kriegserlebnis* novels featured similar violence and suffering, this hardship was expressed as a meaningful sacrifice to the nation. Interestingly, many of the more commercially successful nationalist war novels were produced prior to Remarque's revisionary account, again suggesting that 1928 represented a publicly visible splintering of the war's collective remembrance. In fact, nationalist writers such as Walter Bloem and Franz Schauwecker found some level of popular acclaim immediately following the war.

Much like the antiwar writers, nationalist authors presented industrialized methods of killing such as poison gas as uniquely deadly and horrifying.[33] However, it was Germany's enemies who were responsible for this suffering. No longer presented as entirely of their own mind, weapons such as gas were explicitly controlled by the Allies, and German soldiers were afforded more agency in their narrative counterattacks. Depictions of the gas mask also proved similar to those of the pacifists, reflecting on the mask's unwieldiness and malfunctions. But these technical difficulties were merely another hardship that soldiers had to endure in their national struggle. Certainly, in these more heroic texts, the mask was not conspiring to kill German soldiers or prolong the war.

The importance of *Kameradschaft* for postwar nationalist writers is perhaps most evident in Heinz Grothe's *Der Fronterlebnis*, a 1932 scholarly treatment of war literature. Born in 1912, Grothe was far too young to

[32] Arnold Zweig joined in the publication of pacifist war novels in the mid-1930s, writing both *Erziehung vor Verdun* (1936) and *Einsetzung eines Königs* (1937). Both books maintain discussions of poison gas in World War I, but *Erziehung vor Verdun* features a gas pioneer as the central character. Arnold Zweig, *Education before Verdun* (London: Martin Secker & Warburg, 1936), 395.

[33] In the late 1920s, even the Nazis put on a play entitled *Poison Gas* for the *Kampfbühne*. The content of this play has been lost, but the overall intent of the *Kampfbühne* was to indoctrinate audiences with militaristic and nationalistic themes. Claudia Koonz, *The Nazi Conscience* (Cambridge, MA: Belknap Press, 2003), 79.

serve in World War I, but his nationalist and eventual National Socialist sympathies encouraged him to reflect on the right-wing accounts of the war that had begun appearing in large numbers in the late 1920s.[34] Considering the descriptions of the World War I trenches, Grothe wrote:

> The front is monstrous. Young people dragged out of life, now in fire, gas, and annihilation. The front is an eerie vortex. You can feel its suction power when you are still far away from its center. It attracts men without much resistance.[35]

Here Grothe acknowledged the danger of modern weapons such as poison gas, but he also attributed a certain appeal to their lethality. Without the trials of gas, Grothe argued, there would be no frontline community, no brotherhood forged in the shared experiences of war. It was precisely this brotherhood that formed the foundation of a renewed German nationalism. Grothe wrote:

> There has never been a war fought solely with gas and other treacherous forms of warfare that drowned both the individual and the collective. But the real happenings did not destroy the communal form. It rather welded the people together. People joined together against the daily invisible enemy: death.[36]

Grothe's conception of a frontline soldier community and its ritual baptism in the industrialized weapons of war reflected the ideas of the two most important postwar authors for a broad variety of right-wing political movements, namely Franz Schauwecker and Ernst Jünger.

The war writings of both Schauwecker and Jünger incorporated variations on the nationalist sentiment that appeared in their contemporaries' novels. However, they further reconceptualized the pain and suffering of the war as not only necessary but uniquely positive on an individual level. According to both writers, it was only through ordeals such as gas attacks, that a stronger man could historically emerge. This "New Man" would bring Germany into the modern world, refashioning the nation in the industrial forges of war. A revaluation of human violence encouraged Schauwecker and Jünger to present their readers with particularly graphic depictions of aggression and suffering. These semi-fictionalized scenes were no longer moments of final defeat, as they were for the pacifists, but rather the beginnings of a new German triumph. World War I was merely the prologue to the larger struggle of the twentieth century. Inevitably, some would perish in this chivalric quest

[34] Grothe later joined the National Socialist Bamberg poetry circle and served in the Wehrmacht as a war correspondent.
[35] Heinz Grothe, *Das Fronterlebnis* (Berlin: Joachim Goldstein Verlag, 1932), 54.
[36] Ibid., 29.

for dominance over the modern world, but Schauwecker and Jünger found a certain inevitability in this envisioned future.

After fighting as an enlisted man in World War I, Franz Schauwecker made his breakthrough as a novelist with his 1919 work *Im Todesrachen* (*In the Jaws of Death*). This first-person depiction of trench warfare featured an evocative portrayal of a gas attack:

> In 1918, the opponent had phosgene and something similar to our new, deadly yellow cross [mustard gas]. Invisible and nearly odorless, it sinks heavily into the fields, penetrates into the uniforms like cigar smoke, eats away at things in its liquid form, even sits for days in holes, forests, and valleys, descending, stifling, and poisoning. There is only one salvation: strip out of your uniform and run. Every hesitation is suicide. Like most weapons against which men are powerless, gas is a cruel and deceitful way of fighting that is best characterized by the term "poisoning." Gas! This word creates fear and tightens muscles and tendons, ready to jump.[37]

In this passage, Schauwecker clearly describes gas as an unfair and brutal weapon, and his portrayal of the gas attack is certainly not a straightforward glorification of war. However, in contrast to this passage, the text does later argue that soldiers could survive the suffering that poison gas elicited. This survival required a mix of individual stamina and technological know-how. In the same gas attack scene, Schauwecker wrote, "'Gas!' Here and there heads and bodies rise, some begin vomiting. In no time, the gas masks are pulled out and placed over heads. Then you feel calmer, safer."[38] Similarly, in a passage from his later novel, *Endkampf 1918* (*Final Battle 1918*), Schauwecker described sleep-deprived and flea-bitten soldiers who "snapped their noses in gas masks in the middle of the night and waded in clay to their knees."[39] While Schauwecker portrayed the soldier's pain and fear during these gas attacks, he did not claim that this agony was without meaning. In fact, the calming reassurance of the gas mask offered possible salvation and even rebirth.

After working with Ernst Jünger on a publication that formulated their mutual conception of "New Nationalism" in the mid-1920s, Schauwecker wrote his most widely read novel *Aufbruch der Nation* (*Dawn of the Nation*).[40] Published in 1929, as part of the war novel boom, *Aufbruch der Nation* maintained particularly graphic depictions of

[37] Franz Schauwecker, *Im Todesrachen* (Halle: Heinrich Diekmann Verlag, 1921), 293.
[38] Ibid., 293.
[39] Franz Schauwecker, *Endkampf 1918* (Frankfurt am Main: Verlag Moritz Diesterweg, 1936), 55.
[40] Martin Travers, *Critics of Modernity: The Literature of the Conservative Revolution in Germany, 1890–1933* (New York: Peter Lang, 2001), 98.

maimed soldiers in a military hospital.[41] Schauwecker wrote, "They were wounded. They screamed in pain and moaned in the hospitals. Knives milled in their bodies for the splinters and projectiles. They died in explosions, smoke, gas, mud, and rain. They died in the trenches, buried, and suffocated. They died, eaten alive by gas."[42] Such explicit scenes are placed within the story of a middle-class soldier named Albrecht Urach who loses faith in the war after suffering several severe injuries. However, Urach's exposure to both violence and Bolshevism over the course of the war forces him to recognize the need for a strong German nation that can resist both the lure of communism and the supposedly effeminate weakness of the bourgeoisie. While the war ultimately dismantled the ruling structures of the Second Reich, Schauwecker advanced this destruction as a new beginning for the German people, bursting forth into a modern world pregnant with possibilities.[43] Thus, the explicit pain of the men in the hospital was posited as a necessary trial in the rebirth of the German nation.

Ernst Jünger's war novels play with similar themes of both individual and collective renewal.[44] While Jünger's fictional works, such as his famous *Im Stahlgewitter* (*Storm of Steel*), were not as discernibly nationalistic as Schauwecker's, they featured comparable scenes of building both individual and collective strength through wartime suffering.[45] Pulling on his own experience as a gas officer during the war, Jünger filled his loosely autobiographical novels with chemical weapons, often describing attacks with the flat affect of the consummate technician.[46] For instance, in *Storm of Steel* he wrote:

I leaped over the ramparts of the reserve line, raced forward, and soon found myself enveloped in the gas cloud. A penetrating smell of chlorine confirmed for me that this was indeed fighting gas, and not, as I had briefly thought, artificial fog. I therefore donned my mask, only to tear it off again right away because I'd been running so fast that the mask didn't give me enough air to breathe; also the

[41] Franz Schauwecker, *Aufbruch der Nation* (Berlin: Frundsberg Verlag, 1930), 196.

[42] Ibid., 373.

[43] The Nazis praised Schauwecker's books in the late 1920s and early 1930s. In 1931, Schauwecker joined the Society for the Study of Fascism, an organization that aimed to foster a positive perception of fascism in Weimar Germany. In 1933, under the new Nazi regime, Schauwecker signed the "Vow of Most Faithful Allegiance" and his novels reached their peak sales. Interestingly, Schauwecker never officially joined the Nazi party.

[44] Antoine Bousquet, "Ernst Jünger and the Problem of Nihilism in the Age of Total War," *Thesis Eleven* 132, no. 1 (2016): 17.

[45] In the 1920s, *Storm of Steel* sold 244,000 copies in twenty-six editions. Karl Pruemm, *Die Literatur des Soldatischen Nationalismus der 20er Jahre* (Kronberg: Taunus, 1974), 101.

[46] Jünger was gassed on August 8, 1918. This is one of the fourteen wounds he reportedly suffered during the war. Elliot Y. Neaman, *A Dubious Past: Ernst Jünger and the Politics of Literature after Nazism* (Berkeley: University of California Press, 1999), 26.

goggles misted over in no time, and completely whited out my vision. All of this of course was hardly the stuff of "What To Do in a Gas Attack," which I'd taught so often myself. Since I felt pain in my chest, I tried at least to put the cloud behind me as fast as I could.[47]

In a seemingly odd juxtaposition with scientific and technical language, Jünger used the language of magical realism to enchant modern technology, creating a romantic experience that still lay within mankind's technical relationship to the world. Jünger's fictional soldiers, at first victimized by the technology of modern war, now rode into battle on newly mechanized mythical steeds in order to renew their age-old struggle.[48] As part of this reliance on magical realism, Jünger described poison gas as a creeping cloud and men in gas masks as demons, ghosts, and animals.[49] In one such passage, Jünger wrote,

As I heard the next morning, in that hour in the woods a lot of our men died of poisoning from the clouds of heavy phosgene nestling in the undergrowth. With weeping eyes, I stumbled back to the Vaux woods, plunging from one crater to the next, as I was unable to see anything through the misted visor of my gas mask. With the extent and inhospitableness of its spaces, it was a night of eerie solitude. Each time I blundered into sentries or troops who had lost their way, I had the icy sensation of conversing not with people, but with demons. We were all roving around in an enormous dump somewhere off the edge of the charted world.[50] (Figure 6.1)

By imbuing poison gas and the gas mask with seemingly supernatural qualities, Jünger's war writings seemingly align with passages from Remarque and other antiwar writers. But in pushing the realist narratives of his peers far further, Jünger more explicitly described the sense of unworldly fear that poison gas elicited. However, Jünger's demonic gases could be controlled and conquered by the equally powerful gas mask. In this way, Jünger enchanted the entire technological world and placed it under the control of humanity. Poison gas was no longer faceless and undetectable but rather a formidable weapon that could be directed and employed by the soldiers of the modern battlefield.

Jünger's fictional world was both disturbingly violent and deeply uncaring. While his texts have little stated aim in their depictions of gas warfare, they were clearly part of the author's larger attempt to conceive of World War I as an opportunity for modern rebirth. For Jünger, it was precisely the technological assault on humanity that created a new

[47] Ernst Jünger, *Storm of Steel*, trans. Michael Hofmann (New York: Penguin, 2003), 79.
[48] Ernst Jünger, *Copse 125: A Chronicle from the Trench Warfare of 1918*, trans. Basil Creighton (New York: Zimmerman & Zimmerman, 1985), 8.
[49] Jünger, *Storm of Steel*, 49, 79–80. [50] Ibid., 114.

Figure 6.1 A World War I photograph of a mounted German soldier with a gas mask and a lance. This image, from official German war photography, serves as a visual representation of Jünger's blending of modern warfare with a romantic vision of the medieval knight.
Source. Deutscher Meldereiter mit Gasmaske (IZS) 12674/17

physical environment in which individuals would need to refashion themselves.[51] In the late 1920s and early 1930s, Jünger made this idea more explicit in his nonfiction essays such as *Der Kampf als inneres Erlebnis* (*War as an Inner Experience*, 1922), *Sturm* (*Storm*, 1923), *Die Unvergessenen* (*The Unforgotten*, 1928), and *Über den Schmerz* (*On Pain*, 1934). According to these texts, the war reminded modern humans that they were inherently violent creatures; nineteenth-century civilization was merely a façade for the dark simian conflict that truly propelled human history. Jünger wrote:

> man drives the will to kill through the storms of explosives, iron and steel, and when two people clash in the tumult of battle, so meet two beings, of which only one can exist. These two beings have set themselves in a primal relationship to each other, in the struggle for existence in its most naked form.[52]

[51] Robert Heynen, *Degeneration and Revolution: Radical Cultural Politics and the Body in West Germany* (Boston: Brill, 2015), 265.
[52] Jünger, *Der Kampf als inneres Erlebenis*, 8–9.

Jünger celebrated this reawakening of primitive violence, asserting its importance in the creation of an envisioned "New Man." For Jünger, unbridled violence coupled with the incredible power of modern warfare technology could forge a uniquely hardened individual, a being who could utilize said technology to break through the façade of civilization and reenchant modernity with the more primal gods of aggression, violence, and warfare.[53]

Of course, many soldiers would not survive the ordeal of the trenches. Jünger saw this culling process as essential to the social-Darwinian evolution of his "New Man." The pain elicited by methods of technological warfare such as poison gas were thus recast as a baptism in "fire and blood."[54] Those men who did not endure this ritual were nevertheless essential to Jünger's fashioning of the modern man, the modern community, and the modern German nation.[55] Thus, it was precisely the poison gas attacks and other technologically assisted instances of wartime suffering that gave the war its meaning.[56] For Jünger, to unduly question the importance of this process was a betrayal of the experience itself, a denial of the significance of risk and contingency in the valuation of heroic virtue.[57] Reflecting on the implications of modern warfare technologies such as gas, Jünger wrote that:

> one could see the enhancement of human ability through technology. Compared to the roar of the attack, the clash of weapons and hooves of an earlier time, these [modern weapons] were thousands of times stronger. For that reason, the courage required here greatly exceeded that required of the Homeric heroes.[58]

In passages such as this, Jünger was clearly attempting to create value for chivalric bravery in a form of warfare that many felt to be anonymous and capricious. This rehumanizing of "total war" placed a living operator behind all of the explosives, tanks, airplanes, and gases that defined the Western Front of World War I. For Jünger, gas might have been ghostly and supernatural in a narrative sense, but it was ultimately one of the tools of the "New Man," now encased in rubber and celluloid, who directed and controlled its quasi-mystical power with technical precision.

[53] Klemens von Klemperer, *German Incertitudes, 1914–1945: The Stones and the Cathedral* (Westport, CT: Praeger, 2001), 53.

[54] Ernst Jünger, *On Pain*, trans. David C. Durst (London: Telos, 2008), 1.

[55] Ernst Jünger, *Die Unvergessenen* (Berlin: Wilhelm Andermann Verlag, 1928), 302.

[56] Andres Huyssen, "Fortifying the Heart – Totally: Ernst Jünger's Armored Texts," *New German Critique* 59 (1993): 10.

[57] Thomas Nevis, *Ernst Jünger and Germany: Into the Abyss, 1914–1945* (Durham: Duke University Press, 1996), 85.

[58] Ernst Jünger, *Sturm*, trans. Alexis P. Walker (Candor, NY: Telos Press, 2015), 33.

The Excesses of the Mask: Depictions of Gas Warfare in Visual Art and Film

In her study of poison gas in interwar visual art, historian Doris Kaufmann presents several French and British landscape paintings that attempt to depict poison gas floating over the World War I battlefield.[59] Kaufmann points out that these landscapes struggle to express the "quantum of suffering" that gas inflicted on soldiers. Due to its inherent "lack of pictorialness," poison gas was a particularly difficult weapon to depict through a visual medium. For this very reason, the weapon lent itself to modernist styles such as cubism and futurism, which were able to bend the realities of gas into a subjective experience of horror and pain.[60] In the German context, this is best expressed in Otto Dix's 1917 expressionist painting, *Lichtsignale* (The Flare), in which colored flares light up trenches filled with entangled bodies and barbed wire. The flares are intended to warn of a gas attack and the bodies below wear primal masks.[61] While Dix's painting is a jumble of colors and objects that creates a sense of wartime confusion, its use of the gas mask as a distinct material representation of gas attacks anticipated later developments in German war art.

Dix, a former machine gunner in the German army, turned toward a more realistic style in the 1920s as part of the larger New Objectivity movement. In 1923, he produced "The Trench," another depiction of the World War I battlefield that would be incorporated into a larger *War Triptych* (*Der Krieg*) between 1929 and 1932. Like *Lichtsignale*, "The Trench" relied on a scarred physical landscape, darkened skies, and mutilated bodies to express the destruction of World War I. However, one of the central objects in the painting is a lone soldier in a steel helmet and gas mask. Through the mask, Dix insinuates the presence of poison gas and expresses the toxicity of the entire environment. In this way, the presence of the gas mask is far from coincidental. Much like all masks, the gas mask conceals the human face, making it an object that always maintains/contains excess meaning. Whether worn or not, the mask constantly suggests the obfuscation and suppression of the real. At the same time, this allows the mask to serve as a locus for expanded symbolic significance. For this very reason, it was not uncommon for interwar artists to frequently use the gas mask to represent the possibility, imminence, or aftermath of gas attacks. The ambiguity of temporality in depictions of the gas mask

[59] Kaufmann, "'Gas, Gas, Gaas!'," 171–172. [60] Ibid., 173–174.
[61] For an image of Dix's *Lichtsignale*, see ibid., 173.

further allowed artists to express distinct wartime experiences while pointing to the dangers of a future war.

As a symbolic object, the mask could also be seen as a representation of broader conceptions of modernity. In his posthumously published 1901 work, *The Will to Power*, Friedrich Nietzsche described the masked quality of the triumphant modern man, writing: "He knows that he cannot reveal himself to anybody: he thinks it bad taste to become familiar; and as a rule he is not familiar when people think he is. When he is not talking to his soul, he wears a mask."[62] Likely building on Nietzsche's claims, the interwar philosopher Helmuth Plessner argued in his 1924 work of philosophical anthropology, *Grenzen der Gemeinschaft* (The Limits of Community), that masks constantly mediated modern social life, thereby armoring the individual against social exposure and offering a certain internal freedom in the impersonal.[63] Plessner even described such masks as a form of technology, relying on the more expansive conception of *techne* to include the artifices of human expression. *Grenzen der Gemeinschaft* recognized the value of this protective masking, arguing that it allowed humans to socially exist in a pluralistic society such as the Weimar Republic. Berlin, the city that most fully articulated the promises and pitfalls of modernity to contemporaries, was regularly described as a social arena in which people could now choose their own masks.[64] Of course, this claim referred to both the figurative and literal representation of each city dweller.

The artists of the Dada movement were perhaps the first to pick up on this idea and to use the gas mask as one of their many representations of uncertain identity. For instance, in her 1919 photomontage *Dada-Rundschau*, Hannah Höch placed an ambiguous soldier figure on top of a cloud of poison gas that threatens to envelope the German Minister of Defense, Gustav Noske. The soldier, whose sex appears undefined, is entirely covered by a long trench coat, a helmet, and a gas mask. The combination of images appears to suggest that the new and technologically armored human, as a product of World War I, now threatens the older political order of Germany. Similarly playing with ambiguous identities and threats to the established social world, Rudolf Schlichter's 1920 painting

[62] Friedrich Nietzsche, *The Will to Power: An Attempted Transvaluation of All Values*, trans. Anthony M. Ludovici (Edinburgh: T. N. Foulis, 1910), 367.

[63] Helmuth Plessner, *The Limits of Community: A Critique of Social Radicalism*, trans. Andrew Wallace (Amherst, NY: Humanity Books, 1999), 80; Helmut Lethen, *Cool Conduct: The Culture of Distance in Weimar Germany*, trans. Don Reneau (Berkeley: University of California Press, 2002), 62–63.

[64] Tyler Carrington, *Love at Last Sight: Dating, Intimacy, and Risk in Turn-of-the-Century Berlin* (Oxford: Oxford University Press, 2019), 169.

Figure 6.2 Otto Dix, *Stormtroopers Advance under a Gas Attack*, 1924. © 2023 Artists Rights Society (ARA), New York / VG Bild-Kunst, Bonn.

Dada Dachatelier depicts a number of cyborgs, cobbled together with various common items. At least two of the humanoids have gas masks for faces, representing the ways in which the German military had both figuratively and literally reassembled humans over the course of World War I and into the interwar period.[65]

Returning the gas mask to the battlefield, Otto Dix's work from the mid-1920s focused on the anxiety-producing quality of its inherent anonymity. In Dix's 1924 engraving entitled *Sturmtruppe geht unter Gas* (*Stormtroopers Advance Under a Gas Attack*), the gas mask turns attacking soldiers into violent ghouls, assaulting the viewer through the haze of gas. Like in Schlichter's work, the gas mask appears as the permanent faces of the soldiers, thereby expressing what Dix saw as the horrific mechanical transformations of wartime violence (Figure 6.2).

In the same 1924 portfolio, Dix also included two depictions of gassed soldiers, *Gas-Opfer (Templeux-La-Fosse, August 1916)* and *Die Schlafenden von Fort Vaux (Gas-Tote)*.[66] In both of these works, Dix depicted a row of

[65] Carol Poore, *Disability in Twentieth-Century German Culture* (Ann Arbor: University of Michigan Press, 2007), 34.

[66] Otto Dix, *Gas Victims* (Templeux-La-Fosse, August 1916), 1924; Otto Dix, *Die Schlafenden von Fort Vaux (Gas-Tote)*, 1924.

corpses with darkened skin to express the brutal end effects of asphyxiation under gas. In many ways tied to the image of the advancing stormtroopers, Dix's *Gas Victims* revealed the gas mask's fallibility. Toggling between two competing images of chemical warfare, Dix forced his viewer to contemplate the easy slippage between violent aggressor and lifeless victim, thus raising questions that would animate the intellectual debates over chemical weapons throughout the late 1920s.

While Dix's depictions of the gas mask remained in the trenches of World War I, plenty of other visual artists explicitly brought such military technologies into the social and political context of the interwar period. Reflecting on the ways in which military rearmament threatened Europe in the mid to late 1920s, the gas mask now became a symbol of a broad technological militarism. For instance, in 1924, George Grosz unveiled "Shut Up and Do Your Duty" as part of a larger set of prints entitled *Hintergrund (Background)*. The piece featured a recreation of the crucifixion of Christ in which Jesus is wearing both army boots and a gas mask. Playing on one of the weightiest symbols in European culture, Grosz aggressively politicized his Christ figure for the contemporary moment and reflected on the German soldiers who were killed by the very weapons that they had first employed. Thus, the piece served as a rejection of Germany's stated war aims and the subsequent heroic memorialization of the war. However, the image can also be read to represent German civilians, or broader humanity, in a world where large-scale aero-chemical attacks were possible and normalized. Ultimately, Grosz' provocation achieved its desired effect, and the German state brought criminal charges against both the artist and his publisher for the release of "Shut Up and Do Your Duty." Fighting against accusations of "blasphemy and defamation of the German state," Grosz was now caught up in the longest trial in the history of the Weimar Republic.[67] While Grosz was ultimately acquitted in 1931, his trial most fully revealed the symbolic power of the gas mask and the political importance of national gas protection for both the state and its citizens.

After fleeing the rise of the Nazis in 1933 and settling in the United States, Grosz would return to this visual theme in his 1936 lithograph "Soldier on Horse with Gas Mask." Grosz's horseman is foregrounded against the bombing and gassing of a major city. The cavalryman is given a spectral quality while the skeleton of his zombie horse is visible through its skin. Through these supernatural features, the image reminds the viewer of the pale horseman of a apocalypse, as he rains poison gas down

[67] Herbert Knust, "George Grosz: Literature and Caricature," *Comparative Literature Studies* 12, no. 3 (1975): 235.

on humanity. Serving as a counterpart to Christ in a gas mask, Grosz's masked soldier anticipates the future violence of technologically advanced weaponry. However, the horseman blends both modern and premodern elements, wearing a gas mask alongside a saber and cloak. For this reason, Grosz's piece resembles the World War I photograph of the German cavalryman with a gas mask and spear (see page 247). Expressing the uncanny through this amalgamation, the unrestrained and barbaric violence of an imagined past was brought forward into a more technologically precarious modernity.

For the Dada artists, Otto Dix, and George Grosz, the gas mask undoubtedly served to depersonalize and obscure the human face. For Dix and Grosz, this could serve to represent the fact that violence too had become depersonalized and ever-present in the modern world. However, not all interwar visual artists created such a stark morality tale, instead exploring the complicated politics of violence, complicity, and victimhood that the gas mask seemed to produce. At first glance, Barthel Gilles' 1930 work *Self Portrait with Gas Mask* attempts a straightforward critique of the modern man who perpetually prepares himself for the imminent gas attack. A late product of the "New Objectivity" movement, the painting features the artist standing in front of the nocturnal silhouette of a burning Cologne, presumably just bombed and gassed. Gilles raises a gas mask to his face expectantly, tightening his body in preparation for gas. He gazes intently out at the observer, further expressing a sense of deep trouble and anxiety. In donning the gas mask, Gilles appears aware of the fact that the mask is territorializing his life, forcing him to react to gas attacks in the most intimate and physical way. At the same time, the artist realizes his own complicity in the mask's demand for individual and collective discipline.[68] The scene is frozen at the moment of choice, and the viewer is led to believe that Gilles will most likely put the mask over his face; an overwhelming desire to survive trumps concerns over the systemic structures undergirding this newly violent world (Figure 6.3).

Gilles' self-portrait is clearly the most explicit example of interwar German visual art bringing national defense and poison gas to the attention of those who patronized the often-insular art world. While some Germans may have seen these artworks through reproductions in pacifist literature, the majority of people never had the ability to study their politicized meanings. On the other hand, similar film treatments of poison gas could reach far more German eyes.[69] In his study, *Shell Shock*

[68] Fritzsche, "Machine Dreams," 708–709.

[69] The interwar period saw the breakthrough of film as a mass medium in Germany. In 1910, there were around 1,000 cinemas in Germany with 200,000 seats. By 1928, there were 5,267 cinemas with 1,876,600 seats. By 1930, Germans went to the cinema an

Figure 6.3 Barthel Gilles, *Self Portrait with Gas Mask*, 1930

Cinema, film scholar Anton Kaes has argued that, at least through metaphorical means, some of the best-known Weimar era films tackled the destruction and trauma of World War I. Kaes sees Robert Wiene's 1920 classic, *The Cabinet of Dr. Caligari*, as an attempt to confront the psychic trauma that many German veterans experienced after the war.[70] Similarly, F. W. Murnau's 1922 film *Nosferatu* recreated the narrative of the war by sending young men off to fight a deadly horror, represented by the film's eponymous vampire.[71] Still further, Kaes sets up Fritz Lang's 1924 *Die Nibelungen* as a postwar tale of German national consciousness and its break from an idealistic past.[72] And finally, he argues that Lang's 1927 film *Metropolis* is a reckoning with the scientific, industrial, and military ties that propelled Germany's failed war effort.[73]

Through its depiction of a seemingly utopian society predicated on the power of advanced machinery and the oppression of workers, *Metropolis* can certainly be read in terms of class struggle and Marxist history, with technology playing a central role as a social mediator. Lang's work raises

estimated 6 million times per week. Corey Ross, "Mass Culture and Divided Audiences: Cinema and Social Change in Inter-War Germany," *Past & Present* 193 (November 2006): 160–161.

[70] Anton Kaes, *Shell Shock Cinema: Weimar Culture and the Wounds of War* (Princeton: Princeton University Press, 2009), 48.

[71] Ibid., 101. [72] Ibid., 147. [73] Ibid., 187.

questions about the ways in which technology can deceive and destroy humans and the ways in which humans propagate technologies through an insistence on a totalizing scientific rationality. Utilizing modernist set design, the city of *Metropolis* appears distinctly dystopian due to its hard, cold, technological spaces.

Thus, while Expressionist films such as *Metropolis* had already exposed discerning audiences to questions regarding the destructive powers of technology, Mikhail Dubson's 1929 film *Giftgas* (Poison Gas) expressly treated many of the technological themes surrounding the concurrent poison gas debates. Dubson and his writers based the film on a poorly received 1929 play by Peter Martin Lampel entitled *Giftgas über Berlin* (*Poison Gas Over Berlin*).[74] Pulling on the events of the 1928 Hamburg gas accident, Lampel's play involved a gas leak at a secret military installation in Berlin. The gas subsequently poisons much of the city, creating a desire for national gas protection and leading to the election of General Hans von Seeckt as an emergency military dictator.[75] Only appearing on stage a few times, Lampel's play was banned by Weimar authorities for its supposedly dangerous political message.[76] Nevertheless, Dubson adapted the play for the big screen, telling the story of a young scientist who succeeds in producing a synthetic fertilizer. Clearly referring to the wartime actions of Fritz Haber, the scientist's fertilizer synthesis process also produces poison gas as a byproduct. When the scientist attempts to stop making fertilizer due to his concerns over the production of poison gas, his employer shoots and kills him. However, the pistol fire accidentally unleashes the stores of poison gas, eventually destroying the entire city. The film's final scene features the ghosts of soldiers, who rise from the dead to decry the use of poison gas in World War I. Dubson used gas masks throughout the film to express the insufficiencies of gas protection, and as the gas wafts over the city, civilians fight for scarce masks and leave each other to asphyxiate.[77] Certainly, such violent cinematic scenes forced regular German movie-goers to consider such a future possibility in every German city and every German home.

[74] Werner Hirsch, *Die rote Fahne: Kritik, Theorie, Feuilleton 1918–1933*, ed. Manfred Brauneck (München: Wilhelm Fink Verlag, 1973), 379. Anton Kaes claims that Dubson's *Giftgas* was loosely based on Georg Kaiser's *Gas* trilogy. Kaes, *Shell Shock Cinema*, 187.

[75] Richard Elsner, *Das deutsche Drama: ein Jahrbuch* (Berlin-Pankow: Verlag der Deutschen Nationalbühne, 1929), 233.

[76] Julius Bab, "Giftgas und Zensur," *Die Hilfe* 35 (1929): 170.

[77] Mikhail Dubson, *Giftgas*, 1:10:40.

The Apocalyptic Novel: Dystopian Science Fiction and Visions of Gas Destruction

While *Kriegserlebnis* novels and other cultural production that referenced World War I clearly maintained strong political overtones, their historical setting could potentially serve to depoliticize their claims. Undoubtedly, Germans on the political Right and Left fought over the memorialization of the war through these narratives, but ancillary questions often went unnoticed by those engrossed in the inherent romanticism of the soldier's tale.[78] However, a second body of literature emerged in the Weimar era that presented these postwar questions in a much more politically explicit and future-oriented manner. Apocalyptic and dystopian novels and plays began to hypothesize the role of aerial bombing and gassing in a future war. Pulling on both the experience of modern industrialized killing in World War I and the national aerial defense debates of the 1920s, these proto-science fiction works unambiguously narrated dark visions of the German future. Literary scholar Paul Saint-Amour has described this larger moment in modernist writing as a "perpetual interwar," in which writers were constantly "remembering a past war while awaiting and theorizing a future one."[79]

One of the literary predecessors of this "perpetual interwar" first appeared in 1917. Georg Kaiser's *Die Koralle* (*The Coral*) was the first installment in a trilogy of Expressionist plays that examined the links between industry and warfare that had developed over the course of World War I. Kaiser made this theme more explicit in his second installment, *Gas I*, published in 1918. Set in an industrialized near future, *Gas I* centers on the son of an industrialist billionaire who inherits his father's gas factories. Following his progressive political impulses, the son distributes the factory profits among the workers and gives them fiduciary interest in the business. When one of the factories accidentally explodes, the son wishes to shut down the business, now fully realizing the danger of his products. However, the workers, led by "the Engineer," refuse to abandon the lucrative factories. After a government call for increased national gas production, the son and the Engineer hold a public debate on the future of the gas factories. The Engineer effectively wins, the factories are kept open, and the play ends with the workers chanting for more "Gas!"[80]

The third installment in Kaiser's trilogy, *Gas II* was published in 1920. Set twenty-five years after *Gas I*, the factory workers are now trapped in

[78] Joyce Wexler, "The New Heroism," *Literature & The Arts* 27 (2015): 11.
[79] Saint-Armour, *Tense Future*, 305.
[80] Georg Kaiser, *Gas I* (Potsdam: Gustav Kiepenheuer, 1925), 75.

an endless cycle, unceasingly producing natural gas for their nation's war effort. When the chief engineer synthesizes a new poison gas, he recognizes its value for the battlefield. The current factory owner (and son of the owner in *Gas I*) rejects the chief engineer's proposal to militarize production, instead pleading for universal peace. The factory workers ignore this entreaty due to their strident nationalism and, in an act of anger, the factory owner smashes a vial of poison. Weeks later, when an opposing army approaches the now-abandoned gas factory, the soldiers find the skeletons of the entire workforce. Horrified by the effects of the poison gas, the enemy soldiers commit mass suicide.[81]

It is fairly obvious that, like Dubson's film, Kaiser's plays fictionalize Fritz Haber's production of poison gas in World War I. Reflecting a technological pessimism that grew between 1918 and 1923, Kaiser intended to critique what he saw as a military-industrial complex that had developed during the war.[82] His plays further pulled on Marxist insights to narrate the ways in which capitalist profit-seeking inspired rabid technological advancement. While the tragic characters hold up these advances as a means for achieving utopia, they ultimately lead to human exploitation and greater societal risk. This, Kaiser warned, could produce the apocalyptic scenario that his plays imagined.[83]

As debates over national aerial and gas protection intensified in the late 1920s, Kaiser's plays served as a template for the further proliferation of apocalyptic writing.[84] These fictional accounts took on specific political contexts that expressed distinct German concerns over the roles of France, Britain, the United States, the Soviet Union, and even Japan in a future aero-chemical war. For instance, Peter Martin Lampel's 1929 play *Giftgas über Berlin* posited a future German air war with France that included English and Russian interference.[85] Similarly, in his 1929 novel *Giftküche (The Devil's Workshop)*, Karl August von Laffert imagined a gas war between England and Russia in which Germany suffered in the crossfire. Expressing serious doubts about international diplomacy's ability to avert this future war, Laffert's book left readers with few answers other than personal protective devices such as gas masks.[86]

[81] Georg Kaiser, *Gas II* (New York: Ungar, 1963), 162.

[82] Axel Goodbody, *Nature, Technology and Cultural Change in Twentieth-Century German Literature: The Challenge of Ecocriticism* (New York: Palgrave Macmillan, 2007), 94–95.

[83] Ibid., 108.

[84] Johannes Becher took up Kaiser's themes in his 1925 modernist novel, *(CHCl=CH)3As (Levisite)* while Ernst Toller's plays *Hoppla, We're Alive!* (1927) and *Transfiguration* (1919) both included scenes in which poison gas is satirically justified.

[85] Elsner, *Das deutsche Drama*, 233.

[86] Karl August Laffert, *Giftküche* (Berlin: August Scherl, 1929), 236.

Between 1931 and 1933, authors such as Hans Gobsch, Johann von Leers, Arnold Mehl, and Axel Alexander created similar depictions of aero-chemical warfare between various European states.[87]

Works from the period more obviously classified as science fiction also relied heavily on imaginary apocalyptic scenarios. For example, the engineer-turned-science fiction author Hans Dominik first wrote on the dangers of poison gas in his story *Der Gas-Tod der Großstadt* (The Gas Death of the City). In this then unpublished work, a group of enemy soldiers sit in an underground bunker and direct aerial attacks against a fictionalized capital city. The planes drop gas bombs and phosgene penetrates every building. Dominik wrote:

> The next day, when the sun rose in the east, the last life had died out in the capital of Austrasia. A single gas attack by the superior enemy had turned the thriving city of millions into a cemetery ... The age of the chemical war had dawned. When there were no more fronts, no cannons, no difference between combatants and non-combatants, the whole country of the weaker adversary became a battleground, each one of its inhabitants exposed to the deadly effect of hostile chemical agents.[88]

Clearly influenced by the writings of military theorists like Giulio Douhet, Dominik used artistic license in his portrayal of the dangers of chemical warfare. However, his story also provided a means to combat this apocalyptical vision. Mass chemical death is staved off in the tale after civilians are provided with gas masks. Similarly, in Dominik's later novel *Atomgewicht 500*, scientific and technological solutions succeed in warding off a major chemical and radiological disaster.[89] In this way, Dominik narratively reproduced the arguments of the gas specialists in the late 1920s, claiming that humans could survive in a poisoned atmosphere as long as they received the correct training and equipment. For Dominik, it was only a continued commitment to science and engineering that could possibly save humanity.[90]

Most German authors, however, were not nearly so optimistic, providing no real means of salvation in their fictional recounting of similar events. Pulling on the theoretical insights of the sociologist Ulrich Beck,

[87] Hans Gobsch, *Wahn-Europa 1934* (Hamburg: Fackelreiter-Verlag, 1934); Johann von Leers, *Bomben auf Hamburg* (Leipzig: R. Voigtländers Verlag, 1932); Axel Alexander, *Die Schlacht über Berlin* (Berlin: Verlag "Offene Worte", 1933); Arnold Mehl, *Schatten der aufgehenden Sonne* (Leipzig, Wilhelm Goldmann Verlag, 1935).

[88] Hans Dominik, *Der Gas-Tod der Großstadt* (Dortmund: Synergen Verlag, 2015), 102.

[89] Hans Dominik, *Atomgewicht 500* (Berlin: Verlag Scherl, 1935), 19.

[90] William B. Fischer, *The Empire Strikes Out: Kurd Lasswitz, Hans Dominik, and the Development of German Science Fiction* (Bowling Green: Bowling Green State University Popular Press, 1984), 181.

literary scholar Kai Evers has identified the growth in German apocalyptic writing in the late 1920s and early 1930s as a collective "literature of risk."[91] This focus on human contingency was found not only in war novels and science fiction from the period but also in the writings of Weimar literary luminaries, such as Robert Müsil, Alfred Döblin, Elias Canetti, and Hans Henny Jahnn.[92] As Evers puts it, "Their works recognized society's unpreparedness to imagine the potentially disastrous capacity of otherwise immensely productive and beneficial modern technology."[93] Thus, even *Berlin Alexanderplatz*, Döblin's famous 1929 depiction of Weimar Berlin ends with a short reflection on aero-chemical bombing.[94] Döblin writes:

> Keep awake, keep awake, for there is something happening in the world. The world is not made of sugar. If they drop gas-bombs, I'll have to choke to death; nobody knows why they are dropped, but that's neither here nor there, we had the time to prepare for it.[95]

In this final passage, Döblin presents aero-chemical warfare as a near inevitable condition of modernity. His characters are aware of this condition, but they do nothing to prevent it, reflecting the lack of perceived human agency in the increasingly technologically dangerous world. Just two years after the publication of *Berlin Alexanderplatz*, another soon-to-be widely respected modernist writer, Hermann Broch, similarly used poison gas as a symbol of modernity's tightening noose. In his 1931 literary study of German society, *The Sleepwalkers*, Broch wrote:

> And this age seems to have a capacity for surpassing even the acme of illogicality, of antilogicality: it is as if the monstrous reality of the war had blotted out the reality of the world. Fantasy has become logical reality, but reality evolves the most a-logical phantasmagoria. An age that is softer and more cowardly than a preceding age suffocates in waves of blood and poison-gas.[96]

[91] Kai Evers, "Risking Gas Warfare: Imperceptible Death and the Future of War in Weimar Culture and Literature," *Germanic Review* 89, no. 3 (2014): 271.

[92] This sense of contingency was even present in concurrent newspaper images that attempted to imagine the future aero-chemical war. For instance, the *Berliner Illustrierte Zeitung* printed the image of a lone woman dead in the streets of Berlin in their October 30, 1927, edition. See Kai Evers, "Gassing Europe's Capitals: Planning, Envisioning, and Rethinking Modern Warfare in European Discourses of the 1920s and 1930s," in *Visions of Europe*, 76.

[93] Evers, "Risking Gas Warfare," 272.

[94] Paul K. Saint-Amour, "Air War Prophecy and Interwar Modernism," *Comparative Literature Studies* 42, no. 2 (2005): 149.

[95] Alfred Döblin, *Berlin Alexanderplatz*, trans. Eugene Jolas (New York: Frederick Ungar Publishing Co, 1961), 634.

[96] Broch, *The Sleepwalkers*, 373.

Here, Broch posited the illogical panic of the interwar period as the source of its very downfall. It was the constant concern for future aerochemical warfare, or the mode of "perpetual interwar" writing and discourse, that would inevitably bring the phantasmagoric gas war. And, according to Broch, unlike the soldiers in the trenches of World War I, interwar Europeans were even less prepared to face the deadly new reality that they had invoked.

Technology for the Modern Age: Weimar Intellectuals Enter the Gas Debates

Nearly immediately after his frontline service in World War I had ended, Ernst Jünger began to publish non-fiction writings on the massive technological changes taking place in modern warfare. By 1921, he had written the article "Die Technik in Zukunftsschlacht" ("Technology in the Future War") in which, as the title would suggest, he began to construct his vision of an ever-more technological form of future warfare. As his novels also made clear, Jünger believed that humans still maintained a role in this clash of animated material, for they needed to maintain both the knowledge to direct the violence of their technologies and the discipline to withstand that of their enemy.[97] In his expanded 1922 reflection on the war, entitled *Der Kampf als inneres Erlebnis* (*The War as an Internal Experience*), Jünger continued to consider man's new role on the battlefield, writing:

> The battle of material and trench warfare was fought more recklessly, wilder and more brutally than any other form that man produced. The new soldier was a whole new breed, embodying energy and loaded with the utmost force. Smooth, lean, sinewy bodies, striking faces, eyes petrified in a thousand horrors under the helmet ... Jugglers of Death, masters of explosives and flame, magnificent predators, rushed through the trenches. At the moment of the encounter, they were the epitome of the most combative force in the world, the keenest gathering of the body, the intelligence, the will, and the senses.[98]

In his wartime reflections from 1925, Jünger continued to write about this technological molding of soldiers, often concentrating on the material changes brought on by the use of technologies such as poison gas. He was insistent that gas could never be used on a grand scale due to its unpredictability.[99] Nevertheless, gas had noticeable strategic advantages on the battlefield, allowing it to create impassable terrain and harass entrenched

[97] Ernst Jünger, "Die Technik in der Zukunftsschlacht," *Militär Wochenblatt* 14, October 1, 1921.
[98] Jünger, *Der Kampf als inneres Erlebnis*, 32–33.
[99] Jünger, *Copse 125*, 133.

positions. While gas changed tactics and reconceptualized battlefield space, it also produced physical changes in the common soldier. Jünger claimed that "Since the introduction of gasmasks the long beards that many of [the soldiers] wore ... have disappeared; and certainly the clean-shaven features under the helmet-rim express the strenuous spirit of the man of today a great deal better."[100] For Jünger, it is quite clear that poison gas concurrently altered both the physical appearance and the subjective experience of men on the battlefield.

Intimately aware of the debates surrounding German rearmament and national aero-chemical defense in the late 1920s and early 1930s, Jünger rearticulated his ideas on technology in two intertwined areas of intellectual inquiry. First, Jünger edited and wrote introductory essays for seven photo albums, released between 1928 and 1933.[101] Most of these albums were dedicated to wartime photos, interwar political confrontations, or images of aerial technologies. His final two photo books, *Die gefährliche Augenblick* (*The Dangerous Moment*, 1931) and *Die Veränderte Welt* (*The Altered World*, 1933), attempted to visualize moments of catastrophe and the vast global changes of twentieth-century modernization. As part of this project, Jünger carefully considered the role of photography, a medium that he had previously dismissed as superficial. However, Jünger now began to see photography as a technological weapon that could create carefully constructed narratives of domination.[102] The camera was the tool through which modern man could create an aesthetics of power by curating a seemingly all-encompassing photo narrative.[103] For Jünger, there was no truth outside of the frame of the photograph; the man behind the camera stood in absolute control of reality, in an omniscient position of "stereoscopic vision."[104] This idealized cameraman would be able to shape the course of modernity and enact his own politics.

Granting himself authorial control, Jünger wished to express the revealing qualities of all modern technology. Through his photo albums, Jünger attempted to convey the contingency and risk of the modern world by creating montages of technologically induced death, destruction, and catastrophe. According to Jünger, the industrialized technologies of the

[100] Jünger, *Copse 125*, 40.
[101] Thomas F. Schneider, "Narrating the War in Pictures: German Photo Books on World War I and the Construction of Pictorial War Narratives," *Journal of War and Culture Studies* 4, no. 1 (2011): 31.
[102] Ernst Jünger, "War and Photography," in *New German Critique*, 59 (New York: Telos Press, 1993), 25.
[103] Isabella Gil, "The Visuality of Catastrophe in Ernst Jünger's Der gefährliche Augenblick and Die verärderte Welt," *KulturPoetik* 10, no. 1 (2010): 67.
[104] Carsten Strathausen, "The Return of the Gaze: Stereoscopic Vision in Jünger and Benjamin," *New German Critique* 80 (2000): 131.

early twentieth century did not protect humans from the dangers of the premodern world; rather, they simply masked unavoidable threats, such as disease and natural disasters. In moments of technological failure, these dangers reemerged and revealed the precariousness of modern human existence. In fact, modern technology could serve to intensify the speed and scope of violence in the moment of destruction. Jünger's *Der gefährliche Augenblick* included images of crashed planes, overturning cars, and mine disasters, where human death was predicated on the failure of modern technologies.[105] In *Die Veränderte Welt*, Jünger pointed to poison gas as yet another one of these technologies that could immediately and unexpectedly reveal the proximity of danger. He wrote in the album:

> The increased range and total character of the means of destruction include the possibility of catastrophes, which make planned protection of the population seem ever more necessary. In particular, one also seeks to familiarize the non-combatants with the dangers of gas and air raids.[106]

Die Veränderte Welt also included several images of gas attacks and gas preparation. The photos clearly attempted to express the surprising speed and ethereal character of gas, often depicting people with gas masks as the provisional masters of this new danger. Jünger certainly did not view the technology behind the mask as infallible, but an embrace of this imperfect technological mediation allowed for new forms of action and survival.

While curating and captioning these photo albums, Jünger simultaneously developed similar ideas in his long-form writing. His 1931 essay "Über die Gefahr" ("On Danger") made clear his assertion that increasing levels of danger always lurked behind the façade of modern bourgeois society.[107] Reflecting on technology's role in maintaining this pretense, he wrote, "The history of inventions also raises ever more clearly the question of whether a space of absolute comfort or a space of absolute danger is the final aim concealed in technology."[108] Pointing out that automobiles now killed far more people than warfare, Jünger insisted that technology's concealed endpoint was the destruction and death that awaits all life. While Jünger was clearly attracted to the new and omnipresent dangers of modernity, he remained ambivalent about

[105] Ferdinand Bucholtz, ed., *Der gefährliche Augenblick* (Berlin: Junker und Dünnhaupt Verlag, 1931), 72, 128.
[106] Edmund Schultz, ed., *Die Veränderte Welt* (Breslau: Wilhelm Gottl. Korn Verlag, 1933), 190.
[107] Ernst Jünger, "On Danger," in *New German Critique*, 59 (New York: Telos Press, 1993), 27.
[108] Ibid., 30.

technology's concealing nature and its implications for human freedom.[109] In his 1930 essay "Die totale Mobilmachung," Jünger made it clear that such technology had robbed the professional soldier of his previous glories. By shrinking the space of the battlefield and reducing the prevalance of hand-to-hand combat, technologies such as the tank, airplane, and poison gas had all made military honor, intimate struggle, and genuine heroism practically obsolete.[110] As a proponent of a chivalric militarism, Jünger personally felt this loss while fighting in the trenches of World War I. Nevertheless, he saw the introduction of modern technology to the battlefield as inevitable and diagnosed it as an inherent part of the modern experience. Thus, his intellectual project ultimately attempted to identify and explain the nature of technological warfare in order to carve out a role for the modern warrior.

Jünger most clearly tried to define modern man's role in this complex technological world in his 1932 book *Der Arbeiter* (*The Worker*). In this text, Jünger triumphantly proclaimed that a Gestalt of the modern worker was increasingly intruding upon bourgeois European society.[111] A metaphysical concept, the Worker Gestalt represented an elementary power that was transforming the modern world through revolutionary acts.[112] Jünger enumerated the moments in which the Worker Gestalt had revealed itself and "mobilized the world."[113] Many of these moments of revealing arose through modern, industrialized technology and each individual's embrace and effective use of technology decided his/her relationship to the Worker Gestalt. The heroic human of Jünger's future would thus integrate entirely with his or her technologies, riding into both real and metaphorical battle on steeds of steel and hurling projectiles of fire and gas.

For Jünger, modern weapons of war such as poison gas would literally serve as tools for the revolutionary acts of the Worker.[114] They were already destroying the decadence of the old world and ushering in a new one defined by human discipline and hardness. Unlike the European of

[109] Neil Turnbull, "Heidegger and Jünger on the 'Significance of the Century': Technology as a Theme in Conservative Thought," *Writing Technologies* 2, no. 2 (2009): 12.

[110] Ernst Jünger, *Die totale Mobilmachung* (Berlin: Verlag für Zeitkritik, 1931), 121.

[111] Unlike in Marxist thought, "the Worker" was not a strictly defined socioeconomic category. Jünger rather imagined "the Worker" as a veteran who had come to his own realizations about technology and danger through the experience of World War I. Jünger used the term "Gestalt" to mean a phenomenon that consists of discrete parts or moments but can also be conceived as a whole. Wolf Kittler, "From Gestalt to Ge-Stell: Martin Heidegger Reads Ernst Jünger," *Cultural Critique* 69 (2008): 82.

[112] Ernst Jünger, *The Worker: Dominion and Gestalt*, trans. Dirk Leach (Albany: SUNY Albany Press, 1990), 10.

[113] Ibid., 184. [114] Encke, *Augenblicke der Gefahr*, 194–218.

yesteryear, the Worker could survive in a world inundated by poison gas. By fully integrating into modern technology, he/she would effectively utilize devices such as the gas mask to survive in this chemically uncertain future. In fact, continuing his recurrent interest in facial changes, Jünger postulated that the modern Worker, like the soldier in the trenches of World War I, had even begun to physically resemble his or her own technologies. He wrote:

> First of all, what stands out purely physiognomically is the masklike rigidity of the face which is both acquired and also accentuated and increased by external means such as beardlessness, haircut, and tight headgear. Nor is the role accidental, incidentally, which the mask has begun to play recently in daily life. This becomes apparent in many ways in places where the special work character breaks through – in gasmasks, with which one seeks to arm entire populations.[115]

In Jünger's hands, the employment of poison gas lost its commonly assumed immorality. Warfare technologies such as gas and the gas mask were merely material facts of the modern world. Those who could embrace this reality would be rewarded, but this would certainly not be an easy task. In Jünger's formulation, not entirely unlike Fritz Haber's, modern humans must face their own mortality, passing through the pain of baptismal gas attacks to emerge as technologically augmented masters of the future.[116] A coming mass gassing of German civilians, which Jünger had already hypothesized in his non-fiction writings, would only bring humanity closer to the new truths of the twentieth century. In essence, Jünger attempted to remythologize the modern world, replacing God as absolute arbiter with industrialized technology.[117] In doing so, Jünger also provided humanity with the possibility for substantial control over the dangers of chemical warfare technologies. While control over poison gas could always slip, devices such as the gas mask placed a new human face behind the ethereal gas cloud.

Intellectual works that essentially countered Jünger's developing theory of technology also appeared in the wake of World War I. For instance, in 1918, Karl Kraus published the essay "Das technoromantische Abenteuer" ("The Techno-romantic Adventure") in his literary journal *Die Fackel*. In the essay, Kraus anticipates Jünger's philosophical project, caustically writing, "It must be said, first of all, that especially the German God comes not only in a gas cloud, but also out of the machine; that bravery and honest combat have no part in the chance of a hit of

[115] Jünger, *The Worker*, 142.
[116] Jünger, *On Pain*, 1. This baptismal armoring is the creation of what literary scholar Susan Buck-Morss has called the "anesthetic subject." Buck-Morss, "Aesthetics and Anaesthetics," 32–36.
[117] von Klemperer, *German Incertitudes 1914–1945*, 51–52.

mines, an aerial bomb, or a torpedo."[118] Referring to Germany's war on the Western Front as a "chlorious" affair, Kraus granted technology its god-like status in the modern world, but he disavowed man's ability to control its carnage.[119] Rather, he claimed that human violence and patriotic nationalism set warfare technology in motion, but once this had been done, human agency and heroism were nullified. Thus, modern humanity lost control of its own technology, becoming a victim to its haphazard violence. Kraus wrote:

> Humanity, which has spent its imagination on inventions, can no longer imagine its effectiveness – otherwise it would commit suicide because of regret! But since it has also spent its human dignity on inventions, it lives and dies for all the power that uses such progress against humanity. The inconceivable nature of everyday things, the incompatibility of power and the means to enforce it, that is the condition, and the techno-romantic adventure into which we have engaged will, whatever happens, put an end to the condition.[120]

Not only was Kraus claiming that technology had transcended its human intentions, but he was also insistent that visions such as Jünger's reeked of ethno-national chauvinism. And while Jünger's work did imply that the Germans were poised to effectively control these violent technologies, Oswald Spengler's 1931 work *Man and Technics* was even more explicit in placing modern technologies within the context of a previously predicted world-historical struggle between broad ethno-racially defined civilizations. For Spengler, modern technologies were a manifestation of particular cultural forms and styles. Furthermore, he claimed that certain civilizations were more apt to employ technics, or technological ways of thinking, creating, and being.[121] For instance, Spengler wrote:

> For [the] "colored" peoples ... the Faustian technics are in no way an inward necessity. It is only Faustian man that thinks, feels, and lives in this form. To him it is a spiritual need.[122]

[118] Karl Kraus, "Das technoromantische Abenteuer," *Die Fackel* 474–483 (May 23, 1918).

[119] Kai Evers, *Violent Modernists: The Aesthetics of Destruction in Twentieth-Century German Literature* (Evanston, IL: Northwestern University Press, 2013), 163.

[120] Kraus, "Das technoromantische Abenteuer."

[121] Spengler's desire to claim technological development as part of a uniquely German or Germanic *Kultur* was also picked up by many in the engineering profession during the interwar years. Adelheid Voskuhl, "Engineering Philosophy: Theories of Technology, German Idealism, and Social Order in High-Industrial Germany," *Technology and Culture* 57, no. 4 (2016): 733; Jeffrey Herf, *Reactionary Modernists: Technology, Culture, and Politics in Weimar and the Third Reich* (Cambridge: Cambridge University Press, 1984), 159.

[122] Oswald Spengler, *Man and Technics: A Contribution to the Philosophy of Life*, trans. Charles Francis Atkinson (New York: Greenwood Press, 1976), 51.

For Spengler, this so-called Faustian man who best controlled the powers of technology was a hazy amalgamation of northern European peoples, but for thinkers such as Karl Kraus, such constructions were clearly justifications for an aggressive ethno-racially inflected German nationalism that was tied to militarized technological development.[123]

Picking up on many of Kraus' themes, Dora Sophie Kellner, a journalist with a degree in chemistry and the wife of the philosopher Walter Benjamin, subsequently argued that Germans were unwilling or unable to truly face the deadly realities behind the weapons that they had created.[124] In her 1925 essay "Die Waffen von Morgen" ("The Weapons of Tomorrow"), Kellner pulled on Giulio Douhet's military theories in an attempt to create a "literary montage" that could imagine the real consequences of aero-chemical war.[125] This method of splicing "dialectical images" attempted to jolt the reader out of complacency through evocative imagery, not unlike Jünger's work.[126] For example, Kellner wrote, "The coming war will have a ghostly front. A front that advances over metropolises, into their streets and before every one of their front doors. This war, the gas war from the air ... will be truly 'breathtaking.'"[127] As evidence for this assertion, Kellner evoked the massive interwar chemical arsenals of France and England, claiming that the next war would surely feature the escalation of civilian gassing. She further contemplated the poisoning effects of mustard gas, writing, "By a few 'raids,' the population of the hostile centers is to be filled with unconscious horror in such a way that every appeal for the organization of the defense fails. This terror should approach psychosis."[128] Kellner ended her essay by chastising right-wing Germans who denounced the Treaty of Versailles, and she reminded her reader that there were far more pressing political problems in a world constantly threatened by poison gas.

Further challenging Jünger's early conceptualization of warfare technologies, Walter Benjamin also responded directly to what he proclaimed Jünger's fascist worldview. In his 1930 article, "Theories of German Fascism," Benjamin chided Jünger's attempt to mythologize the technological, claiming that warfare technology would inevitably escape

[123] Ian James Kidd, "Oswald Spengler, Technology, and Human Nature," *The European Legacy* 17, no. 1 (2012): 28.

[124] "The Weapons of Tomorrow" was initially attributed to Walter Benjamin, but it was most likely, at least in large part, written by Dora Sophie Kellner. Kellner would follow the essay with a 1930 antigas novel entitled *Gas gegen Gas (Gas Against Gas).*

[125] Evers, "Risking Gas Warfare," 271.

[126] Strathausen, "The Return of the Gaze," 144.

[127] Dora Sophie Benjamin, "Die Waffen von morgen," *Vossische Zeitung*, no. 303 (1925): 1.

[128] Ibid., 1.

humanity's grasp.[129] Perhaps not surprisingly, Benjamin particularly focused on aero-chemical warfare, writing:

> And we know that there is no adequate defense against gas attacks from the air. Even individual protective devices, gas masks, are of no use against mustard gas and lewisite. Now and then one hears of something "reassuring" such as the invention of a sensitive listening device that registers the whir of propellers at great distances. And a few months later a soundless airplane is invented. Gas warfare will rest upon annihilation records, and will involve an absurd degree of risk.[130]

Alongside these technically overstated but nevertheless meaningful claims, Benjamin went on to rightly argue that Jünger's reappraisal of technology was nothing more than a attempt to recreate an idealized and imagined past in which soldiers could again claim themselves to be the personification of masculinity and honor.[131] According to Benjamin, Jünger's desire to enforce his prescriptive relationship on humanity, nature, and technology was, in fact, the "face of fascism." On a practical level, Benjamin claimed that Jünger's personal philosophy was a serious threat to humanity because it only took a single pilot to drop gas bombs on millions of people. If lone reactionaries attempted to fulfill Jünger's vision, Benjamin warned, "... millions of human bodies will indeed inevitably be chopped to pieces and chewed up by iron and gas."[132]

In his famous 1935 essay, "The Work of Art in the Age of Its Technological Reproducibility," Benjamin reiterated his critique of so-called New Nationalists such as Jünger. He claimed that the glorification of war and the technologization of the human body were attempts to aestheticize politics. For Benjamin, this was a fascist endeavor aimed at giving the masses a cultural voice while denying them tangible rights.[133] Only war could mobilize people and diffuse their productive energies enough to pacify them under a fascist state.[134] Benjamin's essay was not only a condemnation of Jünger's politics but also his aesthetic means of expression. For Benjamin, Jünger's earlier photo albums were an attempt to render the human body mechanically armored and numbed.[135] Indeed, Benjamin claimed that Jünger pulled his viewer outside of

[129] Walter Benjamin, "Theories of German Fascism: On the Collection of Essays War and Warrior," *New German Critique* 17, no. Special Walter Benjamin Issue (Spring 1979): 120.
[130] Ibid., 121. [131] Ibid., 126. [132] Ibid., 128.
[133] Siegfried Kracauer was one of the first to diagnose Jünger's attempt to subsume politics into the aesthetic. Strathausen, "The Return of the Gaze," 148.
[134] Walter Benjamin, *The Work of Art in the Age of Its Technological Reproducibility and Other Writings on Media*, eds. Michael Jennings, Brigid Doherty, and Thomas Levin (Cambridge, MA: Harvard University Press, 2008), 42.
[135] Buck-Morss, "Aesthetics and Anaesthetics," 38.

history, forcing the spectator to become overwhelmed and passive in the face of Jünger's graphic images and thus granting the author dictatorial power over historical narrative.[136] Metaphorically speaking, one could say that Jünger was pulling the gas mask over his reader's face. However, Jünger's theory of photography grasped not only the viewer but also the photographer in its optic regime.[137] According to Benjamin, Jünger and his Worker Gestalt were trapped in their own authoritarian aesthetics, unable to theorize beyond the borders of their visual narratives and unable to see the lack of human control over poison gas. They too were thus encircled by what the communist writer Otto Biha contemporaneously described as the "dangerous poisonous gas deployed on the cultural front."[138]

The intellectual dispute between Jünger and Kellner/Benjamin essentially reflected the larger air and gas debates that wracked the Weimar Republic. Escalating in the late 1920s and early 1930s, the concern over aero-chemical attack thus permeated everything from the most theoretical discussions to the literary and cinematic consumption of the average German citizen. This, in turn, only increased the sense of risk and contingency that defined much of the Weimar era and forced the German government to enter into discussions of national security with greater urgency. In his personal letters, the novelist Stefan Zweig attempted to express the dizzying nature of what seemed like a newly dangerous modern world through a reflection on the French philosopher Michel de Montaigne. Zweig wrote that de Montaigne maintained an admirably level head, for "in an equally dirty time like ours [he] tried to stay independent and to think clearly even under the gas mask."[139] The gas mask's ability to distort and fog the vision of its wearer serves as an apt metaphor for the inability of interwar Germans to diagnose and describe the problem of aero-chemical defense. Contemporaries remained at odds over who or what controlled poison gas, how best to avoid future chemical warfare, and whether Germany could successfully harness its powers for either national survival or rebirth.

[136] Strathausen, "The Return of the Gaze," 145.

[137] Gil, "The Visuality of Catastrophe in Ernst Jünger's Der gefährliche Augenblick and Die verärderte Welt," 69.

[138] Otto Biha, "Der proletarische Massenroman. Eine neue Eine-Mark-Serie des 'Internationalen Arbeiterverlages,'" *Die Rote Fahne* 178 (August 2, 1930).

[139] Stefan Zweig, *Stefan Zweig Briefe 1932–1942*, eds. Knut Beck and Jeffrey B. Berlin (Frankfurt am Main: S. Fischer Verlag, 2005), 334.

7 Synthesizing the "Nazi Chemical Subject": Gas Masks, Personal Armoring, and Vestiary Discipline in the Third Reich

In the spring of 1932, the Weimar government created six air raid education units to provide air defense demonstrations and training sessions across Germany.[1] Nine more units were added that summer when gas raid exercises confirmed the worst fears of air protection activists. Reports from the early drills claimed that civilians were unable to follow simple orders; they panicked in attack simulations and exercises devolved into chaos. In particular, the reports claimed that women and children became too excited, while members of the working class actively disobeyed orders.[2] For many air protection specialists who reflected on these results, the atomized individual proved the greatest danger in the event of an aero-chemical raid. Civilians, they argued, would have to work selflessly and cooperatively in order to ensure the safety of their entire community.[3] Successful air protection was not just a matter of technical knowledge but also one of mass-psychological composure.[4] As such, air protection specialists began to wonder if the problems that they encountered in these exercises were related to the dissolution of broader German social ties. Thus, they asked themselves if perhaps the inability to work collectively was a problem inherent to democratic equality. Foreshadowing future developments, one observer of an air raid exercise in East Prussia claimed that the unruly civilians urgently needed a strong-willed leader.[5]

In the early 1930s, demands for better air and gas protection continued to feed into the larger concerns for national security that helped to define the unstable German political landscape. Gas experts and air protection advocates helped to fan the flames of this debate. As historian

[1] Reichsluftschutzbund, *Luftschutz ist Selbstbehauptungswille: Aufgaben und Erfahrungen über die Ausbildung im zivilen Luftschutz* (Berlin: Reichsluftschutzbund RLB, 1936), 5.
[2] Fritzsche, "Machine Dreams," 698.
[3] Ritter von Mittelberger, in *Gasschutz und Luftschutz*, 2 (August 1932): 170.
[4] Fritzsche, "Machine Dreams," 697.
[5] "Die zivile Luftschutzübung in Ostpreussen vom 23–25 Juni 1932: Kritische Betrachtung und Auswertung ihre Ergebnisse," *Gasschutz und Luftschutz* 2 (August 1932): 179–180.

Andreas Linhardt has pointed out, the desire for national gas and air protection was often just a political cloak for general German rearmament or the implementation of more authoritarian state structures.[6] A 1932 letter petitioning the state authorities of Baden-Württemburg to create better air and gas defense measures reveals this form of political pressure. In the letter, an unnamed German engineer wrote that "[aero-chemical war] is the most important issue of all, as aircraft pose a real mortal danger to all those people who are quietly asleep, trusting in the effectiveness of international treaties, which *will* fail to protect them."[7] The engineer went on to boldly (and falsely) assert that French aerial attacks could kill 600,000 Germans in a single day. The German authorities knew this, he claimed, but chose to do nothing.

While many economic, social, and political events contributed to the rise of the Nazis in the early 1930s, anxieties over national security certainly assisted the party in soliciting German votes in both the 1930 and 1932 federal elections. With the perceived threat of an aero-chemical attack looming (most likely from a chemically armed Soviet Russia) and political street violence increasing, the democratic governing parties lost significant credibility in their attempts to quell violence and secure German borders.[8] The Nazis, on the other hand, conveyed a sense of ruthless order and national steadfastness in the face of perceived crisis.[9] In a country with vastly reduced armed forces, the SA brownshirts appeared as a potential auxiliary military force, while Hitler's calls for a strong and proud Germany resonated with those concerned with geopolitical vulnerability.[10] Historian Michael Wildt has argued that many accepted the authoritarian Nazi vision in 1933 because "Germans hoped for the end of party squabbles and political bickering … [they] wanted peace and order, [to] re-establish the state's authority and alleviate Germany's economic problems."[11]

[6] Andreas Linhardt, "Die Fachzeitschrift 'Gasschutz und Luftschutz' unter dem Einfluss des Nationalsozialismus," n.d., p. 3, www.bbk.bund.de/SharedDocs/Downloads/BBK/DE/FIS/DownloadsDigitalisierteMedien/FachzeitschriftNotfallvorsorge/GasschutzundLuftschutz/Ausarbeitung_Dr_Linhardt_PDF.pdf?__blob=publicationFile.

[7] Gesuch des Oberingenieurs Karl Leuprecht in Karlsruhe Luft und Gasschutz betreffend, 1932 (LBWK) 231, 6516.

[8] Michael Wildt, *Hitler's Volksgemeinschaft and the Dynamics of Racial Exclusion: Violence against Jews in Provincial Germany, 1919–1939*, trans. Bernhard Heise (New York: Berghahn Books, 2012), 268.

[9] Richard J. Evans, *The Coming of the Third Reich* (New York: Penguin, 2003), 293.

[10] Jerome G. Kerwin, "The German Reichstag Elections of July 31, 1932," *The American Political Science Review* 26, no. 5 (1932): 932.

[11] Michael Wildt, "Self-Reassurance in Troubled Times: German Diaries during the Upheavals of 1933," in *Everyday Life in Mass Dictatorship: Collusion and Evasion*, ed. Alf Lüdtke (New York: Palgrave Macmillan, 2016), 56.

Just four days after Hitler had risen to the position of chancellor on January 30, 1933, he organized a private meeting with his senior military commanders and outlined his plan to rearm Germany and invade Eastern Europe. Publicly, the issue of national security took center stage during the Reichstag fire on February 27 of that same year. Once the communist arson suspect was rounded up, Hitler was able to pass the "Decree for the Protection of the German People," which allowed the Nazis to suspend civil liberties in the name of "public safety and security."[12] The subsequent suppression of the communists cleared the way for yet another set of elections and the passing of the Enabling Act, which then granted Hitler dictatorial powers on March 23, 1933.[13]

Once in power, the Nazis began to fully recognize the importance of air and gas protection in the minds of many Germans. Political solutions could not only allay anxieties over aero-chemical attacks but also forge the kind of social ties that the Nazis envisioned for the German nation. Technical experts had already expressed the significance of creating a seamless air and gas protection community predicated on collective cooperation and submission to authority. During the 1932 air raid exercises, an early incarnation of this functional air raid community was said to best be found in the small towns where political divisions were not particularly pronounced and local club culture created intimate ties between neighbors.[14] The task for the Nazis, then, was to replicate this homogenized and hierarchical society on the national level.

The first step was the centralization of air and gas protection services within the state. Rather than rely on private organizations like in the Weimar era, the Nazis charged Josef Seydel, Bruno Loerzer, and Erich Hampe with quickly setting up the *Reichsluftschutzbund* (Reich Air Protection League or RLB) for national air and gas protection.[15] On April 29, 1933, the *Reichsluftschutzbund* became operational with the Flak artillery officer Hugo Grimme serving as president, but ultimate control lying in the hands of the Minister of Aviation, Hermann Göring. As early as the summer of that year, the league launched a massive

[12] Benjamin Carter Hett, *Burning the Reichstag: An Investigating into the Third Reich's Enduring Mystery* (Oxford: Oxford University Press, 2014), 71.

[13] Not coincidentally, a traveling air defense exhibition opened in Munich only days before the announcement of the Enabling Act. Süß, *Death from the Skies*, 35.

[14] Kurt Kottenberg, "Zum Luftschutzproblem in Mittel- und Kleinstädten," *Gasschutz und Luftschutz* 2 (June 1932): 127–129.

[15] Josef Seydel was an elected member of the Reichstag with a background in civilian air defense. Bruno Loerzer was the Chairman of the Association for German Air Defense. Erich Hampe was the head of the *Technische Nothilfe*. Nicole Kramer, *Volksgenossinnen an der Heimatfront: Mobilisierung, Verhalten, Erinnerung* (Göttingen: Vandenhoeck & Ruprecht, 2011), 107.

campaign to teach air readiness to all German citizens. This campaign was inaugurated on June 24 with the dropping of leaflets over Berlin. In an attempt to express the seriousness of air protection, these leaflets stressed that only paper was currently falling on the capital city, but soon it would be real bombs.[16] By the end of 1933, the voluntary *Reichsluftschutzbund* enjoyed a membership of 1 million Germans. By the following summer, 2.5 million more had joined the ranks.[17] The public presence of the *Reichsluftschutzbund* and its summer campaign of 1933 made it abundantly clear that the Nazis would afford greater importance to air and gas protection than the Weimar government could ever muster.

Alongside the creation of a formal institutional structure for air and gas protection, the Nazis also looked for new ways to instill the work ethic and selflessness that were deemed necessary for wartime survival. Months before Hitler's seizure of power, the vice chief of the *Technische Nothilfe*, Erich Hampe, began to question the prevailing orthodoxy of Weimar-era civilian aero-chemical protection. Since World War I, experts had assumed that women, children, the infirm, and all other supposedly weaker members of society would never be able to conduct air raid exercises. These drills required strenuous manual labor, an overriding concern for others, and most importantly, a cool and calm demeanor amid danger. However, reflecting on the failures of previous air raid drills, Hampe now argued that *all* civilians should be given air and gas protection duties, regardless of their age or gender. According to Hampe, a sense of responsibility and collectivity could fortify all Germans and enter them into a social fabric that could best withstand aero-chemical attacks.[18] This argument further aligned with the Nazis' creation of the centralized but voluntary *Reichsluftschutzbund* and the more general Nazi desire to create a self-sufficient and socially integrated German subject.

Distinguishing Nazi air and gas protection protocol from those of other European countries, no German citizen would now receive special treatment in the event of an aero-chemical attack. Much like the soldiers of 1918, German civilians would be forced to face the perilous realities of a chemically toxic world. Following the logic of former soldiers such as Ernst Jünger, the Nazis saw this deadly adversity as a way to forge a

[16] Peter Fritzsche, *A Nation of Fliers: German Aviation and the Popular Imagination* (Cambridge, MA: Harvard University Press, 1992), 208.
[17] Ibid., 211.
[18] Erich Hampe, "Technische Nothilfe und Luftschutz," *Gasschutz und Luftschutz* 1 (September 1931): 43.

national community of crisis in this presumed era of "total mobilization." It was thus an individual citizen's responsibility to confront deadly technologies as an inherent part of the greater national struggle.[19]

While *Selbstschutz* (self-protection) within a collective air raid community became the catchword of air and gas defense under the Nazis, specific technical questions remained open-ended. The gas specialists, many of whom easily translated their work into the authoritarian terms of National Socialism, were still unsure whether gas masks were necessary for each German civilian. Nevertheless, regardless of its real effectiveness, the RLB ultimately chose to use the gas mask as a material tool in the attempted creation of a chemically minded German subject. Accordingly, the RLB ordered millions of the newly designed civilian gas masks to be sold at a discounted rate to Germans across the Reich. With masks donned, German civilians now appeared as technologically augmented soldiers in the presumed struggle for national survival. In the eyes of the *Reichsluftschutzbund*, the mask created a physically homogenized society that could now survive, if not thrive, in a modernity defined by its atmospheric toxicity. By affixing their masks, German civilians could potentially take on the new face of the impervious master of the modern world, thereby becoming complicit in Nazi civic mobilization and implicitly accepting a chemically precarious future.[20]

The Big Business of Chemical Defense under the Nazis

Alongside the creation of the *Reichsluftschutzbund*, the Nazi seizure of power brought a renewed focus on chemical production and its subsidiary industries.[21] While inconsistent with some of the party's foundational rhetoric, the Nazis quickly expressed their willingness to allow large corporations such as IG Farben to pursue nationwide capitalist enterprise. As historian Peter Hayes has shown, the giant chemical concern achieved significant profits through the Nazis' state-sponsored economic programs while simultaneously respecting Nazi policies that discouraged the growth of foreign markets and/or instituted ethno-racial discrimination. Summing up this relationship, Hayes writes, "Nazi economic policy rested on the recognition that so long as a state displays its

[19] Wirth, "Die Neuordnung des Selbstschutzes," *Gasschutz und Luftschutz* 8 (August 1938).
[20] Fritzsche, "Machine Dreams," 698.
[21] Florian Schmaltz, "Chemical Weapons Research on Soldiers and Concentration Camp Inmates in Nazi Germany," in *One Hundred Years of Chemical Warfare: Research, Deployment, Consequences*, 230.

determination but permits businessmen to make money, they will let themselves be manipulated as to how."[22]

Prior to 1933, Carl Bosch, the de-facto leader of the IG concern, criticized the Nazis' pursuit of autarky and the ways in which their racial ideology would impinge on the autonomy of international science.[23] However, the Nazis' ability to put down labor disputes and their lucrative contracts for IG fuel, synthetic rubber, and other chemical products eventually quieted Bosch's initial concerns. Over the course of 1933, the IG Farben group gave over 4.5 million Reichsmark to the Nazi Party and gradually moved toward domestic accommodation of the new regime.[24] For many like-minded industrialists, the Nazi regime represented potential economic stability after the tumultuous experience of the Great Depression.[25]

While early Nazi economic programs did little to help IG Farben's bottom line, the pursuit of German autarky presented the possibility that chemical armaments might again be produced and sold.[26] Thus, the IG group cozied up to the Nazi regime with hopes of future military contracts, the absorption of foreign chemical companies, and general political sway in the Reich.[27] The open announcement of German rearmament on March 16, 1935, and the introduction of the Nazi Four Year Plan on September 9, 1936, began to make this relationship pay off. While IG Farben's net profits had stagnated between 1933 and 1935, they doubled in 1936.[28] These financial successes were most fully felt in the early 1940s, when Nazi military conquest finally produced significant growth in IG Farben's sales, raising yearly profits to 300 million Reichsmark.[29]

[22] Hayes, *Industry and Ideology*, 379.

[23] Peter Hayes, "Carl Bosch and Carl Krauch: Chemistry and the Political Economy of Germany, 1925–1945," *The Journal of Economic History* 47, no. 2 (1987): 357.

[24] However, many of IG Farben's Jewish researchers were not removed from their positions until 1938. While this may appear as if the company was protecting its employees due to a moral stance, it was more likely a practical consideration based on the importance of their research. Ultimately, keeping Jewish researchers in Germany until 1938 put them in a dangerous situation. Stephan H. Lindner, *Inside IG Farben: Hoechst during the Third Reich* (Cambridge: Cambridge University Press, 2008), 154–159.

[25] Werner Plumpe, *German Economic and Business History in the 19th and 20th Centuries* (London: Palgrave Macmillan, 2016), 225–226.

[26] Hayes, *Industry and Ideology*, 114.

[27] Diarmuid Jeffreys, *Hell's Cartel: IG Farben and the Making of Hitler's War Machine* (New York: Metropolitan Books, 2008), 280.

[28] By November 1936, IG Farben was encouraging the Wehrmacht to stockpile chemical weapons. Ian Kershaw, *The Nazi Dictatorship: Problems and Perspectives of Interpretation* (London: Edward Arnold, 1985), 59.

[29] These rising profits were partly predicated on the absorption of foreign chemical competitors and the use of slave labor. Hayes, *Industry and Ideology*, 325.

As part of IG Farben's ongoing research into pesticides, the chemical plant at Leverkusen assigned the chemist Gerhard Schrader to study organic fluoro-phosphorine compounds. On December 23, 1936, Schrader synthesized a nerve agent that was later named Tabun. After accidentally poisoning himself with the gas, Schrader realized Tabun's ability to induce asthma and painfully constrict the pupils of the eye.[30] The IG Concern reported the discovery of Tabun to the Army Ordinance Office, which expressed great interest in its potential military applications.[31] Alongside the production of Tabun at Bayer, BASF chemical plants began to again manufacture chemical weapons such as phosgene and mustard gas under the direction of Dr. Otto Ambros.[32] In 1938, Schrader then isolated the more toxic nerve gas Sarin, which alongside Tabun, could penetrate all known gas mask filters.[33] In September of the following year, the German military contracted IG Farben to construct a Tabun factory at Dyhernfurth. At its peak, the factory produced 1,000 tons of Tabun per month.[34] A second IG plant at Falkenhagen raised this production capacity in 1943. Estimates suggest that the Germans produced up to 30,000 tons of Tabun before the end of World War II.[35] Undoubtedly, the Nazis hoped that nerve gas would prove to be the ultimate wonder weapon in the future war that they imagined.[36]

Chemical research institutes experienced similar changes under the Nazi regime. The 1934 Civil Service Law forced many Jewish chemical researchers to abandon their positions. This included Fritz Haber, who resigned as the director of the Kaiser Wilhelm Institute for Physical and Electrochemistry on October 1, 1933, after facing anti-Semitic attacks in the German *Journal for Natural History*, regardless of his earlier

[30] Florian Schmaltz, "Neurosciences and Research on Chemical Weapons of Mass Destruction in Nazi Germany," trans. Stefani Ross, *Journal of the History of the Neurosciences* 15, no. 3 (2006): 188–189.

[31] Rolf-Dieter Müller, "World Power Status through the Use of Poison Gas? German Preparations for Chemical Warfare, 1919–1945," in *The German Military in the Age of Total War*, 184–185.

[32] In a 1948 Nuremberg trial, Otto Ambros was sentenced to eight years in prison for war crimes and the use of slave labor. Eleven other IG officials were also given prison sentences. IG AG Kampfstoffe P 70/I Diverse Ausarbeitungen zu Kampfstoffen im Allgemeinen Dyhernfurth Gendorf finden sich in IG A 866/1 (BASF) IG A 866/1.

[33] Monatsmeldungen an das Oberkommando des Heeres uber die Produktion von Sprengstoffen und Vorprodukten sowie den Rohstoffverbrauch Dez, 1939–1941 (BASF) Pertinenzbestand P.7./6.

[34] Albert Speer, *Inside the Third Reich* (New York: Simon and Schuster, 1970), 413.

[35] For the production numbers of all nerve agents in the Third Reich, see Rolf-Dieter Müller, "Die deutschen Gaskriegsvorbereitungen 1919–45: Mit Giftgas zur Weltmacht?," *Militärgeschichtliche Zeitschrift* 27, no. 1 (1980): 46.

[36] Friedrich Reck, *Diary of a Man in Despair*, trans. Paul Rubens (New York: New York Review of Books, 2013), 135, 148.

conversion to Protestantism.[37] To the end, Haber held on to his allegiance to Wilhelmine Germany, writing in his resignation letter: "You will not expect from a man in his sixty-fifth year a change of mental outlook, which in the preceding thirty-nine years directed his academic life, and you will understand that the pride with which he has served his German homeland for his entire life prescribes this request for retirement."[38] However, in one of his final acts as a leading figure at the KWI, Haber refused the German military's request to transition his laboratories back to chemical warfare research. Furthermore, he began to organize the emigration of other German-Jewish scientists and secured a spot for himself at a Cambridge University laboratory.[39]

Suffering from insomnia and anxiety, Haber did not find Cambridge to be a particularly comfortable academic home. He certainly had his international admirers and supporters, but there were still plenty of people who resented his role in chemical warfare. For instance, in what has become a historically symbolic anecdote, the renowned physicist Ernest Rutherford reportedly refused to shake Haber's hand due to the latter's willingness to work for the German state. Thus, the conflict between the international ideals of science and distinct national pride haunted Haber to the end of his life. Choosing to leave his position in England, he subsequently accepted a spot at the Sieff Research Institute in Palestine. While traveling through Switzerland on his way to take up his new position, Haber succumbed to his continuing illnesses and died of heart failure on January 29, 1934. A German patriot to the end, Haber died in exile from his beloved country, leaving behind a complicated legacy that includes both the development of uniquely treacherous weapons of war and the synthesis of a chemical element that allowed for the subsequent Green Revolution.[40] But perhaps even more important than his chemical products, which were later used in various ways by various people, Haber was an unrepentant promoter and builder of the kinds of large-scale nationalized scientific projects that would come to dominate the mid-twentieth century.

[37] Dietrich Stoltzenberg, *Fritz Haber: Chemist, Novel Laureate, German, Jew* (Philadelphia: Heritage Press, 2004), 285–286.

[38] Stern, *Einstein's German World*, 155.

[39] Haber notably helped Chaim Weizmann emigrate from Germany. Haber found his own laboratory position at Cambridge through Sir William Pope, a chemist who developed mustard gas for the British during World War I. Ibid., 158, 161.

[40] The physicist Max von Laue organized a commemorative service for Fritz Haber despite assured Nazi censure. At the service, Max Planck and Otto Hahn spoke to a large crowd of industry scientists, military men, and diplomats. This funeral is often cited as one of the German physics community's few acts of open resistance to the Nazi regime. Lewin Sime, *Lise Meitner*, 156.

The removal of Jews like Haber from the ranks of the Kaiser Wilhelm Society initially strained all of the state-sponsored research institutes. It was only through connections to corporate interests and a fairly insular bureaucratic structure that the Society, now led by the physicist Max Planck, was able to remain focused on its research agenda without much state interference in the early years of the Third Reich.[41] However, with the announcement of the Nazi Four Year Plan, the Society increasingly fell under the control of the Reich Research Council and the Military Weapons Office. Over the first six years of the Nazi regime, the Kaiser Wilhelm Society's budget doubled, continuing to climb well into World War II.[42] By 1942, Society research, now monitored by the state minister Albert Speer, was again almost completely beholden to military interests.[43] Of the over forty Kaiser Wilhelm Institutes active in the Third Reich, seven worked on projects related to offensive or defensive chemical warfare.[44] Further research was conducted at the Institute of Medical Research at Heidelberg, where the biochemist Richard Kuhn synthesized the exceedingly lethal nerve agent Soman in 1944.[45] Even more deadly than Tabun and Sarin, a single drop of Soman induced vomiting, extreme muscle pain, and eventual nervous system failure. IG Farben produced 70 kilograms of Soman before the end of the war.[46]

University and smaller private chemical research laboratories often maintained even less independence from Nazi directives. Indeed, the small size and decentralization of laboratories such as Ferdinand Flury's at the University of Würzburg was ideal for keeping research low-profile in a supposedly disarmed Germany.[47] In the early years of

[41] The Kaiser Wilhelm Society managed to protect the positions of several Jewish researchers in the early 1930s. Max Planck met with Hitler on May 16, 1933, to intercede on behalf of Fritz Haber. After Hitler reportedly worked himself into a lather, Planck left and Haber's position at the KWI remained in jeopardy. David Cassidy, *Uncertainty: The Life and Science of Werner Heisenberg* (New York: W. H. Freeman and Company, 1992), 307.

[42] Kristie Macrakis, *Surviving the Swastika: Scientific Research in Nazi Germany* (New York: Oxford University Press, 1993), 5.

[43] Ibid., 95.

[44] Florian Schmaltz, "Chemical Weapons Research in National Socialism: The Collaborations of the Kaiser Wilhelm Institutes with the Military and Industry," in *The Kaiser Wilhelm Society under National Socialism*, eds. Susanne Heim, Carola Sachse, and Mark Walker (Cambridge: Cambridge University Press, 2009), 313–315.

[45] Richard Kuhn was also the president of the *Deutsche Chemische Gesellschaft* (DChG) and a departmental head for the Reich Research Council. Like the *Verein Deutscher Chemiker*, the DChG also involved itself in chemical weapons research under the Nazis. The DChG, composed mostly of academic chemists, predominantly worked on offensive weapons production in military funded labs. Maier, *Chemiker im "Dritten Reich"*, 265.

[46] Schmaltz, "Neurosciences and Research on Chemical Weapons," 198–199.

[47] Ibid., 186–187.

the Third Reich, this decentralization allowed for a variety of different chemical warfare research agendas, encouraging specialization and problem-driven approaches.[48] This included the continued testing of many of the gases developed during World War I in university laboratories at Marburg, Münster, Giessen, Würzburg, Greifswald, and Danzig.[49]

As part of this decentralized chemical weapons research, the German military set up and maintained three separate production and testing facilities that employed upwards of 1,000 scientists and support staff. The largest site was the Army Proving Grounds north of Münster, where scientists started to produce Tabun in 1937. The second was the Army Gas Protection Laboratory at the Spandau Citadel, which began testing the effects of nerve gasses on animals. And the final site was the Gas Protection Center of the Army Ordnance Office in Berlin. Besides these three large chemical testing sites, the Military Medical Academy in Berlin produced valuable supplemental research.

Set up in 1934, the Academy was largely intended to train sanitary officers in the Wehrmacht. However, through the Academy's Department of Pharmacology and Military Toxicology, directed by the gas specialists Wolfgang Wirth and Otto Muntsch, scientists began testing Tabun on both themselves and other volunteer officers.[50] By 1939, the testing at these state-directed sites, as well as at IG factories, KWI institutes, and university laboratories provided the Nazis with a substantial national advantage in state-of-the-art chemical weapons research. After Germany invaded Poland in 1939 and Hitler declared Germany's official "Preparations for the Gas War," IG Farben and the Wehrmacht were well positioned to begin large-scale production.[51]

While a comprehensive picture of poison gas research in the Third Reich still remains difficult to piece together, evidence of gas protection research is abundant. Private research groups, the military, and manufacturers conducted studies on everything from new gas mask satchels to hand-held gas alarms.[52] Unsurprisingly, the gas mask received the most attention. Through studies on cardiac and respiratory activity under the gas mask, scientists and engineers hoped to create masks and filters that

[48] Maier, *Chemiker im "Dritten Reich"*, 265.

[49] Schmaltz, "Chemical Weapons Research on Soldiers and Concentration Camp Inmates in Nazi Germany," 232.

[50] The Military Medical Academy's Department of Pharmacology and Military Toxicology coordinated many of the disparate university laboratories that worked on chemical weapons research. Schmaltz, "Neurosciences and Research on Chemical Weapons," 189–190.

[51] Besprechung bei Herrn Generalmajor Thomas am Sept 5, 1939 (BA-MA) RH 12-9/25.

[52] Sondergerat fur Gasschutz, 1931–1936 (BA-MA) RH/12/4; Beschaffung der Tragetaschen zür Gasmaske 24, 1932 (BArch-B).

could last longer, provide a more comfortable fit, and be worn easily during daily life.[53] This research appeared particularly imperative because the most recent Auergesellschaft gas mask (the Gas Mask 30) often had problems with its gas-tight fit.[54] And, at the time, most German soldiers were still training with defective or broken gas masks from World War I.[55]

To address these problems, the Auergesellschaft and the Drägerwerks began working on both a new military gas mask and a mask that would be affordable enough for every German citizen. After barely surviving the Great Depression, the Dräger company was reluctant to recommit itself to this kind of military-specific manufacturing.[56] Under Heinrich Dräger, the company attempted to continue to manufacture about 50 percent of its products for international markets, focusing primarily on medical and mining equipment. Nevertheless, the Nazi state's demands for the production of civilian gas masks required the acquisition of new factories and raw materials suppliers.[57] Due to the Nazis' apparent concern for air and gas defense, military contracts shot up to about 65 percent of all Dräger production in 1935. Between 1933 and 1939, Dräger's final profits increased by more than forty times, sitting at 6 million Reichsmark by the end of the decade. In 1939 alone, the Dräger company received contracts worth over 6 million Reichsmark for civilian gas masks.[58] This windfall slowed between 1940 and 1943, and Dräger cut back its civilian gas mask production by 50 percent due to wartime rubber shortages.[59]

The Auergesellschaft also navigated murky financial waters after 1933. The Jewish partners Leopold Koppel and his son Albert Leopold were removed as chairmen, forcing the remaining board to convert Auer into a public limited company.[60] In 1934, the Degussa company bought significant shares of the Auergesellschaft, allowing Auer to continue its

[53] Karl Graus aus Schmechten, *Elektrokardiographie Untersuchungen bei Belastun älterer Personen unter der Gasmaske* (Höxter: Huxaria Druckerei und Verlagsanstalt, 1939), 5.

[54] Neue Gasmaske, 1933 (BA-MA) RH/12/4.

[55] Verwaltung, Lagerung, Nachschub, Unterbringung, Besichtigung des Gasschutzgerats bei der Truppe, 1933–1936 (BA-MA) RH/12/4.

[56] The Drägerwerks maintained an often-contentious relationship with the Nazis. Heinrich Dräger protected several of his Jewish employees from Nazi persecution. However, the company also utilized 1,200 forced laborers from Eastern Europe and 50 prisoners of war in the 1940s. Marc Buggeln, *Slave Labor in Nazi Concentration Camps*, trans. Paul Cohen (Oxford: Oxford University Press, 2014), 132–133.

[57] Knoll, *Die Drägerwerke im Dritten Reich*, 10–12.

[58] Lorentz, *Industrieeltie und Wirtschaftspolitik 1928–1950*, 173.

[59] Knoll, *Die Drägerwerke im Dritten Reich*, 11–12.

[60] After the removal of the Koppels, Dr. Karl Quasebart took on a greater leadership role in the company. In 1937, Quasebart was made chairman of the board and head liaison to

production of incandescent lamps and gas masks.[61] Much like for Dräger, contracts for civilian gas masks skyrocketed in the mid-1930s, eventually making up 90 percent of all Auer profits (64 million Reichsmark in 1939). This lack of market diversity practically forced Auer to seek even more military contracts that could be quickly fulfilled in Auer's large Oranienburg factory, conveniently located just outside Berlin. Competition with Dräger further spurred costly expansion, drawing concern from Auer's parent company Degussa. However, there was little that Degussa could do to rein in Auer's dependence on government contracts due to the company's increasingly intimate relationship with the now-powerful *Reichsluftschutzbund* and the Wehrmacht.[62] In 1933, Auer received 10 million Reichsmark to produce about 500,000 gas masks for the military.[63] After the implementation of the Four Year Plan, the company made about 2.8 million civilian gas masks per year.[64] Reflecting on this increased demand in its 1936 product catalog, the Auer company claimed, "The implementation of the Four Year Plan requires the application of all the scientific, technical, and economic energies of our people."[65] However, due to increasing investment in assets and slow payment from government contracts, Auer lost 2.5 million Reichsmark in the 1937 fiscal year, and by 1938, Auer owed its creditors about 7 million Reichsmark.[66] Degussa reeled in Auer's proposed expansions, now only allowing the company to acquire stock in a radium and uranium firm in the Sudetenland.[67]

The boom in gas mask production even inspired Hugo Stoltzenberg to attempt to reenter the German chemical protection market. Stoltzenberg spent about 90,000 Reichsmark on the development of a new gas mask

the German military. Auergesellschaft, 1939 (BArch-B) NS/19/ Sig 3446 Ort 51 Haus 901/EG.

[61] Hayes, *From Cooperation to Complicity*, 128.

[62] Auer's desire to continually expand production partly inspired their decision to utilize 2,000 female Sachsenhausen concentration camp prisoners for gas mask production at Oranienburg after 1944. Vorermittlungsverfahren Tatbestand: NS-Verbrechen in Nebenlagern des Konzentrationslagers Sachsenhausen Tatort: Nebenlager Auerwerk-Oranienburg, 1944–1945 (LBWL) EL 48/2 I.

[63] This was three times the total number of masks that they had sold in the previous year. Degea Aktiengesellschaft (Auergesellschaft) Berichte, Korrespondenz, 1933 (EIA) Degussa IW 24.5./2-3.

[64] Hayes, *From Cooperation to Complicity*, 130–131.

[65] Degea A.G., *Farbspritzen und Gasschutz* (Berlin: Degea A.G., 1936).

[66] Degussa, on the other hand, experienced lasting profit increases from the Nazi Four Year Plan. Like IG Farben, Degussa stood to gain even more from Nazi warfare and conquest in the 1940s. Peter Hayes, "The Chemistry of Business-State Relations in the Third Reich," in *Business and Industry in Nazi Germany* (New York: Berghahn Books, 2004), 69.

[67] Otto Frisch, *What Little I Remember* (Cambridge: Cambridge University Press, 1979), 89.

that featured a fully transparent face shield. Writing to the Ministry of Aviation, Stoltzenberg boasted, "The construction of this mask is based solely on German inventions, while the previous mask types were invented and developed primarily by Jews."[68] In a second letter, Stoltzenberg claimed that the Auergesellschaft was a Jewish monopoly that was now driving his Aryan factory out of business.[69] This attempt to curry favor with the Nazis ultimately failed, and Stoltzenberg's mask was never widely adopted by state services. His effort to drive competitors out of business by utilizing the rhetoric of Nazi "Aryanization" was a common practice.[70] Ever the opportunist, Stoltzenberg did not give up his attempts to court the Nazis and he joined the party in 1941, but he struggled to create lasting business relationships with important members of the party. In fact, his continued international business deals made him unpopular with the regime, and he was even detained for a month by the Gestapo under the suspicion of treason.

Aero-Chemical Rearmament and Mobilization in the Third Reich

The 1933 Nazi takeover also brought swift measures for the rearmament of the Reichswehr. Minister of the Interior Wilhelm Frick and Minister of Economics Hjalmar Schacht introduced several convert financial structures that could effectively funnel state money into the German armed forces. For instance, in May 1933, Schacht set up the dummy company MEFO, which issued promissory notes used to contract the major German armaments manufacturers to produce arms. Other seemingly civil organizations such as the *Deutsche Verkehrsfliegerschule* (German Air Transport School) operated as covert military academies. Poison gas training was part of this renewed military spending, and Reichswehr soldiers learned to both employ and defend against the gases of World War I. With their timeworn gas masks, soldiers entered pitch-black training rooms that would be subsequently filled with tear gas. Testing composure in an imagined gas attack, the army expected men to be able to don their gas masks in fifteen seconds while holding their breath.[71] Like the German regulars of 1918, the Reichswehr soldiers of the early 1930s

[68] Geratetechnisches, Konstruktion, Erfindungen, Versuche, Formveranderungen usw., 1933–1936 (BA-MA) RH/12/4.
[69] Gasschutz-Volk, 1934–1936 (BA-MA) RH/12/4.
[70] Hayes, *From Cooperation to Complicity*, 74.
[71] Gas und Entseuchungsraume, 1931–1936 (BA-MA) RH/12/4.

also learned to properly fit and clean their masks, reporting to their NCO if their gas masks malfunctioned or exhibited any holes.[72]

By March 16, 1935, the Nazis openly disobeyed the Treaty of Versailles and publicly announced that they were rearming Germany.[73] They described their plans for rearmament as an assurance of peace, gesturing toward the dangers of the Soviet Union as an obvious justification for a restored German military.[74] The best practices for poison gas training now became an open topic for the commanders of the newly renamed Wehrmacht. Understandably, new guidelines still drew heavily on the experience of chemical warfare in World War I. For instance, Major General Friedrich von Tempelhoff reiterated that success in chemical warfare relied on the gas discipline of each soldier.[75] As such, soldiers who learned to live among toxic gases were best prepared to leverage the offensive capabilities of chemical weapons.[76] Tempelhoff wrote that "progress depends not only on the technology, but also on how the soldier understands how to use both gas and gas protection devices."[77]

The technology behind gas protection had progressed slowly but significantly since World War I. After the public announcement of German rearmament, Wehrmacht soldiers began to train with full-body rubberized gas suits that could resist vesicants such as mustard gas. Their Gas Mask 30 filters were designed to resist a greater number of gas variants for longer periods of time, and their auxiliary protective devices such as gas blankets and gas alarms were designed for mobile combat.[78] Nevertheless, with gas masks affixed, both visibility and audibility remained major problems for troop cohesion.[79] Understandably, the soldiers' inability to hear their group leaders still greatly concerned military commanders. The direction of a strong and decisive leader was seen as essential for both the completion of any mission and basic survival in gas attacks. One military report stated:

[72] Anweisung für die Handhabung der Gasmaske 30; mit Abbildungen, 1936 (LBWH) M 635/1 Nr 1461.

[73] On March 7, 1936, the Nazis reoccupied the Rhineland. Hitler avoided confrontation with Great Britain and France by crafting a vague Peace Plan that made overtures to détente policies and claimed adherence to international limits on bombers and poison gas. Richard Overy, *The Bombing War: Europe 1939–1945* (London: Penguin, 2013), 31.

[74] *Gasschutz und Luftschutz* Jahrgang 7 Heft 1, 1937.

[75] Friedrich von Tempelhoff, *Gaswaffe und Gasabwehr: Einführung in die Gastaktik* (Berlin: Verlag von E.S. Mittler & Sohn, 1937), vi.

[76] Gasschutz Vorschriften und andere Vorschriften, 1938 (BA-MA) RH/12/9.

[77] von Tempelhoff, *Gaswaffe und Gasabwehr*, 210.

[78] RH/12/9 Sig 4, 1938 (BA-MA).

[79] Gasabwehrubung, 1937 (BA-MA) RH/12/4.

There is a danger that the troops, who are not used to this fighting, will lose their nerve during the first fierce aero-chemical attack and that those of morally weak natures will take this opportunity to flee the battlefield. In this way, the attack power of whole units may be called into question before the force has even come into the area of fire.[80]

On the whole, German military theorists assumed that aero-chemical attack would produce some of the greatest damage on the battlefields of the future. According to the Army Office, traditional chemical weapons could no longer surprise an unprotected enemy. Therefore, battlefield successes would only come from new gases that could penetrate current gas protection devices or attacks that could cripple army supply lines. Aero-chemical attacks seemed particularly valuable for shutting down enemy depots and arms manufacturing. A 1937 military report argued that the Wehrmacht should throw "all available chemical weapons" at the enemy to shut down military production. The report claimed that a yearly manufacture of at least 10,000 tons of chlorine was necessary to achieve such a goal.[81] Simultaneously, commanders drew up hundreds of detailed plans for protection against enemy aerial gassing. At the Wehrmacht Academy, they regularly discussed and delivered these plans through lectures on "gas fighting" and "gas protection."[82]

While the military required the manufacture of millions of gas masks as part of the Nazis' rearmament program, directives from Hermann Göring and the *Reichsluftschutzbund* spurred subsidiary gas mask production in the mid-1930s. As early as the spring of 1933, the Hitler Youth and the SA began handing out air protection literature and rattling tins to solicit funds for the recently formed *Reichsluftschutzbund*.[83] As the RLB grew, it engaged in an increasing number of air and gas protection activities. Dramatizing the German nation's vulnerability to aero-chemical attack, the RLB set off sirens and stopped traffic to stage sensationalized bomb diffusions in the middle of crowded city streets.[84] They further dropped paper bombs filled with sand to express the looming possibility of aerial attacks against civilians. And at an even more omnipresent level, the RLB put up posters, handed out pamphlets, and erected massive dummy bombs in public spaces. One American observer of these activities claimed, "Military preparedness … pervades every aspect of life in modern Germany … Every house bears the placards of an organization for defense

[80] Gasabwehrubungen, 1938 (BA-MA) RH/12/9.
[81] Gaskampf. Grundlegende Erlasse und Arbeitspläne II, 1937 (BA-MA) RH/12/9.
[82] RH/12/9 Sig 22 (BA-MA) S. [83] Süß, *Death from the Skies*, 36–37.
[84] Peter Fritzsche, *Life and Death in the Third Reich* (Cambridge, MA: Belknap Press, 2008), 55.

against air attack. The language of the street, the press, the radio, the newsstand and even the library and classroom smacks of war."[85]

RLB members even conducted air defense shows where firefighters, members of the *Technische Nothilfe*, and Red Cross workers would douse fires and rescue victims from small-scale wooden cities. The finale of one specific show entitled, "Kiel in Flames," featured the firing of tear gas into the crowd of spectators.[86] While the RLB was certainly aware that these measures could potentially lead to civilian panic, prevailing logic claimed that any visible measure was better than the inaction of the Weimar government. As such, practically no method was off limits if it could impart the seriousness of aero-chemical warfare to German civilians.[87]

A combination of these scare tactics and Nazi pressure to participate in civic duties motivated millions of German civilians to join the *Reichsluftschutzbund*. By 1936, the RLB boasted 8.2 million members in over 7,000 local branches, making it one of the largest civic organizations in Nazi Germany. Putting its membership into perspective, historian Peter Fritzsche writes that the total number "included one out of every six Berliners and over 10 percent of Germans living in Hesse, the Palatine, and Baden."[88] By the mid-1930s, the RLB was conducting large-scale air raid exercises across Germany, forcing civilians to enter air raid shelters, don gas masks, and turn off lights on a regular basis. In 1934 and 1935 alone, an impressive 28 percent of all Germans in twenty-one select cities visited an air defense show. In cities and towns close to German borders, these percentages increased drastically. For instance, 40 percent of the population of Heidelberg, 46.7 percent of Frankfurt an der Oder, and 48 percent of Dusseldorf spent at least one day learning the basics of aero-chemical defense.[89]

Theoretically, all RLB activities could have been carried out by the Weimar government's Ministry of the Interior. In fact, the league incorporated all of the state services that had been previously tapped by Weimar leadership to serve in national air and gas protection. Thus, there were

[85] Ibid., 55. [86] *Internationales Luftfahrt Archiv*, 6 October 1933.

[87] The use of tear gas in air defense shows was eventually prohibited by the Ministry of Aviation because the regime feared that it would deter civilians from participating in defense exhibitions. Torben Möbius, "World War II Aerial Bombings of Germany: Fear as Subject of National Socialist Governmental Practices," *Storicamente* 11, no. 21 (2015): 12.

[88] Fritzsche, *A Nation of Fliers*, 211. The RLB also ran 5,088 air raid protection schools with 28,000 air protection teachers. Generalmajor a.D. von Tempelhoff, "Chemische Kampfmittel und Gasabwehr in der 'Russischen Felddienstordnung!'," *Gasschutz und Luftschutz* 1 (August 1938).

[89] *Die Sirene*, no. 6 (March 1936).

few structural differences between the RLB and the Weimar institutional apparatus. Admittedly, the RLB consolidated the work of the private air and gas protection organizations, but it placed itself precisely where the Ministry of the Interior had previously sat atop the administrative structure. It was not the nominal changes to institutional structures that made the RLB appear different but rather its grassroots activities and groundswell support. When bombastically praising the ways in which the RLB effectively harnessed local activism in 1933, Hermann Göring wrote, "Under the leadership of the National Socialist state, the period of individualized air protection is no longer compatible with the modern age. With all acknowledgement of the work done by the air defense advisors in a difficult time, I ask you to dissolve private local air defense groups immediately."[90]

While the private air defense organizations did disappear, state services and ministries such as the *Technische Nothilfe*, Ministry of Transportation, German Railroad, Ministry of Work, Ministry of Agriculture, Postal Service, Ministry of Science, Ministry of Finance, German Red Cross, Ministry of Propaganda, consulting VDCh chemists, and local police and fire departments all continued to serve underneath Göring's RLB in the push to prepare German civilians for aerochemical warfare.[91] Part of the decision to keep these smaller bureaucratic structures in place was purely practical. Not only did they already maintain air protection experience from Weimar-era exercises, but it was also apparent that local authorities fared the best in leading air raid drills. Through intimate knowledge of a specific town or area, local police and firefighters could most quickly diagnose regional air defense problems and build the civic community structures that could best respond.[92]

At the same time, the localized RLB structure created the kind of disjointed polycracy that often-defined Nazi government.[93] While Hermann Göring served as the face of the larger organization, much of the real work was carried out by the undersecretary of the German Air Ministry and the RLB director of Civilian Air Protection, Kurt Knipfer.

[90] Einzel-u. Sammelschutz, 1933–1936 (BA-MA) RH/12/4.
[91] Richard Bauer, *Fliegeralarm: Luftangriffe auf München 1940–1945* (München: Heinrich Hugendubel Verlag, 1997), 10.
[92] Fritzsche, "Machine Dreams," 698.
[93] Jurisdictional conflicts pervaded air defense in the Reich. While the RLB oversaw all civilian preparations, Himmler's Order Police struggled to maintain control of the state air defense services over the Reich Interior Ministry. In 1938, the RLB agreed to a comprise, stating that the RLB would maintain responsibility for all air defense while Himmler's Order Police would run any emergency operation in the case of a real aerial attack. Jörn Brinkhaus, *Luftschutz und Versorgungspolitik: Regionen und Gemeinden in NS-Staat, 1942–1944/45* (Bielefeld: Verlag für Regionalgeschichte, 2010), 16.

Knipfer, a former air force pilot and Weimar bureaucrat in the Prussian Ministry of Commerce, was responsible for promoting the RLB's push for a militarized national community of air protection. Already in conversation with many of the gas specialists through the pages of *Gasschutz und Luftschutz*, Knipfer picked up many of his organizational ideas from theorists such as Erich Hampe. Accordingly, Knipfer and the RLB demanded that each German locality create an authoritarian structure that reflected the larger national regime. This necessitated strong leaders and independent services in each neighborhood, but it did not always foster inter-community cooperation.[94] While daily conditions in each locality might differ, the RLB's rhetorical insistence on individual responsibility to a militarized and collectivized national community permeated the entire endeavor.

The Air Defense Act of 1935 placed the RLB under the newly established air force and restructured the organization.[95] The RLB was subsequently divided into fifteen regional groups that oversaw approximately 66,300 local district departments and 644,000 public officials.[96] Local RLB leaders each controlled their own "air-defense sector" and were expected to direct police and fire brigades in the event of an attack. Furthermore, each city apartment building elected a "house warden," who would then assign a *Hausefeuerwehr* (fire detail) for the individual building. House wardens reported to elected *Blockwarte* (block wardens), who sent information up an extensive chain of command that theoretically ended with Göring himself.[97] The editor of *Gasschutz und Luftschutz*, Heinrich Paetsch, described these mini-dictatorships of air and gas protection:

> Decisions to be taken are for the good of the entire population; there cannot be any debate, any conflicts of opinion, or any half-hearted resolutions. Only the man who demonstrates that he is a leader, who has iron nerves, who is cold-blooded and fearless, who has a clear and quiet aspect, and shows endurance, will serve in the air defenses.[98]

[94] Einzel-u. Sammelschutz, 1933–36 (BA-MA) RH/12/4.

[95] In 1939, Kurt Knipfer's Air Protection Department and the RLB were placed under Air Force Inspectorate 13, but the Luftwaffe interfered very little with civilian measures and Knipfer retained real power over the RLB.

[96] Generalmajor a.D. von Tempelhoff, "Chemische Kampfmittel und Gasabwehr in der 'Russischen Felddienstordnung!'," *Gasschutz und Luftschutz* 1 (August 1938).

[97] The RLB also assigned *Melder* who would relay local messages and *Laeinhelfer* who were responsible for caring for the elderly, young, and wounded.

[98] Heinrich Paetsch, "Oertliche Führung im Luftschutz," *Gasschutz und Luftschutz* 2 (May 1932), 97–100.

While the bureaucratic structure under the RLB remained complicated, the league placated air protection advocates by involving them in its aggressive programs and daily visual presence. Even if state protective services were still difficult to organize at the granular level, German civilians were now very much aware of the dangers of aero-chemical attack. Clearly, Göring and the RLB had no problem with using scare tactics and sensational propaganda to make air and gas protection an omnipresent existential concern in the minds of German citizens.[99]

Scientific Research or Authoritarian Spectacle: The Gas Specialists in the Third Reich

While the Nazis disbanded the private air and gas protection organizations and integrated their members into the RLB, the most prominent poison gas scientists and engineers continued their efforts to raise air and gas protection awareness through public lectures and their various professional journals. The journals *Gasschutz und Luftschutz* (*Gas and Air Protection*) and *Beruf und Stand* (*Profession and Status*) specialized in questions of aero-chemical protection, while *Luftschutz-Rundschau* (*Air Protection Review*) focused on broader questions of national air protection.[100] The Auergesellschaft's *Die Gasmaske* (*The Gas Mask*) and the Dräger company's *Drägerheft* (*Dräger Booklet*) continued to feature scientific articles alongside purchasing catalogs. In addition to these more established journals, *Die Sirene* (*The Siren*) first appeared in late 1933 as the RLB's monthly publication.

After the Nazi takeover, Heinrich Paetsch and Rudolf Hanslian remained the editors of *Gasschutz und Luftschutz*. Most of the contributors, a veritable "who's who list" of powerful men in the police, fire brigade, *Technische Nothilfe*, Wehrmacht, and chemical research laboratories continued to write for the journal under the Nazi regime. In the early months of 1933, the journal remained largely focused on technical and scientific issues, although articles frequently lauded the Nazis' insistence on the importance of air and gas defense. But in June of that same year, Kurt Knipfer wrote of the Nazis' political rise as a "national revolution," specifically pointing to the significance of the centralization of air and gas protection under the *Reichsluftschutzbund*.[101] By the fall, contributors were cheering the end of international armament limitations

[99] Westermann, *Flak*, 36.
[100] *Beruf und Stand* was controlled and published by the *Verein Deutscher Chemiker*.
[101] Linhardt, "Die Fachzeitschrift 'Gasschutz und Luftschutz' unter dem Einfluss des Nationalsozialismus," 4.

and quoting sections of *Mein Kampf*.[102] In this same vein, the 1935 - reestablishment of the Wehrmacht inspired a long paean to the Nazis' tireless work. According to the article, now every German man, woman, and child was "genuinely indebted to carry out the tasks assigned to them to the best of their knowledge and ability."[103] While *Gasschutz und Luftschutz* was never officially controlled by the Nazi party, the use of Nazi insignia and propaganda increased over the course of the 1930s.[104] Both managing publishers, Dr. August Schrimpff and Dr. Franz Ebeling, joined the Nazi Party, and Ebeling went so far as to become a member of the SS in 1933.[105]

Through their own publications, the RLB supported the gas specialists in their continued political struggle against opponents who claimed that air and gas defense was merely an indirect form of militarism. Articles, books, and pamphlets by the RLB labeled such pacifists as enemies of the state, while simultaneously legitimating the arguments of gas specialists such as Rudolf Hanslian and Hermann Büscher by championing their ideas and referring to them as *Fachleute* (technical experts).[106] Air protection literature continued to deny any overtly political aim, repeatedly referring to "facts" and "reality" in an attempt to simply "educate" and "help the German people."[107] Air and gas specialists were insistent that education and the technical application of knowledge was Germany's best hope for salvation in the newly dangerous modern world.

Given that they had already been providing public lectures in the late 1920s, members of the *Verein Deutscher Chemiker* (VDCh) continued their consulting role under the RLB. In fact, the chemist Heinrich Remy attempted to formalize this position, demanding the institution of a state examination to qualify "air-raid chemists."[108] From June 8 to 11, 1933, over 700 of these chemists met in Würzburg, where members such as Hugo Stoltzenberg gave lectures on poison gas protection to a packed room. The Würzburg conference ended with a massive air raid

[102] Vgl. K. Sckerl, "Deutschland vor einer Revolution der Bauwirtschaft," *Gasschutz und Luftschutz* 3 (October 1933): 256–260.
[103] *Gasschutz und Luftschutz* no. 1 (January 1936).
[104] *Gasschutz und Luftschutz* no. 2 (1938).
[105] Linhardt, "Die Fachzeitschrift 'Gasschutz und Luftschutz' unter dem Einfluss des Nationalsozialismus," 8.
[106] Status titles and insignia were an important method by which the RLB brought the gas specialists and the German public into the fold. Detlev Peukert, *Inside Nazi Germany: Conformity, Opposition, and Racism in Everyday Life*, trans. Richard Deveson (New Haven: Yale University Press, 1987), 73.
[107] Ulrich Müller-Kiel, *Die Chemische Waffe: Im Weltkrieg und Jetzt* (Berlin: Verlag Chemie, GMBH, 1932), 1.
[108] Maier, *Chemiker im "Dritten Reich"*, 271.

demonstration that simulated a chemical attack on the front lawn of the conference center.[109]

While couched as civilian protection, such public displays revealed the *Verein Deutscher Chemiker*'s familiarity with military concerns. This was further revealed through the 1934 awarding of the association's Liebig Medal for excellence in chemistry to Ferdinand Flury. In his acceptance speech, Flury proclaimed:

> The threat posed by chemical warfare agents, which only previously threatened front-line soldiers in the war, can become an existential question for all of us, for the entire German homeland, tomorrow. This creates a serious duty for all of us to prepare and expand air defense ... [this honor] is an outward sign of the gratitude that the German chemists of today have for their wartime peers, who silently sacrificed their lives in the scientific laboratories, on the test fields, and in the chemical warfare companies, just like the countless unknown soldiers of the World War.[110]

In their writings from the mid-1930s, gas specialists claimed poison gas as a real existential threat in their own geopolitical terms. Contributors to *Gasschutz und Luftschutz* sustained their denouncement of antiair protection pacifists by continually pointing to the dangers posed by the chemical weapons of the United States and the Soviet Union.[111] One article warned, "The Russian air force is powerful, considering both the quality of their weapons and the training of their crews. The Soviets are in the process of creating an air force that will be terrible to behold in the future."[112] Another mocked the Women's League for International Peace and Freedom, describing the female pacifists' commitment to international diplomacy as dangerously naïve.[113] At the same time, however, the gas specialists continued to discredit many of the apocalyptic visions of a future gas war that had become so popular in the Weimar Republic.[114] One Dr. Curt Rosten wrote:

> The stupid idea [that the gas threat is beyond protection] has been rife for a long time. The press has done its part to perpetuate the gas danger, endlessly tending towards a sentimental fear of poisons and intoxicants ... but the fear of chemical

[109] A subsequent meeting of "air raid chemists" took place at the Auergesellschaft gas school at Oranienburg in September 1933.

[110] Maier, *Chemiker im "Dritten Reich"*, 278.

[111] Erich Hampe, *Der Mensch und die Gase: Einführung in die Gaskunde und Anleitung zum Gasschutz* (Berlin: Räder-Verlag, 1932), 63.

[112] "Der Gasaufklärungs-, Gasbeobachtungs- und Gasalarmdienst in der Roten Armee," *Gasschutz und Luftschutz* no 11 (May 1936).

[113] Heinrich Hunke, *Luftgefahr und Luftschutz: Mit Besonderer Berücksichtigung des Deutschen Luftschutzes* (Berlin: Verlag von E.S. Mittler & Sohn, 1935), 62.

[114] "Gas War Danger Exaggerated" in the May 22, 1938 edition of *The Observer*. RH/12/9 Sig 71, 1937–1938 (BA-MA).

warfare agents is only justified if one is not aware of the measures that can be taken to avert this danger.[115]

In the sometimes-contradictory minds of the gas specialists, the continual reminder of a serious but measured anticipatory threat best spurred the German people to participate in air and gas protection. Undoubtedly, with the rise of the Nazis and the creation of the RLB, the gas specialists had essentially won their political debate with the proponents of international treaties.[116] However, they had to continue to walk a fine line between stoking public anxiety to inspire aero-chemical preparation and creating a sense of political paralysis or widespread panic.[117] Thus, they sought a form of careful emotional influence that could flex and shift according to the public's expectations, and later experiences, of aerial attack.[118]

Regardless of the increased attention and esteem that the RLB garnered for the air and gas specialists and their aligned interest in the buildup of national defense, many of the gas specialists remained critical of some of the RLB's specific guidelines for aero-chemical protection in the early 1930s. Some experts were still unsure whether civilians could remain calm and perform air raid activities under the threat of aerial bombing. The chaotic failures of several 1930s air raid exercises underscored these doubts.[119] An article in the Auergesellschaft's *Die Gasmaske* pointed out that civilian air and gas protection was a psychological matter and that correct guidelines could be instilled only after all fear and doubt had been banished from the collective civilian mind.[120] A military report from 1938 imagined how an improperly trained citizenry might act:

Even against the civilian population – the poison gas will be used as the weapon of war, and here too, its unpredictable and uncanny effects will create long-term panic and lead to a complete stalling of life in a besieged city. The feelings of insecurity among the population will grow, industry and traffic will be suspended for a long time, the psychosis of poisoning will cause the hospitals to overflow, and the sanitary and detoxification measures will fail.[121]

[115] Curt Rosten, *Was man vom Luftschutz wissen muss* (Berlin: Verlag Deutsche Kultur-Wacht, 1935), 75.
[116] The Nazis would eventually ban and burn Waldus Nestler's and Gertrud Woker's texts.
[117] Westermann, *Flak*, 36.
[118] Möbius, "World War II Aerial Bombings of Germany," 3–4.
[119] Bernd Lemke, *Luft- und Zivilschutz in Deutschland im 20. Jahrhundert* (Potsdam: Militärgeschichtliches Forschungsamt, 2007), 85.
[120] "Übungen mit Gasmasken von Dilp-lng Paul Maack, Führer des Luftschutzdienst der Technische Nothilfe Hamburg," *Die Gasmaske* (January 1936): 13.
[121] Ausnutzung der deutschen chemischen Industrie fur eine entscheidungsuchende Kriegfuhrung, 1938 (BArch-B) R/3113/.

Reflecting further on these problems, some gas specialists prioritized civilian evacuation over defense. One expert wrote, "The difficulties faced by the problem of accommodation in evacuation are very great; but they do not seem to be insurmountable. Even if only small successes can be achieved, they are better than none and worth the work."[122] Evacuation was certainly a contentious issue, but the most common concern among gas specialists remained the gas mask. While the RLB extolled the importance of the device in civilian air and gas protection, technical experts were unsure whether civilians could properly handle the mask or if masks would even be necessary in the war of the future.[123] If gas masks were not entirely essential, then their frightening appearance might simply serve to sow anxiety and panic among the German populace.

Undoubtedly, providing gas masks for millions of German civilians would also prove incredibly costly, and the RLB struggled to deliver enough masks throughout the decade.[124] Reflecting on these problems, one gas specialist wrote:

> With regard to gas protection, it is important to say that gas masks are not enough to provide all of the population with total protection. They are prohibitively expensive and they are of limited use, assuming they receive careful maintenance and are in constant readiness.[125]

In 1934, the newly designed Auer and Dräger S-Mask cost twenty-two Reichsmark, a one-third reduction in price from the previous year.[126] Given its pared-down simplicity, the S-Mask was not the highest-performing device, but it still cost far too much for public distribution.[127] For this reason, only certain industrial workers and members of the air and gas protection services received S-Masks.[128] Beyond the problems of price and availability, experts further claimed that some civilians would certainly forget their masks or fail to put them on fast enough.[129] Still others worried that civilians would be outfitted with the wrong type of filter if an enemy dropped an unknown or unusual gas. The pages of *Gasschutz und Luftschutz* teemed with these and other technical questions throughout the 1930s.

[122] *Gasschutz und Luftschutz* no. 10 (October 1935).
[123] Viktorit-Antiplyn, eds., *Was ein jeder über die Gasmaske wissen soll und muss* (Brünn: Viktoria Gummiindustrie, 1933), 1.
[124] Dräger Geräte Gesellschaft, 1933 (TMA) III.2 08746.
[125] Hunke, *Luftgefahr und Luftschutz*, 185.
[126] "Die S-Maske," *Gasschutz und Luftschutz* no. 4 (April 1934).
[127] "Ober Kritik an der S-Maske," *Gasschutz und Luftschutz* no. 4 (April 1935).
[128] Luftschutzdienst der Gruppe Karupfalz, 1935–1937 (LBWK) 465 c Zugang 1991–1997, 83.
[129] Henry F. Thuillier, *Gas im nächsten Krieg* (Berlin: Albert Nauck & Co, 1939), xi.

However, toward the end of the decade, Erich Hampe, a regular contributor to *Gasschutz und Luftschutz*, began to strongly affirm that German civilians could in fact conduct successful air protection drills. In a passage that is demonstrative of his broader stance, he wrote:

> Man can protect himself against the danger of gas. He will overcome it if he has acquired the necessary skills and has made the necessary preparations. As the fatherland demands of every soldier that he sell his life as dearly as possible, so the newly united German people demand the same from every citizen, man or woman. This duty is imposed on them, not just so that they may live as long as possible, but also so that Germany lives![130]

Hampe's arguments aligned with the RLB's belief in collective protection, encouraging the organization to attempt to solve universal gas mask distribution. In 1937, the RLB took a major step in this direction with the introduction of the *Volksgasmaske* (the People's Gas Mask).[131] The *Volksgasmaske* only cost five Reichsmark, and RLB propaganda exhorted all Germans to buy one as part of a national safety campaign.[132] The editors of *Gasschutz und Luftschutz* responded to the introduction of the *Volksgasmaske* by triumphantly proclaiming:

> It can be said that the Reich Minister for Aviation has taken a step here which is of the utmost importance to German civil air defense and has introduced a new phase in its development ... In a short time, they have managed to make millions of copies of this device at a reasonable price.[133]

This heroic declaration, however, ignored the fact that both the Auergesellschaft and Drägerwerks were struggling to produce the millions of masks that total civilian defense required. Furthermore, in order to make the mask affordable, manufacturers significantly reduced the quality of construction.[134] These basic *Volksgasmasken* often sat awkwardly on the head, and their filters could only withstand the gases used in World War I. For this reason, the RLB further required propaganda

[130] Erich Hampe, *Der Mensch und die Gasgefahr: Einführung in die Gaskunde und Anleitung zum Gasschutz* (Berlin: Räder-Verlag, 1940), 117.

[131] It is important to note that Hampe joined the Nazi party in the spring of 1933, and he maintained his position as vice chief of the *Technische Nothilfe* until 1941 when he was reassigned to the technical emergency corps of the Wehrmacht. This would suggest that Hampe had personal and political reasons to promote a form of civilian aero-chemical defense that aligned with the RLB's platform.

[132] Jedem Volksgenossen seine Volksgasmaske! 1937 (SBB) Einbl. 1937, 008 kl.

[133] *Gasschutz und Luftschutz* no. 6 (June 1937).

[134] Even Adolf Hitler admitted that the *Volksgasmaske* was a "product of two weeks labor" and that its production was "all a mess." Helmut Heiber and David M. Glantz, eds., *Hitler and His Generals: Military Conferences 1942–1945* (New York: Enigma, 2003), 431–432.

campaigns that could inspire high levels of civilian confidence in such a flimsy piece of technology.

The distribution of civilian gas masks most clearly represented the RLB's increasing disregard for the technical debates of the gas specialists. In the mid-1930s, gas specialists began to complain about the lack of specialist consultants in influential RLB roles. For instance, the air raid chemist Kurt Stantien decried the RLB's spotty employment of what he called "handmaid chemists."[135] Indeed, between 1934 and 1936 the number of official *Verein Deutscher Chemiker* lectures on air and gas protection decreased significantly.[136] After 1937, the RLB clearly controlled the direction of German air and gas defense through propaganda campaigns that extolled the strength of the Reich's air and gas defense measures. As long as they were willing to promote the party line, gas specialists were welcome in the implementation of these measures. And, not surprisingly, the gas specialists who maintained close ties with the Nazi Party tended to back the RLB.[137]

While less ideologically or politically committed gas specialists may have expressed greater disagreement over policy choices, it is worth considering the broader commitments of these engineers and applied scientists in order to understand their general willingness to continue to toe the party line. The gas specialists were certainly dedicated to the then popular ideals of technocratic efficiency, but they also tended to support obedience to authority and a general concern for public safety.[138] Undoubtedly, most, if not all, gas specialists saw their work as a pressing matter of national security. Furthermore, they now belonged to a scientific field that was granted national importance and corresponding financial support.[139] This could be seen in the gas mask company profits from

[135] Maier, *Chemiker im "Dritten Reich"*, 278.

[136] In 1934, there were twenty-two official air and gas defense lectures, but by 1936, there were two. Maier, *Chemiker im "Dritten Reich"*, 279.

[137] Heinrich Remy and Wilhelm Haase-Lampe joined the Nazi Party in 1933 while Ferdinand Flury joined in 1937. Hans-Georg von Tempelhoff and Otto Muntsch held positions in the Wehrmacht. Maier, *Chemiker im "Dritten Reich"*, 272.

[138] Carl Mitcham, *Steps toward a Philosophy of Engineering: Historico-Philosophical and Critical Essays* (London: Rowman & Littlefield International, 2020), 266. See also Jessica M. Smith, *Extracting Accountability: Engineers and Corporate Social Responsibility* (Cambridge, MA: MIT Press, 2021), 24–25.

[139] Local air protection officials contributed articles to *Gasschutz und Luftschutz*, *Die Gasmaske*, and *Die Sirene*, but the same fourteen-odd men from the Weimar period dominated the pages of these journals and the specialist field more broadly. They also published standalone works, often with the aid of the RLB or the Nazi Party. It was largely through this state-approved publishing that the German public was now directly exposed to the ideas of the gas specialists. Indeed, the RLB churned out countless books, posters, and magazines throughout the 1930s. In 1941 alone, the RLB printed 2.5 million copies of its magazine, *Die Sirene.*

the 1930s and it bears mentioning that many of the gas specialists had a personal financial interest in gas mask sales.

At the same time, however, the gas specialists did continue to debate the specific details of the RLB's protection plans in the pages of *Gasschutz und Luftschutz* and to apply scientific research to new air and gas protection questions. For instance, some argued that only German civilians living in major cities should be provided with a gas mask, while others pointed out that the physical difficulties of wearing a gas mask could hamper all air and gas protection activities.[140] But overall, the decision to transform German civilians into armored members of an air and gas protection community was no longer a site of meaningful debate. Now, the most pressing question was how to best achieve this transformation without creating civilian panic.[141] While the *Volksgasmaske* might not provide much real protection against aerial bombs and gas, both the RLB and its allied gas specialists realized its value as a tool for creating a disciplined and chemically minded German subject. For precisely this reason, an article in the July 1938 edition of *Die Sirene* sported the dramatic title "The Gas Mask: The Symbol of Fate for Both the Army and the Homeland."[142]

The *Reichsluftschutzbund* and Aero-Chemical Defense in Daily Life

The *Reichsluftschutzbund* made its most immediate mark in the daily lives of Germans through air and gas defense education. The RLB director of Civilian Air Protection, Kurt Knipfer, pronounced the importance of educational programs, writing that:

> Aerial attack threatens every house, every family and every business. All people must collectively protect themselves against the threat. The German people have been enlightened in the recent years by books, lectures and meetings on air protection and have been educated to a large extent through courses.[143]

[140] General a.D. von Tempelhoff, "Welchen Einfluß haben die angelegten Gasmasken auf die Marsch- und Gefechsfähigkeit der Truppe?" *Gasschutz und Luftschutz* no. 9 (September 1938).

[141] A shift in tone of *Gasschutz und Luftschutz* publications can best be seen in a 1937 addendum to an article on civilian evacuation by General Hugo Grimme. Grimme claimed that French forms of evacuation were not an option for the German people, who would never show such cowardice. Julia S. Torrie, *"For Their Own Good": Civilian Evacuations in Germany and France, 1939–1945* (New York: Berghahn Books, 2010), 19.

[142] "Die Gasmaske, das Symbol gleichen Schicksals von Heer und Heimat," *Die Sirene* no. 14 (July 1938).

[143] Kurt Knipfer and Werner Burkhardt, *Luftschutz in Bildern: Eine gemeinverständliche Darstellung des gesamten Luftschutzes für jeden Volksgenossen* (Berlin: Landsmann, 1935).

As early as 1935, air and gas protection courses were mandatory in every German school. In such courses, teachers instructed their students on the basic chemical structures of the most common poison gases and the types of weapons that could deploy them. Furthermore, lessons about the dangers of aerial attack were integrated across curricula.[144] For instance, common math questions asked students to calculate the trajectories and explosive power of various bombs and bullets, while writing assignments asked students to reflect on the necessity of blackout measures.[145] One aero-chemical specialist wrote of these lessons: "Enlightenment begins best in school. It is the patriotic task of the German teaching staff … [to present] chemical experiments and the weapons of aircraft squadrons."[146] To aid in this patriotic task, the Auergesellschaft offered both films and slides that teachers could use to expose students to the dire consequences of aero-chemical attack.[147] Teachers could also attend RLB courses, where they would learn the necessary lessons.[148]

A significant portion of aerial defense class sessions was dedicated to teaching children how to best protect themselves in the event of an attack.[149] However, the effective protection of children proved to be both technically and conceptually difficult. Children's small faces did not fit the three sizes of either the S-Mask or the *Volksgasmaske*. To account for this, Auer and Dräger produced specific children's sizes. Teachers or RLB members would measure each child's head and try to find an appropriate mask.[150] Sometimes this would be done as part of a school-wide air and gas protection course where large numbers of children would line up to be fitted.

Younger children presented even greater difficulties. In order to protect toddlers and infants, the gas mask companies produced bag-like hoods and *Gaskettchen* (airtight cribs) in which mothers could place their children.[151] Foot pumps then allowed parents to pump air into the sealed

[144] Aerial defense was also integrated into common children's games like *Fliegeralarm* (Air Raid Warning) and *Marsch in den Luftschutzraum* (March to the Air Raid Shelter).
[145] Nicholas Stargardt, *Witnesses of War: Children's Lives under the Nazis* (New York: Alfred A. Knopf, 2005), 24.
[146] K. Schütt, *Die chemischen und physikalischen Grundlagen des Luftschutzes in der Schule* (Berlin: C.J.E Volckmann Nachf., 1935), 6.
[147] Auer-Gesellschaft, *Schulungs- und Lehrmittel für den Gasschutz* (Berlin: Auergesellschaft Aktiengesellschaft, 1941), 13.
[148] Reichsluftschutzbund Unterrichts-Anweisung (TMA) II.Kleine 1906.
[149] W. Mieienz, "Gasschutz für Säuglinge und Kleinkinder," *Gasschutz und Luftschutz* no. 1 (January 1939).
[150] Karl Wollin und Pepp Seidl, "Gesichter und Gasmasken," *Die Gasmaske* no. 5 (September 1933).
[151] Judith Sumner, *Plants Go to War: A Botanical History of World War II* (Jefferson: McFarland & Company, 2019), 270.

chamber. The April 13, 1939 edition of the *Berliner Illustrierte Zeitung* enthusiastically cheered these technological solutions, publishing a cover photo with a mother and child in gas masks that triumphantly stated: "The Gas Mask for the German Child Is Here!"[152]

Beyond the fact that child gas masks were often as flimsy in construction as the adult *Volksgasmaske*, experts continued to doubt whether children could properly put them on. Their apparent lack of coordination and shorter attention spans worried both parents and air protection theorists alike. Even if children did manage to properly don their masks, experts assumed that they were liable to take them off at the first moment of discomfort. For this reason, gas mask education was intended to convey the seriousness of aero-chemical attack and to normalize the process of donning a mask.[153] One gas protection publication summed up this process, writing, "The child must be taught to overcome the fear of the gas mask."[154]

In an attempt to make gas masks more approachable for children in the 1930s and 1940s, some British and American gas masks were designed to loosely resemble the faces of animals or cartoon characters. German children, on the other hand, could only hope for a black leather *Volksgasmaske*.[155] RLB illustrations tried to make the utilitarian gas mask appear friendly and dependable through anthropomorphic cartoons, but German mask shortages and the inability of the already-strained manufacturers to even consider customization ensured that all German masks were essentially identical.[156]

At the same time, this uniformity served the purpose of entering German children into the national community of air and gas protection. Looking just like any adult, children wearing "People's Gas Masks" were expected to bear the same responsibilities in protecting Nazi Germany from aero-chemical attack. Air and gas protection courses through the Hitler Youth and the League of German Girls further attempted to provide children with the knowledge and skills necessary to serve as

[152] *Berliner Illustrierte Zeitung* (April 13, 1939): cover page.

[153] Hans Helbig and E. Sellien, *Der Luftschutz in Schulen und Hochschulen: eine Erläuterung der LDv 755/2 mit einer Sammlung der einschlägigen Gesetze, Verordnungen, Erlasse* (Berlin: Otto Stollberg, 1942), 13.

[154] Olmütz-Luttein, *Die Gasmaske : Auszüge aus den in den Gas- u. Luftschutzkursen in Olmütz-Luttein Vorgetragenen Vorträgen* (Olmütz-Luttein: Olmütz-Luttein Verlag Chema, 1936), 9.

[155] Gabriel Moshenka, "Gas Masks: Material Culture, Memory, and the Senses," *The Journal of the Royal Anthropological Institute* 16, no. 3 (2010): 616.

[156] Gasmaske Plakatt (IZM) PS 152.

auxiliaries for the RLB.[157] Thus, the idealized Nazi child was extremely familiar with the gas mask, trusting that it would help them perform their duties for the Reich. In this case, air and gas protection training produced at least some of its desired effect, and many children did, in fact, become accustomed to the mask. For instance, a report from a school in Munich claimed that the children were proud to bring their masks to school and that those who did not own one pleaded with their parents to buy them a mask.[158]

Of course, air and gas protection courses were not solely geared toward children. Three thousand four hundred permanent air and gas protection schools as well as local courses in first aid, firefighting, and gas detection provided similar instruction for adult citizens.[159] Still more of this training was conducted through RLB exhibitions or adult education centers.[160] Like German children, adults were expected to at least learn how to correctly don a gas mask and file into an air raid shelter. An SA air and gas protection training course claimed that:

> The goal of the training is to create feelings of safety and confidence in the mask. Further, the principles of gas combat allow civilians to master the ability to do any service under the mask for prolonged periods of time. Basic knowledge in the field of civil air defense is required.[161]

Both the RLB and the gas specialists insisted that chemical knowledge, alongside a collective spirit, would prevent communal panic in the event of an aero-chemical attack. Indeed, one gas training instructor wrote that the best protection against gas was "the disciplined behavior of the population and the most comprehensive explanation of the nature of aerial warfare in every school and house."[162] Thus, through the RLB, millions of civilians learned about the various types of poison gases and regularly practiced wearing their gas masks.[163]

German civilians could buy exactly one *Volksgasmaske* from distribution centers or at RLB exhibitions. Given that extra gas masks were rarely available due to production shortages, the RLB encouraged Germans to

[157] Michael H. Kater, *Hitler Youth* (Cambridge, MA: Harvard University Press, 2004), 176; "Annemarie under der Volksgasmaske," *Die Sirene* no. 19 (September 1938); Georg Erich Griesbach, *Lost wird Luftschutzhauswart* (Leipzig: A. Anton & Co, 1934).

[158] Bauer, *Fliegeralarm*, 14.

[159] R. J. Overy, *The Bombers and The Bombed: Allied Air War over Europe 1940–1945* (New York: Viking, 2013), 235.

[160] E. Meyer, E. Sellien, and Pol-Major Borowietz, *Schule und Luftschutz* (München: Verlag von R. Oldenbourg, 1934).

[161] Organisation (Luftschutz), 1934 (LBWK) 465 c Zugang 1991–1997, 14.

[162] Wilhelm Lichtenberg, *Luft- und Gasschutz-Fibel als Instruktion für die Hitler-Jugend und B.D.M.* (Berlin: Reichsjugendführung, 1934), 20.

[163] Vorschriften uber Gasschutz, 1935–1936 (BA-MA) RH/12/4.

carry their gas mask at all times.[164] To this end, the RLB sold a carrying case that could be worn as a satchel bag over the course of a day's regular activities. In the climate of aero-chemical fear that continually hung over the nation, Germans now lined up in droves to buy their own *Volksgasmaske*.[165] Reflecting on the popularity of masks in the late 1930s, the pacifist Waldus Nestler acerbically wrote:

> Thousands of fathers and mothers have bought this indispensable piece of equipment for young and old alike, and last Christmas the gas mask was successfully pitched as the gift of the year. Even the little marzipan angels and the Santa Clauses in the shop windows had gas masks and gas suits. It is understandable then that in those days, just about every second person was wearing a gas mask. So instead of asking about someone's health, they would ask "Is your gas mask in working order?"[166]

Much like the children at school, German adults would line up at a distribution center to have their face measured by a member of the RLB.[167] After receiving one of three sizes, they paid their five Reichsmark and took home the device that they were supposed to wear, in one way or another, at all moments of their waking lives. In this way, the gas mask became one of the omnipresent artifacts of the Third Reich, one that, according to the historian Peter Fritzsche, was most emblematic of "the new standards of cooperation, training, and expertise to which members of the people's community expected one another to adhere."[168] Alongside their new gas masks, Germans also received a user manual that described the best practices for mask care and use. In its dire tone, the pamphlet warned, "Take care of your gas mask even better than you take care of your clothes, because the gas mask could possibly save your life!"[169]

Aside from RLB and municipal service workers, only industrial workers might be able to procure more than one gas mask after 1937.[170] If considered essential to the Reich, industrial workers could receive an S-Mask that remained at their place of work.[171] Depending on the industry, factories might also provide rubberized gas suits and the necessary cleaning tools

[164] Reichsfinanz-ministerium, 1935 (BArch-B).
[165] C. E. Brigham, "Der chemischer Krieg der Zukunft," *Die Gasmaske* no. 6 (November/December 1937).
[166] Pehnke, *Der Hamburger Schulreformer Wilhelm Lamszus*, 131.
[167] Karl Wollin, "Gesichter und Gasmasken," *Die Gasmaske* no. 5 (September 1934).
[168] Fritzsche, *Life and Death in the Third Reich*, 52.
[169] Die deutsche Volksgasmaske (TMA) III.2 08754.
[170] Luftschütz, 1939–1945 (GStA) I HA Rep 178 B Nr 795.
[171] Industrial gas protection was directed by the *Werkluftschutz* (WLS), a subsection of the *Reichsluftschutzbund*. Each industrial plant in Germany was assigned a WLS warden, who was responsible for all aero-chemical defense training and adherence to national guidelines. *Gasschutz und Luftschutz* no. 1 (January 1935).

to detoxify the factory floor.[172] Not surprisingly, chemical factories took extra precautions against both aerial attack and accidental gas leaks. Aside from receiving personal protective equipment, workers at chemical factories under the IG Farben umbrella, for instance, were required to regularly participate in additional gas drills such as performing knee bends under the gas mask.[173]

Public air raid exercises among city-dwelling civilians involved less programmatic calisthenics, but they were not necessarily any less strenuous. Given that the Nazis endeavored to turn every German citizen into a member of an idealized national air raid community, people of all ages, genders, and physical constitutions were required to take part in aero-chemical defense exercises. At the same time, the RLB attempted to create a distinct hierarchy of civilian responsibility in which each German would help to protect themselves, their family, and the community at large. This organic model was best represented through gas mask protocol, which dictated that each person was to put on their own mask before they assisted their children and other family members.[174] After taking care of their family, citizens were then expected to help their friends and neighbors, especially the elderly and infirm. And all of the necessary training to create this hierarchy of responsibility would be done under the benevolent eyes of the RLB air wardens.[175]

As Germany rushed towards war in the late 1930s, increasing numbers of able-bodied German men either served in the expanded Wehrmacht, worked in what were deemed essential industries, or belonged to a municipal service such as the police corps or fire brigade. The preoccupation of so many men essentially forced the RLB to rely on women and teenagers to serve as local air wardens. The Third Reich's employment of women in air and gas protection services contradicted the Nazis' stated position on the role of women in society as well as much of the rhetoric that air and gas protection proponents had used to encourage national defense since the end of World War I.[176] From 1918 on, the gas specialists and other advocates of national defense had repeatedly asked Germans how women could possibly survive an aero-chemical attack. They assumed that the

[172] Werkluftschutz-Entgiftung, 1933–1941 (BASF) BASF PB/A.8.7.3./9 Pertinenzbestand.
[173] Direktions Abteilung Luft- und Gasschutz, 1934–1940 (BA) 329-0350-02.
[174] Möbius, "World War II Aerial Bombings of Germany," 6.
[175] Hugo Stoltzenberg, *Experimente und Demonstrationen zum Luftschutz* (Hamburg: Hugo Stoltzenberg Fabrik, 1933), 59.
[176] Broadly speaking, the Nazi Party believed that women's work should be restricted to the home. As mothers of the German race, they were essential to the rearing of Nazi children. While necessity required the acceptance of women in the air protection services, propaganda continued to present acceptable female roles. One RLB poster read "The woman in a soldier's post, but still a woman."

supposedly weak female constitution and the allegedly disordered female mind would never be able to effectively conduct air raid drills.[177] Women, they claimed, were not psychologically composed enough to protect themselves, let alone others. Thus, advocates of aero-chemical defense used the necessary protection of German women to drum up a highly gendered support for national defense measures (Figure 7.1).

In need of serviceable bodies, however, the RLB turned this logic on its head. With men serving in other state-required roles, approximately 280,000 women under the age of sixty now became the heroines of local air and gas protection.[178] This number made up about 34 percent of the RLB's officials, and in some localities such as Bad Lauterbach, women comprised up to 50 percent.[179] As block wardens, members of local fire brigades, and apartment building caretakers, these women helped to train the rest of the civilian population in air and gas protection practices.[180] The head of RLB propaganda, Otto Teetzman, correspondingly wrote, "[A] very important part of the most important self-protection tasks must rest on the shoulders of women. Accordingly, every German woman who is physically fit, must be incorporated and trained [in the RLB]."[181] While Teetzman's inclusion of women in the RLB was largely motivated by necessity, it also served the ideological purpose of integrating all Germans into the community of national defense. If gas bombs did not discriminate between genders, nor would the RLB.[182]

To justify the value of female air wardens, RLB officials and their allies now amassed new forms of propaganda. For instance, a publication from the Dräger company claimed that women were more psychologically apt to instinctively feel than to rationally think. For this reason, Dräger claimed, women asked fewer questions about the necessity of air protection and they learned to trust protective technologies more completely. Due to their inherent docility, the article asserted, they were even better at maintaining an even pattern of breathing while wearing the

[177] Peter Thompson, "Wardens of the Toxic World: German Women's Encounters with the Gas Mask, 1915–1945," *German Studies Review* 43, no. 2 (2020): 359.
[178] Jill Stephenson, *The Nazi Organization of Women* (Totowa, NJ: Croom Helm, 1981), 206.
[179] Nicole Kramer, "'Kämpfende Mütter,' und 'gefallene Heldinnen' – Frauen im Luftschutz," in *Deutschland im Luftkrieg* (München: Oldenbourg Verlag, 2007), 88–89.
[180] "Luftschutz-die Dienstpflicht der Frau," *Die Sirene* no. 11 (June 1935).
[181] Otto A. Teetzmann, *Der Luftschutz Leitfaden für alle* (Berlin: Verlag des Reichsluftschutzbundes, 1935), 94.
[182] Karl Hummel and Friderich Schweinsberg, *Erste Hilfe im täglichen Leben und im Luftschutz* (München: Dr. Friedrich Schwinsberg, 1936), 59.

Figure 7.1 A depiction of a male Red Cross member saving a defenseless woman from poison gas fumes. The image comes from the front cover of a 1930 edition of the *Auergesellschaft* trade journal, *Die Gasmaske*.
Source. Die Gasmaske no. 6 (1930)

mask.[183] Furthermore, RLB officials argued that mothers were especially capable in helping both children and the elderly. According to this logic, the firm and authoritarian mother could quickly direct her family into an air raid shelter while keeping them calm with her matronly presence.[184] On the most intimate level, the imagined German mother would provide the stoicism and diligence that the then president of the RLB, Hugo Grimme, deemed necessary for all civilians in the fight against aero-chemical threats.[185]

Collectivity and Responsibility: The National Community of the Air Raid Shelter

At the height of its power, the RLB coordinated public air raid exercises as part of both air defense exhibitions and city-wide preparations. Large-scale drills usually required preparatory planning so that citizens knew when to don gas masks, head to public air raid shelters, or turn out their house lights.[186] For instance, during a special 1933 air defense exhibition in Karlsruhe, marked planes from the local flying club flew over the city in six separate waves. Civilians were told that these planes dropped both explosives and gas bombs. Once the aircraft were detected from the ground, various preselected industrial firms set off their air raid sirens. Before filing into apartment-building basements, the citizens of Karlsruhe closed their windows and shop doors to protect against imaginary shrapnel and splinters. Police officers ushered people out of the street and into public shelters as the planes began to drop red smoke canisters down on the city. Even the hospital prepared for the drill, creating a separate ward for imaginary patients exposed to either phosgene or mustard gas. The hospital staff wore gas masks to protect themselves from secondary contamination while they provided chlorinated lime, warming blankets, and hot water bottles to the imagined gas victims (Figure 7.2).[187]

In urban areas such as Karlsruhe, air raid authorities designated large underground structures, such as subway stations, as public air raid shelters.[188] The RLB sometimes conducted their drills at night to

[183] Dräger, *Gasschutz im Luftschutz*, 252.
[184] Kramer, "'Kämpende Mütter' und 'gefallene Heldinnen'," 91.
[185] Hugo Grimme, *Der Reichsluftschutzbund: Ziele, Leistungen und Organisation* (Berlin: Junker und Dünnhaupt Verlag, 1936), 9.
[186] Luftschutzubungen mit Nebel, 1931–1936 (BA-MA) RH/12/4.
[187] Major a.D. von Laer, "Schutz der Heimat: Die erste Lufschutzübung der Stadt Karlsruhe," *Luftschutz-Rundschau* no. 6 (June 1933).
[188] These public shelters were meant to protect only those citizens who did not have access to their own shelter. There was a continual shortage of public shelters throughout the 1930s and, in 1939, the Reich calculated that it would cost 3 billion Reichsmark to

Figure 7.2 An air raid exhibition in Bremen. Emergency workers carry off supposedly injured children as smoke billows in the background. *Source. Dräger Gasschutz Im Luftschutz*, vol. 2 (Lübeck: Kommissionsverlag H.G. Rahtgens, 1933), 152

simulate an evening air raid. Darkness, of course, made it particularly difficult for German civilians to find their way to shelters or to descend into basements. At the same time, civilians also had to contend with the visual and auditory impairment of the gas mask. Just as soldiers in World War I struggled to see through the lenses and to hear the commands of their officers, civilians found it difficult to coordinate themselves under the rubberized mask.[189] Regardless of their level of familiarity with the mask, these sensory difficulties made wearing it extremely unpleasant. A young girl in Dresden described her experience with the gas mask, writing:

An air protection kit next to your bed – Quiet horror oozes from the gas masks nearby. Their handling had to be practiced regularly; they were drawn like bathing caps taut over the face, creating a feeling of suffocation and filling the nose with a penetrating rubber odor. The eyes sitting behind the glass eye pieces and the snout-like ventilation filter make people appear like monstrosities from

provide shelter for every German citizen. David Crew, *Imagining the Bombing of Germany, 1945 to the Present* (Ann Arbor: University of Michigan Press, 2017), 26.

[189] Heinz Denckler, ed., *Gasschutz und Gasmaske* (Berlin: Heinz Denckler Verlag, 1935), 10.

another world. We sat together silently because we could not speak, listening anxiously to the dull drones of the bomber squadrons above our heads, destined to drop their deadly load on German cities.[190]

German air protection authorities were well aware of the difficulties and fear associated with wearing a gas mask. They hoped that by demanding careful preparation and by providing step-by-step instructions for an aero-chemical attack, they could inspire civilians to effectively protect themselves and buy into the national endeavor. This kind of rote learning could not be accomplished through sporadic, or even regularly scheduled, large-scale air raid exercises. Thus, the apartment building or apartment block air raid drill became the primary site for molding civilians into armored combatants in the struggle for air and gas protection. Starting in 1933, the RLB attempted to assign a caretaker to every residential building in Germany.[191] The apartment caretakers and the local air wardens would then conduct regular air and gas exercises for their residents and neighbors. On the more intimate level of the *Blockgemeinschaft* (block community), wardens could teach civilians to prepare food and water in case of attack, to turn off lights and ovens once the air sirens blared, and to keep a well-serviced gas mask nearby at all times.[192]

During local air raid practice, air wardens also served to lead residents from their individual apartments to the basement of their building. Most apartment buildings constructed in the 1930s featured air raid shelters that maintained supplies and emergency equipment.[193] While the Nazi building authorities tried to retrofit older buildings with purposefully designed air raid shelters, the prohibitive cost of this construction forced most urban residents to rely on ground-floor storage rooms or cellars.[194] Even though the structural integrity and gas-proofing of many of these impromptu air raid shelters were dubious, they did provide the RLB with a means of collective mental comfort. While civilians might be lacking personal equipment such as a gas mask, they could still mentally rely on the perceived strength of both their building's air raid shelter and their local community. An RLB official described this psychological value, writing:

[190] Karin Husmann, Es bleibt unsagbar... Zwei Kinder erleben das Inferno Dresdens 1945 und die Kriegs- und Nachkriegsjahre, 1999 (DTA).

[191] There were over 800,000 building caretakers in the Third Reich. "Gasschutz tut not," *Neuer Vorwärts* no. 69, 3 (1934).

[192] Luftschütz in der Familie (TMA) II.Kleine 1907.

[193] Schutzraum (TMA) II.Kleine 1656.

[194] Schutzmassnahmen in Unterkunften, 1935–1936 (BA-MA) RH/12/4.

> The air raid shelters do not only protect the life of the civilian population and help to maintain the spiritual resilience of the homeland ... their presence is still one of the most important prerequisites for the material capabilities of the people.[195]

Once huddled together in an air raid shelter, civilians would be ostensibly surrounded by family members and neighbors. However, with gas masks affixed, these normally intimate companions could resemble a collection of militarized ghosts, automatons, or aliens. Like for soldiers in World War I, gas masks created a foreign lifeworld in which the familiar features of the human face were replaced by technologically altered forms. Ultimately, the RLB believed that civilians could and would learn to accept this fact, given that the gas mask might mean the difference between life and death. For civilians, one of the best means for normalizing this odd sensory experience was humor.[196] In passing jokes and newspaper comics from the 1930s, Germans regularly laughed about the fact that they resembled elephants or Carnival revelers. German women joked that the mask was the newest and most modern trend in fashion, while men quipped that at least their wives could not look any worse.[197]

Propaganda, Ideology, and the Attempted Creation of a Nazi Gas Protection Community

Propaganda and ideology were undoubtedly essential components of the RLB's attempt to ready Germany for aero-chemical war. In fact, a 1933 *Gasschutz und Luftschutz* article claimed that "propaganda and the press is the solid foundation for the entire air defense program. The propaganda must do the necessary work to create a willing und knowledgeable population."[198] Reflecting on the difficulties that the Weimar government faced in creating national air and gas protection services, RLB publicists recognized the tensions between realist preparation and apocalyptic predictions.[199] Many pacifists in the Weimar Republic wished to warn Germans of the dangers of aero-chemical warfare by illustrating the

[195] Hans Frommhold, *Luftschutzraum-Bauweisen* (Berlin: Verlag Gasschutz und Luftschutz, 1939), 5.

[196] Erhard Schütz, "Wahn-Europa: Mediale Gas-Luftkrieg Szenarien Der Zwischenkriegszeit," in *Krieg in Den Medien* (Amsterdam: Rodopi, 2005), 141. For examples of French gas mask humor, see Susan Grayzel, "'Macabre and Hilarious', The Emotional Life of the Civilian Gas Mask in France during and after the First World War," in *Total War. An Emotional History*, eds. Claire Langhammer, Lucy Noakes, and Claudia Siebrecht (Oxford: Oxford University Press, 2020), 58.

[197] *Die Sirene* no 26 (December 1938).

[198] "Propaganda und Presse," *Gasschutz und Luftschutz* no. 8 (1933).

[199] Bernhard Rieger, *Technology and the Culture of Modernity in Britain and Germany 1890–1945* (Cambridge: Cambridge University Press, 2005), 252.

horrors of a potentially violent future. But this tactic clearly backfired, encouraging the German public to demand greater national defense. Influenced by the writings of gas specialists, the RLB further realized that the technological solutions available for national defense could never provide complete protection. Gas masks and air raid shelters were reactive technologies, and they could not prevent bombers from flying over German cities and dropping explosives or gas, nor could they stop enemy scientists from developing new chemical weapons. Even more aggressive technologies such as antiaircraft guns or a strengthened Luftwaffe still could not guarantee national protection due to the potential for human error and gaps in any system of defense.

For these reasons, the RLB believed that, in coordination with specific protective technologies, an ideology of national collectivity was required to make Germans psychologically resistant to aero-chemical attacks. The rhetorical creation of a national air raid community, populated by *Volksgenossen* (national comrades), would inspire Germans to fully face the threat posed by aircraft and poison gas.[200] In this new world of imminent risk and contingency, civilians could never fully rely on government services or protective technologies to shield them from danger. As such, they would have to mobilize and educate themselves if they were to survive. As Ernst Jünger had prophesized in the early 1930s, the distinction between civilian and soldier was no longer viewed as tenable in a world defined by technologically amplified violence.[201] Just like soldiers in the Wehrmacht, members of this envisioned national air raid community (i.e., all German civilians) would have to accept the possibility of sacrificial death in order to ensure that Germany would survive.[202] The further intention behind militarizing the nation was to create a civilian form of "war fatalism" that would, according to the Nazis, prevent the home-front dissension of World War I and force Germans into a life-or-death struggle for national supremacy.[203]

Historian Peter Fritzsche has described this ideological campaign as a quest to create a German civilian who could exist between "the lords in the air and the victims of the lords of the air," the two sociological categories that dominated military thinking in the 1930s.[204] While ascribing to such

[200] Peter Fritzsche, "Nazi Modern," *Modernism/Modernity* 3, no. 1 (1996): 7.
[201] See also Erich Ludendorff, *Der totale Krieg* (München: Ludendorffs Verlag, 1935), 5.
[202] Michael Geyer, "There Is a Land Where Everything Is Pure: Its Name Is Land of Death: Some Observations on Catastrophic Nationalism," in *Sacrifice and National Belonging in Twentieth-Century Germany* (College Station: Texas A&M Press, 2002), 121–122.
[203] Fritzsche, "Machine Dreams," 705.
[204] Peter Fritzsche, *An Iron Wind: Europe Under Hitler* (New York: Basic Books, 2016), 42–43.

assumptions, RLB air protection propaganda also revealed its geopolitical beliefs by continuing the gas specialists' denunciation of international chemical weapons treaties.[205] These treaties, the RLB insisted, could not possibly protect Germans from future aerial warfare. In fact, they were Allied or Jewish schemes that aimed to weaken Germany's military defenses in preparation for the coming future war.[206] The RLB supported this claim by pointing to the continued development of chemical weapons and long-range bombers in Great Britain, France, the Soviet Union, and the United States.[207] In the broader Nazi conception of zero-sum geopolitics, where national struggle was an inevitable fact of life, arms production could only lead to eventual war. One RLB official reiterated this claim, writing:

> Air protection is necessary! In the deep sleep lies the German nation, probably dreaming of the everlasting peace. But outside with a busy hand, they continually forge new weapons. Ten thousand airmen are ready to start! Gas and bombs bring death. Wake up, you German people, already time is pressing! Wake up, my people, air defense is needed![208]

Under such stark geopolitical terms, the RLB felt that they could utilize their own form of carefully constructed alarmism. Thus, in their pamphlets and posters, they made it clear to the German people that the Americans and Russians were producing new and more deadly forms of gas.[209] Furthermore, the RLB routinely pointed out Germany's weak geographic position at the center of Europe, claiming that air raids could come from any cardinal direction.[210] Reflecting on the apparent perspicuity of such geopolitical visions, one air and gas protection enthusiast stated that:

> If one sees all around the huge preparations for a gas war, then the demand of Germany for equal rights to re-arm becomes ever more urgent, and the tasks of passive air protection for our defenseless country become fully comprehensible.[211]

[205] The gas specialist Hermann Büscher explicitly argued that pacifist calls for German disarmament were part of larger Jewish conspiracies and a culture of defeatism. Hermann Büscher, *Giftgas und wir?: Die Welt der Giftgase Wesen und Wirkung, Hilfe und Heilung* (Leipzig: Johann Ambrosius Barth, 1937), x.

[206] Müller-Kiel, *Die Chemische Waffe*, 5.

[207] Orts-Gruppe Heidenheim des Reichsluftschutzbundes, *Die Luftschutzgrundschule: Ein Laitfaden für die Schulung im Luftschutz* (Stuttgart: W. Kohlhammer Verlag, 1937), 1.

[208] Rosten, *Was man vom Luftschutz wissen muss*, frontispiece.

[209] "Russland," *Gasschutz und Luftschutz* no. 3 (1935); RH/12/9 Sig 70, 1934 (BA-MA).

[210] Josef Weber, *Das neue Handbuch für den Luftschutz: Ein Buch für Jedermann!* (Berlin: J. Weber Buch-Verlag, 1935), 10.

[211] Deutsche Luftschutz und Gasschutz Korrespondenz, 1933 (BArch-B) R/36/.

Further still, the creation of a sense of imminent danger afforded the Nazis with a justification for both their chemical weapons program and a reevaluation of the history of chemical warfare. According to this carefully crafted story, the Nazi production of poisons such as phosgene and mustard gas and the official creation of the Luftwaffe were merely assurances of peace.[212] The Nazis further claimed that Germans did not need to feel guilty for the deployment of poison gas in World War I. Riffing on the arguments of men like Fritz Haber, they assured their subjects that the French had first used gas grenades and that the Germans had always fought defensively.[213] This argument guided the publication of Rudolf Hanslian's well-received history of the 1915 German gas attacks at Ypres. Published in 1934, Hanslian's text claimed that the French were undoubtedly the first to use chemical weapons and that postwar attempts to ban Germany's chemical arsenal were merely Allied efforts to gain a future military advantage.[214] Popular historical accounts now treated the gas pioneers of World War I as national heroes and the anniversary of the Second Battle of Ypres was reinstated as a national holiday. On the twenty-fifth anniversary of the battle, both the Drägerwerks and the Auergesellschaft donated money to construct large monuments to fallen gas pioneers while the living veterans held a celebration in Berlin that ended with three "Sieg Heils" in praise of the Führer.[215]

While national rearmament may have eased some minds, the Nazis' imagined aero-chemical war still ran the risk of sending Germans into a panic. To counter this, RLB propaganda walked a fine line, relying on descriptions of collective resiliency to bolster the psychological resistance of German civilians. RLB literature frequently described the "iron will" of the German people and their undeniable desire to survive.[216] Hermann Göring claimed that "a people that carries the iron will of self-preservation, will successfully defy the dangers of the air attacks."[217] This vision of *Lebenswille* or "will to survive" created a level of rhetorical agency for the German people that interwar apocalyptic visions tended to lack. According to the Nazis and the RLB, the Germans could indeed survive the future gas war if they had the fortitude to resist its attacks.

[212] Ernst Ohlinger, *Bomben auf Kohlenstadt* (Oldenburg: Gerhard Stalling Verlag, 1935), 128.

[213] Deutscher Jägerbund, 1934 (TMA) I.4.040 08039; Reichsluftschutzbund betr. Luftschutz, Vortragsmaterial, 1944 (TMA) KE 1656.

[214] Gas und Nebel im Volkerrecht, 1933–1936 (BA-MA) RH/12/4.

[215] Rundschreiben an die Mitglieder der Kameradschaft der Offiziere der ehemaligan Gestruppen, 1940 (MPG).

[216] Walther Doering, *Luftschutz und Gasschutz: Markbuch für den Wehrfähigen* Vol. 3 (Breslau: Völkischer Verlag Walter Uttikal, 1934), 8.

[217] Weber, *Das neue Handbuch für den Luftschutz*, frontispiece.

Erich Hampe bolstered this idea in a 1937 air protection publication, writing:

> Never before in history has a danger so determined the future of a people. What will be decisive is how the people as a whole react to this danger. In this same way, every German can contribute to shaping the future of his people. He has the duty to do so before the judge of history.[218]

Here, Hampe argued that individuals certainly could protect themselves if they were able to bind themselves together in a community of responsibility. Fellow propagandists also used eschatological language such as the "judge of history" or the "will of God" to express the world-historic quality of the German struggle.[219] However, by positing God's will as open-ended or tied to the success of the nation, they expressed the earthly agency of Germany in shaping its own historical outcome. At the same time, the Germans were said to be entirely alone in this fight for survival, thus forcing them to rely solely on their national comrades. For instance, an article calling for diligent air and gas protection in *Die Gasmaske* quoted Johann Gottlieb Fichte in claiming that:

> It can be rigorously proven, and we will find this out in our own time, that no man, no god, and no event of any kind in the realm of luck can help us. We must alone help ourselves if we are to be helped.[220]

Following this same logic, the RLB frequently used the slogan "*Luftschutz ist Selbstschutz*" or "air protection is self-protection" to express the importance of personal responsibility in an effective national air and gas protection program.[221] Another popular motto, "*Ein Volk hilft sich selbst,*" or "a people helps itself" was emblazoned on most RLB publications and posters.[222]

While the RLB was willing to acknowledge that no technological solution could entirely protect the population, they and the larger Nazi party clearly did not reject the use of technology in the furtherance of national ideology.[223] In fact, a military report from 1938 claimed that chemical weapons technologies were inherently German in nature because they "corresponded to the special scientific-technical talent of

[218] Kurt Knipfer and Erich Hampe, "*Der Zivile Luftschutz*" *Ein Sammelwerk Über Alle Fragen Des Luftschutzes* (Berlin: Verlagsanstalt Otto Stollberg, 1937), 131.
[219] Friedrich Krüger, *Der Luftschutz für Jugend und Schule* (Leipzig: Verlag der Dürr'schen Buchhandlung, 1933), 28.
[220] Fritz Geisler, "Schutz der Zivilbevölkerung und die Abrustungskonferenz," *Die Gasmaske* no. 5/6 (October 1932).
[221] Luftschutz ist Selbstschutz Plakatt, 1939 (IZM) 11/ZGa 035.005.
[222] *Die Gasmaske*, 1937–1938 (SBB) 4 Ona 30/98-9/10.
[223] Rieger, *Technology and the Culture of Modernity in Britain and Germany*, 250.

the Germans."[224] According to such a view, the disciplined German constitution, coupled with technological augmentation, would allow Germans to live, if not thrive, in a world permeated by poison gases. Responding to interwar fears over the inherent uncontrollability of poison gas, the RLB insisted that it was the Germans who were destined to dominate a modern world soaked in toxic chemicals. IG Farben similarly averred that "chemical weapons are the weapons of superior intelligence and superior technical-scientific thinking. As such they are able to play a decisive role for Germany."[225]

This kind of chauvinistic rhetoric was essentially a return to the attempted creation of a "chemical subject" during World War I. But now this reawakened social engineering project would not be exclusively applied to soldiers. Under the direction of the Nazis and the RLB, all Germans would be recast as the masters of a new Nazi "chemical modernity." Men, women, children, and the elderly could become "Nazi chemical subjects," learning to live with the gas mask permanently affixed to their faces. For this reason, the *Kameradschaft* (community of comrades) of the World War I trenches was conceptually morphed into the *Volksgemeinschaft* (people's community) of the nation, and the technically exclusive S-Maske was similarly recast as the egalitarian *Volksgasmaske* (People's Gas Mask).

Indeed, the gas mask proved to be the ideal tool in the creation of this *Volksgemeinschaft* of chemically minded civilians. With the mask on, Germans visually resembled a new race of chemically immune cyborgs. Of course, the mask could always elicit fear but, according to insistent RLB propaganda, it provided a greater sense of comfort and heightened the human will to live. Germans in masks could thus tangibly see and feel the steps that their regime was taking to protect the population. Feelings of safety in a world defined by risk could presumably justify or obscure the authoritarian discipline that air and gas protection devices demanded. And the continually sensationalized reminder that enemy planes were waiting on the borders of Germany created the social myth that ultimately did encourage so many Germans to don the mask and join this imagined community of defense.

Furthermore, gas masks obscured the individual facial distinctions of the Germans who populated this community, thereby supposedly amplifying the sense of national collectivity and encouraging cooperation and compliance. In this sense, the mask came to serve as the new uniform of

[224] Ausnutzung der deutschen chemischen Industrie fur eine entscheidungsuchende Kriegfuhrung, 1938 (BArch-B) R/3113/.
[225] Müller, "World Power Status through the Use of Poison Gas?," 186.

Ernst Jünger's "totally mobilized" and "totally militarized" society.[226] By simply attempting to protect themselves, countless Germans entered into the Nazis' envisioned community of national defense. In the moment of pulling the mask over their heads, even those who did not join organizations such as the RLB or the Hitler Youth gave a certain level of implicit consent to the violent world of Nazi geopolitics.[227] In turn, the RLB could at least claim that this process of collective vestiary discipline, driven by anxieties over future chemical warfare, was entirely voluntary and there is little doubt that public anxieties led to a certain level of self-policing within German society. A citizen who failed to douse their lights during a mandatory blackout or refused to wear a gas mask at designated times could easily be marked as a traitor to the larger community. Thus, in 1935, the director of the RLB, Kurt Knipfer, was able to plausibly pronounce that, "the German people, with their *own* idealism and their *own* self-sacrifice, have done an outstanding job [in air and gas protection] through *voluntary* action."[228]

The Munich Crisis and the Stillborn Fate of the Nazi "Chemical Subject"

On the whole, Nazi preparations for aero-chemical warfare did, in fact, provide Germans with emergency plans more detailed than those of the French and English. While other western European nations had continued to expand both their air forces and chemical weapons programs throughout the 1920s and 1930s, they did less to formally prepare their civilian populations for aero-chemical attack. The French and British only began to provide civilian gas masks in the late 1930s, in response to increased Nazi aggression in central Europe.[229] However, once these two nations began to ready their citizens for the impending war, they sought more comprehensive technological solutions.[230]

[226] Moshenka, "Gas Masks," 611.

[227] Buck-Morss, "Aesthetics and Anaesthetics," 38–41.

[228] Kurt Knipfer, "Zum neuen Luftschutzgesetz," *Gasschutz und Luftschutz* no. 7 (1935) (emphasis added). International experts viewed German aero-chemical defense as the gold standard. In the late 1930s, commissions from nations like Great Britain and Japan traveled to Nazi Germany to examine their defensive measures. Germany agreed to this because it provided information on other national programs. Sheldon Garon, "On the Transnational Destruction of Cities: What Japan and the United States Learned from the Bombing of Britain and Germany in the Second World War," *Past & Present* 247, 1 (2020): 257.

[229] Fritzsche, "Machine Dreams," 705.

[230] Susan R. Grayzel, *The Age of the Gas Mask: How British Civilians Faced the Terrors of Total War* (Cambridge: Cambridge University Press, 2022), 89. See also: Jordan

In 1937, at the behest of Francisco Franco, the Nazi Luftwaffe and the Fascist Italian Aviazione Legionaria bombed civilians in the Basque town of Guernica. This notorious act was then followed in September 1938 by the Munich Crisis, in which the Nazis expressed their clear intention to invade Czechoslovakia.[231] In an oft-criticized historical moment, the British and French chose to appease Hitler by granting him the predominantly German-speaking parts of Czechoslovakia. This stopgap political solution was largely intended to avoid a major war, but it did not stop the British and French from recognizing the need to prepare for expected aero-chemical attacks. In fact, in this heightened moment of geopolitical anxiety, the British began distributing more than 35 million gas masks to civilians.[232]

Seeing the mask as a necessary evil in a Europe threatened by fascism, the British government soon succeeded in providing a mask for every English adult.[233] Tracking the usage of these masks in a large-scale market research project, the British government noticed that significant percentages of citizens carried their masks with them at all times during moments of political or military tensions, such as the aforementioned Munich Crisis or the evacuation of Dunkirk.[234] Like the Germans, the British certainly attempted to create a national civil air raid community based on the ideal of cooperative service. And in moments of real or perceived national danger, British air and gas protection directives could appear similarly authoritarian.[235] However, the British government never desired civilian resignation to the possibility of violence and death quite like the Nazis.[236] For this reason, British best practices during an aero-

Malfoy, "Britain Can Take It: Chemical Warfare and the Origins of Civil Defense in Great Britain, 1915–1945" (Dissertation, Florida International University, 2018).

231 German newspapers attempted to justify the Nazi invasion of Czechoslovakia by claiming that the Czechs had threatened ethnic Germans in the Sudetenland with gas attacks. Richard Evans, *The Third Reich in Power: 1933–1939* (New York: Penguin, 2005), 671.

232 Moshenka, "Gas Masks," 609.

233 Many British citizens received their mask on what the press called APR Sunday. Vans with loudspeakers rolled through London, encouraging civilians to be fitted for a gas mask. Brett Holman, *The Next War in the Air: Britain's Fear of the Bomber, 1908–1941* (London: Routledge, 2014), 207–208.

234 At any one time, the highest percentage of British citizens carrying gas masks in the street was about 75 percent. This number significantly fluctuated based on the sense of imminent danger across the nation, and by the end of the war, very few civilians still carried a gas mask. Moshenka, "Gas Masks," 624.

235 Lemke, *Luftschutz in Grossbritannien und Deutschland*, 91–95.

236 Furthermore, and perhaps even more importantly, the British national air raid community was not predicated on stark ethno-racial distinctions. Dietmar Süss, "Wartime Societies and Shelter Politics in National Socialist Germany and Britain,"

chemical attack remained a relatively open topic for experts throughout World War II.[237]

Historian Roxanne Panchasi has explored similar events in the French context. Panchasi claims that the gas mask served as both a literal and figurative barrier between the French body and the dangers of the chemically impregnated modern world.[238] However, the gas mask was not the only French defense against aero-chemical warfare. Fearing civilian panic during the Munich Crisis, France planned and implemented large population transfers away from its eastern border with Germany. While these transfers were not obligatory and male heads of households would ultimately decide if their family moved, mass evacuation remained a viable option in the event of a German gas bombing.[239] In Paris, too, as many as 3 million civilians fled the city during the Munich Crisis due, in part, to gas mask shortages.[240] Ultimately, these evacuation plans proved disadvantageous during the 1940 German invasion of France, but the willingness to move civilians rather than expose them to aerial bombardment represented a significant difference from Nazi air protection plans.

According to the Nazis and the RLB, the personal responsibilities and the supposed fortitude of the members of the militarized air raid community precluded Germans from leaving their city or town. Beyond the fact that this became a national directive, rhetorical claims insisted that there was no escape in a nation that was supposedly surrounded by enemy long-range bombers.[241] In many ways, the Munich Crisis proved to the Nazis that their methods of air protection were superior. Hearing of the desperate preparations of the French and British, the RLB appeared comparatively calm and collected.[242] The cloud of fear that hung over France and Britain was reported to be omnipresent and stifling, with British eyewitnesses noting the "anxiety, tension, uncertainty; throngs of people in the streets; the indignity and black humor of being fitted with gas-masks."[243]

in *Bombing, States and Peoples in Western Europe, 1940–1945* (London: Continuum, 2011), 43.

[237] Historian Susan Grayzel claims that an important distinction between the British and German air and gas protection practices was Britain's ability to "domesticate the air raid," rather than militarize the civilian. Susan R. Grayzel, *At Home and Under Fire: Air Raids and Culture in Britain from the Great War to the Blitz* (Cambridge: Cambridge University Press, 2012), 223.

[238] Roxanne Panchasi, *Future Tense: The Culture of Anticipation in France between the Wars* (Ithaca: Cornell University Press, 2009), 96.

[239] Torrie, *"For Their Own Good"*, 32.

[240] Fritzsche, *An Iron Wind*, 53.

[241] Torrie, *"For Their Own Good"*, 24–25.

[242] *Gasschutz und Luftschutz* no. 1 (1939).

[243] Michael Graham Fry, "The British Dominions and the Munich Crisis," in *The Munich Crisis, 1938 Prelude to World War II* (London: Frank Cass, 1999), 307.

Through the first week of August 1938, the imminent threat of aerochemical war continued to waft through the air and penetrate to the bones of British and French civilians.[244] Gas mask distribution centers stayed open at all hours, and citizens were instructed to stay near air raid shelters. In the national media, conjecture over war preparations spread like wildfire, leading one Briton to write that English civilians "had been told that the devastation of an air attack would be beyond all imagination. They had been led to expect civilian casualties on a colossal scale. They knew, in their hearts, that our military preparations were feeble and inadequate."[245]

In Germany, on the other hand, civilians repeatedly lauded the RLB and other air protection agencies from 1938 until late 1944.[246] This was particularly striking, given the Nazis' refusal to consider evacuation during these moments of heightened international tension. Furthermore, many German civilians (mostly in rural areas) still lacked gas masks and/or effective air raid shelters.[247] The price to produce gas masks for every German citizen remained too high and as both chemical weapons and gas mask technologies continued to evolve, civilians needed new gas masks and filters that the Reich could not possibly hope to provide.[248] Even if the RLB could somehow afford such an expense, the gas mask manufacturers did not have the production capacity to generate so many masks. It therefore seems clear that it was RLB propaganda and Nazi ideology that allowed for larger gaps and flaws in technological preparation. Their vision of a collective German air raid community populated by "Nazi chemical subjects" clearly did provide a certain level of determination and confidence. This collective resolve only grew in importance over the course of World War II as greater financial difficulties and reduced industrial capacity led to even greater shortages in gas mask production.[249]

Yet while the RLB publicly insisted that a collective resolve created efficient and effective air and gas protection over the course of World War II, daily German experience for both soldiers and civilians was a mixed

[244] William L. Shirer, *Berlin Diary: The Journal of a Foreign Correspondent, 1934–1941* (Baltimore: Johns Hopkins University Press, 2002), 129.

[245] Faber, *Munich, 1938*, 357.

[246] This public approval is impressive given the amount of leadership turnover in the RLB. RLB leadership was not considered a form of active duty and, as more generals were required for World War II, RLB presidents were shuffled off to artillery and Luftwaffe positions. Fritzsche, "Machine Dreams," 706.

[247] Between 1938 and 1940, the RLB continued in their quest to provide a gas mask for every German civilian. However, only 12 million masks were distributed outside of major industrial centers. Militar und Kriegssachen: Luftschutzmassnahmen in Heil- und Pflegeanstalten, 1933–1945 (SF) G 1215/3, 711.

[248] *Die Sirene* no. 16 (1942): 218.

[249] Knoll, *Die Drägerwerke im Dritten Reich*, 11–12.

bag. For the Wehrmacht soldier, the gas mask was an essential part of military training throughout the 1930s. Men jogged in gas masks, conducted target practice in gas masks, and sat in tear-gas-filled chambers for extended periods of time.[250] At the onset of World War II, Wehrmacht soldiers, or at least their commanders, were convinced that this training would be put to the test against the French and British.[251] Two years later, the German invasion of the Soviet Union gave this assumption even more credibility. A military report from 1941 claimed that "the gas mask is not an arbitrarily interchangeable piece of equipment, but it is rather inseparable from the soldier ... the fact that chemical weapons have not yet been used should not lead to laxity in this important training."[252] Even as late as 1944, the German military demanded that soldiers maintain gas discipline and train in their gas masks, despite the fact that chemical weapons had never been intentionally employed on the battlefields of World War II.[253]

Given both the discomfort of wearing the gas mask and the weight that it added to their packs, German soldiers increasingly resented these orders that required them to perpetually carry and train with the mask. Ultimately, many soldiers ended up throwing their gas masks away during highly mobile operations or desperate retreats.[254] According to one mildly amusing anecdote, a soldier received praised from his officers for always having his gas mask carrying case nearby. However, the officers did not know that this soldier had thrown out his gas mask months ago, now using his case to carry a camera.[255] In a war that increasingly appeared as if it would not feature chemical weapons, such stories became more common and soldiers simply no longer saw the value of the gas mask.

Gas discipline also frequently broke down among German civilians. Even in the mid- and late-1930s, plenty of civilians disobeyed air raid orders and failed to carry their gas masks in public. Like for many soldiers in World War I, the mask became a physical and mental burden to civilians. The carrying case constantly hugged the body, thus reminding

[250] Hedda Kalshoven, *Between Two Homelands: Letters across the Borders of Nazi Germany*, trans. Peter Fritzsche and Hester Velmans (Urbana: University of Illinois Press, 2014), 146.

[251] Daniel Uziel, *The Propaganda Warriors: The Wehrmacht and the Consolidation of the German Home Front* (Bern: Peter Lang AG, 2008), 212.

[252] Anhaltspunkte fur die Prufung des Standes der Ausbildung in der Gasabwehr bei Besichtigungen Ubungen usw, 1940 (BA-MA) RH/11/IV.

[253] RH/12/9 Sig 72, 1944 (BA-MA).

[254] Ben H. Shephard, *War in the Wild East: The German Army and Soviet Partisans* (Cambridge, MA: Harvard University Press, 2004), 146.

[255] Wolfgang Sannwald, *Schiefertafel, Gasmaske und Petticoat: Erlebte Dinge und Erinnerungen aus dem Landkreis Tübingen* (Gomaringen und Tübigen: Gomaringer Verlag und Schwäbisches Tagblatt, 1994), 78.

Germans of the imminence of aero-chemical attack and its violent disruption of daily life. The mask itself stifled breathing and smelled of synthetic rubber, seemingly choking out the life that it was meant to protect. Regular feelings of discomfort, apathy, and even indifference often meant that Germans would comply with the regime at one moment and break with its rules and regulations in the next.[256]

To combat air raid laxity, RLB propaganda claimed that poor individual gas discipline would both help the enemy and lead to a violent death.[257] They supplemented these claims with the threat of authoritarian punishment. Civilians could be fined or imprisoned for not adhering to air defense regulations.[258] Infractions included failing to turn out lights during a mandated blackout, not partaking in an air raid drill, and forgetting or refusing to carry one's gas mask.[259] These punitive measures were apparently enough to encourage most German civilians to passively follow air and gas protection guidelines, while others actively joined the RLB's idealized air protection community. By 1943, the RLB boasted 22 million members, or roughly 33 percent of the national population.[260]

The proclaimed successes of the RLB began to fade with the arrival of the bombing campaigns of World War II. As is commonly known, the Allies did not drop gas bombs, but instead released millions of tons of explosives and firebombs on German cities. The sheer destruction of these air raids forced the Nazis to consider, and eventually conduct, civilian evacuations from western cities.[261] Given the lack of initial planning, the idea of massive evacuations of a city the size of Cologne now appeared daunting if not impossible.[262] Daily and nightly life in many German cities increasingly featured near-constant air raid sirens and explosions, numbing many Germans to the seriousness of each individual moment of danger.[263]

256 This would suggest that, at any given moment, based on their ability or desire to follow the guidelines of the national air raid community, German civilians could alternate between complicity with, and opposition to, the regime. For this reason, it is useful to carefully consider the emotive categories of "apathy" and "indifference" in German relationships to the Nazi regime. John F. Sweets, *Choices in Vichy France: The French under Nazi Occupation* (New York: Oxford University Press, 1986), 147.

257 Ludwig Hohlwein, *Kannst du das Verantworten? Du hilfst dem Feind!*, 1940.

258 Overy, *The Bombing War*, 414.

259 Peter Fritzsche, *The Turbulent World of Franz Göll: An Ordinary Berliner Writes the Twentieth Century* (Cambridge, MA: Harvard University Press, 2011), 170.

260 Garon, "Defending Civilians against Aerial Bombardment."

261 By 1943, German civilians began to flee the cities in large numbers.

262 Eva Schwedheem, Kriegsjahre in Köln, 1943/44, 1943–1944 (DTA).

263 Unnamed Private Tagebucher, 1941, 1942, 1943 (TMA) II.Kleine 2090.

For many civilians, it was precisely the absence of gas bombs that challenged much of the RLB training that they had previously received. For instance, one civilian wrote:

> We were never told against what types of gases the masks would protect us and what we should do in the event of a gas attack. Could one smell the lethal gas? When would one put the device on and when would one take it off? Was there to a be a public warning like the siren indicating enemy aircraft approaching? The planning was sketchy at best and the saving grace was that, fortunately, there was never a need.[264]

By the early 1940s, air raid questions like these flooded the offices of *Die Sirene*, forcing the journal to redirect all inquiries to local RLB offices.[265] Regardless of the apparent confusion, the RLB continued to demand that civilians stay vigilant for gas attacks by carrying their gas masks and conducting gas drills.[266] But this demand became increasingly unreasonable as the Reich repeatedly failed to provide enough gas masks over the course of the war.[267] One anti-Nazi pamphlet from 1939 pointedly asked German citizens if they truly had enough gas masks for their families. Claiming that French and British citizens as well as Nazi government officials all received state-of-the-art gas masks, the pamphlet asserted that the *Volksgasmaske* would be "useless after a few weeks."[268]

To many Germans, the dream of an effective national gas protection community now appeared hollow. To counter this growing sentiment, the RLB directive of *Selbstschutz* (self-protection) was increasingly couched as a form of self-preservation rather than an act of individual heroism. At the same time, as part of the Nazi party's broader attempt to take on a more paternalistic role in assisting bombed-out civilians, party members took over many of the RLB's functions in air and gas protection.[269] While air defense certainly remained important through 1945, the social myth of collective national gas protection began to disappear. With attention and materials turned toward combatting the traditional bombing raids that took 600,000 German lives and destroyed 3.5 million homes, gas protection journals were left unread and unsupported.[270]

264 Knell, *To Destroy a City*, 289.
265 Carolyn Birdsall, *Nazi Soundscapes: Sound, Technology and Urban Space in Germany, 1933–1945* (Amsterdam: Amsterdam University Press, 2012), 121.
266 NS/6/ 1943 (BArch-B).
267 Luftschutz hier: Versorgung der Bevolkerung mit Gasmasken, 1944–1945 (SF) B 702/ 1, 4852.
268 Hast du eine Gasmaske Volksgenosse? 1939 (SBB) Einbl. 1939/45, 4422. H-J.
269 Möbius, "World War II Aerial Bombings of Germany," 15–16.
270 *Die Sirene* (1944); *Die Gasmaske* (October/December 1940). Only *Gasschutz und Luftschutz* remained in publication up to 1945, but in 1943, it shifted its sole focus to firebomb protection. In 1945, the editors stopped the regular publication of *Gasschutz*

In the final years of the war, aggressive Allied fire-bombing intensified the feelings of deadly irony that now tainted poison gas protection. Incendiary bombs set fire to large swaths of German cities, creating firestorms that could quickly suck the oxygen from the surrounding area. Because the *Volksgasmaske* was entirely ineffective without exterior oxygen and it provided no protection against carbon monoxide, death awaited even those few Germans who had time to don their masks during a fire bombing.[271] Author Jörg Friedrich estimates that 70 to 80 percent of all casualties in the bombings of Kassel and Hamburg were due to carbon monoxide inhalation.[272]

Formal German air protection services remained active until 1944 and the Nazis continued to pitch air raid activities as a form of national struggle with some success. But the bomber's ability to overwhelm German air and gas protection services forced many civilians to reconsider the supposed strength of their national air raid community.[273] With broken supply lines and inadequate protective technologies, some civilians now felt that they had been left to fend for themselves.[274] Reflecting on their individual fortitudes, Germans were again forced to consider if they could survive, let alone thrive, in this newly violent modernity. A similar reckoning is both predicted and given voice in Irmgard Keun's 1937 novel *After Midnight*:

Old Pütz is a pensioner, leading a quiet, peaceful life on his own. He has nicely brushed white hair and walks with neat, tottery little footsteps. Aunt Adelheid made him come to air raid drill. That day we had to put on gas masks, which practically smothered you, and then run up a staircase. Old Pütz stood in a dark corner, all shaky, holding the gas mask in his thin little hands and no doubt hoping nobody would notice him. But Aunt Adelheid's beady black eyes noticed him all right. He had to put his gas mask on, and Aunt Adelheid chased him up the staircase ahead of her. Up in the loft he collapsed. Putz's crumpled body lay there on the floor in his one good, dark blue Sunday suit, and we could hear him breathe stertorously inside his mask. Aunt Adelheid had put the mask on him wrong, and it was difficult getting his head out again.[275]

und Luftschutz. The journal was reintroduced in 1952 as *Ziviler Luftschutz*, now focusing on issues of nuclear protection. *Gasschutz und Luftschutz* no. 15 (1945).

[271] David Crew, *Bodies and Ruins: Imagining the Bombing of Germany, 1945 to the Present* (Ann Arbor: University of Michigan Press, 2017), 154.

[272] Friedrich, *The Fire*, 273, 331.

[273] Nazi Germany's civilian air defense was remarkably resilient until late in World War II. Lemke, *Luftschutz in Grossbritannien und Deutschland*, 209.

[274] Knell, *To Destroy a City*, 278.

[275] Irmgard Keun, *After Midnight*, trans. Anthea Bell (Brooklyn: Melville House, 2011), 10.

After seeing Old Pütz nearly die from exhaustion, Keun's young female narrator reflects further: "I mean, it's pure chance that poison gas isn't eating my body away right now. The Führer doesn't mind taking risks. He can say the word and declare war tomorrow and kill the lot of us. We're all in his hands."[276]

[276] Ibid., 32.

8 Prophets of Poison: Industrialized Murder in the Gas Chambers of the Holocaust

Tracing the various German visions of a distinct "chemical modernity" provides us with a unique cultural history of scientific and technological development. From 1915 to 1945, the German nation was inextricably tied to the advancement of modern militarized chemistry. Undoubtedly, this relationship was largely the result of the well-documented national strength in chemical research and production that Germany enjoyed in the late nineteenth and early twentieth centuries. Yet Fritz Haber's use of chlorine gas at the Second Battle of Ypres mobilized the full power of chemical firms such as Bayer, BASF, and Hoechst for imperial German military endeavors. In doing so, however, Haber failed to provide a breakthrough for the German army, and his chlorine gas instead created an arms race that would increase the lethality of both the gases themselves and the means of delivery.

Following World War I, foreign national chemical arsenals would haunt the German nation, making its citizens fully aware of both their military losses and their weakened geopolitical position throughout the 1920s and 1930s. Feelings of national weakness and uncertainty provided German ultra-nationalists and militarists with one of their central political messages. According to men such as Hitler and the other leaders of the National Socialist Party, the Germans had an undeniable right to national security in a world defined by imminent conflict. Chemical weapons, they claimed, represented German ingenuity and ruthlessness in this newly modern world where even the atmosphere could be immediately weaponized. In this way, a populist message of rearmament and the reassertion of national pride masked a willingness to play on the public fears of total chemical war. Regardless of whether the Nazis would ultimately elect to use poison gas on the battlefield, German chemical rearmament and defense was an assumed necessity for national survival. Ironically, while IG Farben produced nerve gases such as Tabun, Sarin, and Soman in the name of reassurance, the continued buildup of chemical weapons again contributed, if less directly this time, to a war that would ultimately bring about a German defeat in 1945.

With a ten-year head start in nerve gas production, Germany maintained stocks that far outnumbered those of the Allies during World War II.[1] Between 1942 and 1943 alone, the production site at Dyhernfurth produced enough Tabun to kill approximately 7.5 billion humans.[2] Hitler apparently knew about these stocks by December 2, 1941, at the latest, which has subsequently led historians and political scientists to try to explain why Hitler and the Nazis never chose to employ these weapons on the battlefield. Some have claimed that Hitler's reserve was related to his personal experience as a victim of gas in World War I. Serving as a message-runner for the 16th Bavarian Reserve Infantry Regiment, Hitler apparently witnessed the utterly demoralizing effect that gas had on German troops.[3] He also had first-hand experience with gas masks, since he spent February of 1918 testing old masks for three Reichsmark a day.[4] On October 14, 1918, Hitler himself was apparently exposed to mustard gas outside of Ypres, sending him to a military hospital in Pasewalk and effectively ending his military service. Recently, however, scholars have exposed possible embellishments in Hitler's personal story, with some claiming that he was only exposed to trace amounts of mustard gas and then suffered subsequent psychosomatic blindness.[5] But regardless of the medical specifics resulting from Hitler's gassing, the Pasewalk story would certainly support the idea that Hitler harbored a personal distaste for chemical weapons. Perhaps more importantly, it reveals the salience of gas in the decades following World War I. Indeed, Hitler's trial by gas became a highly symbolic story in Nazi Germany, and members of the Hitler Youth were forced to memorize this heroic part of the Hitler mythology.[6]

If Hitler did in fact become averse to poison gas in World War I, then this position was certainly not unique, and one can find denouncements of gas in ego documents from all segments of postwar German society.[7] This would suggest, as some political scientists have postulated, that chemical weapons might naturally disgust the human mind. As World

[1] The Nazis reached a monthly production capacity of 12,000 tons of all war gases: von Hippel, *The Chemical Age*, 232.

[2] Following the defeat of the Nazis in 1945, the Allies began the destruction of over 75,000 tons of German chemical weapons. 2 Weltkrieg Chemische Waffen, 1933–1945 (BA) 204-006.

[3] Thomas Weber, *Hitler's First War: Adolf Hitler, The Men of the List Regiment, and the First World War* (Oxford: Oxford University Press, 2010), 196.

[4] Ian Kershaw, *Hitler, 1889–1936: Hubris* (London: Allen Lane, 1998), 117.

[5] Alan Kramer, *Dynamic of Destruction: Culture and Mass Killing in the First World War* (Oxford: Oxford University Press, 2007), 324.

[6] Frederic C. Tubach, *German Voices: Memories of Life during Hitler's Third Reich* (Berkeley: University of California Press, 2011), 44–45.

[7] See Eva Schwedheem, Kriegsjahre in Köln, 1943/44 (DTA); Unnamed Private, Tagebucher, 1941, 1942, 1943 (TMA) II.Kleine 2090.

War I soldiers came to realize, the weapon destroyed the idea of fairness and chance in war, leading to regular descriptions of gas as perfidious, treacherous, barbarous, and inhumane.[8] Thus, such an argument maintains that humans are intrinsically appalled by gas' pervasive form of atmo-terrorism and the subsequent horrors of asphyxiation. However, this supposedly "natural" aversion did not stop German soldiers in World War I from cheering on their own chemical weapons when they terrorized the British and French. It also did not stop Hitler and the Wehrmacht from stockpiling nerve agents and mustard gas up until 1945.[9] This apparent willingness to deal in chemical weapons casts doubt on the claim that Germany's ultimate decision not to use gas in World War II was purely based on either moral considerations or a general aversion toward the weapon.

Political scientist Jeffrey Legro has stressed that it was the Wehrmacht that ultimately chose to forgo the use of gas. This was because the speed of *Blitzkrieg* tactics could best be achieved without a weapon that traditionally hampered mobility and delineated the battlefield.[10] Furthermore, as historian Florian Schmaltz has made clear, the Nazis actively feared the possibility that the Allies would gas both German soldiers and civilians as a reprisal for their own chemical attacks.[11] This threat of equal retribution was only effective because the Germans erroneously believed that the Americans and Russians had the ability to produce their own nerve agents early in the war. Trusting that the Allies now had the advantage in chemical weaponry, Hitler ordered an increase in nerve gas production in 1943 but also became increasingly hesitant to use his supposed super weapons.[12] The Nazi Minister of Armaments and War Production,

[8] James W. Hammond, *Poison Gas: The Myths versus Reality* (Westport, CT: Greenwood Press, 1999), x.

[9] In December 1941, Wehrmacht commanders seriously considered using mustard gas to take the city of Leningrad. They viewed gas as an ideal method for clearing urban spaces and demoralizing/killing Russian civilians. Adrian E. Wettstein, "Urban Warfare Doctrine on the Eastern Front," in *Nazi Policy on the Eastern Front, 1941: Total War, Genocide, and Racialization*, eds. Alex J. Kay, Jeff Rutherford, and David Stahel (Rochester: University of Rochester Press, 2012), 63–64.

[10] This argument does not address the fact that poison gas had been effectively integrated into mobile combat at the end of World War I. Jeffrey W. Legro, *Cooperation under Fire: Anglo-German Restraint during World War II* (Ithaca: Cornell University Press, 1995), 180.

[11] Churchill made it clear to the Nazis that if they used gas on the Russians, then the RAF would drop gas on German cities. Florian Schmaltz, *Kampfstoff-Forschung im Nationalsozialismus: Zur Kooperation von Kaiser-Weilhelm Instituten, Militär Und Industrie* (Göttingen: Wallstein Verlag, 2005), 25–28; Luftkrieg und Gaskrieg (SBB) Einbl. 1939/45, 4319.125; 125Vor; 125.2.Ex.

[12] The Nazis invested over 200 million Reichsmark in chemical weapons production over the course of World War II. IG AG Kampfstoffe P 70/I Diverse Ausarbeitungen zu

Albert Speer, claimed that Hitler considered using nerve gas against the Russians once the Wehrmacht was already in retreat, but by then, the hesitancy of Wehrmacht generals and the destruction of German chemical production sites made this plan infeasible.[13]

Of course, Hitler did order a scorched earth policy at the very end of the war with the intention of slowing down the American and Russian advances. It is quite curious that this directive did not include the use of chemical weapons, which would have been ideal for this purpose. It was for this reason that political scientist Richard Price added to any normative or rationalist claims about Nazi restraint by arguing that the Nazis' refusal to use chemical weapons was part of a shared "cultural taboo." Price writes:

> Chemical weapons appeared at a germinal moment that witnessed an emerging awareness that a civilization based upon progress in technology might be turned upon itself. This category of weapons was impregnated with institutionalized symbolic value. Chemical weapons portended that humankind might already have reached the moment of ultimate self-mastery implied by technology – the very capability to decide the fate of the species itself.[14]

Much like its successor, the atomic bomb, poison gas thus served as a "litmus test and symbol of unease" for humanity's ability to control its own destructive technologies in the twentieth century.[15] According to Price, the restraint that this symbol created was essentially irrational when weighed against the tactical value of gas on the battlefield. Nevertheless, a combination of irrational restraint, individual abhorrence, and rational calculations appear to have protected soldiers and civilians alike from an unrestrained chemical war in the 1940s.[16]

At the same time, however, none of these theories account for the fact that the Nazis did actually employ chemical weapons against former citizens and denizens of their expanded Third Reich. Price and others can attempt to explain the reticence to use chemical weapons under the

Kampfstoffen im Allgemeinen Dyhernfurth Gendorf finden sich in IG A 866/1 (BASF) IG A 866/1.

[13] Speer, *Inside the Third Reich*, 413.

[14] Richard Price, *The Chemical Weapons Taboo* (Ithaca: Cornell University Press, 1997), 166.

[15] Interwar anxieties over poison gas tend to resonate with subsequent atomic fears. In many ways, the discourses surrounding poison gas, especially cultural visions of aerochemical attack, were formative for the earliest imaginings of nuclear apocalypse. Peter Thompson, "From Gas Hysteria to Nuclear Fear: A Historical Synthesis of Chemical and Atomic Weapons," *Historical Studies in the Natural Sciences* 52, no. 2 (2022): 223–224.

[16] Richard Price, "A Genealogy of the Chemical Weapons Taboo," *International Organization* 49, no. 1 (Winter 1995): 102–103.

precepts of formalized warfare between sovereign nations, but these rational calculations and cultural taboos did not hinder the Nazi use of poison gas on relatively powerless internal populations. The mentally and physically disabled, Roma, Slavs, and Jews were never part of the imagined *Volksgemeinschaft* that defined German citizenship under the Nazis. For this reason, they were also excluded from the chemical protections that the RLB's national air raid community claimed to afford its citizens. This Social Darwinian and often ethno-racial demarcation between the people who would survive the envisioned aero-chemical war and those who were meant to perish was central to the Nazi conception of their desired "chemical subjects." Accordingly, as early as 1934, Jews and *Mischlinge* (people of mixed Jewish and Aryan ancestry) were barred from all state air protection activities.[17] Only Jewish World War I veterans could serve as air wardens for buildings inhabited solely by other Jews. The Nazis then formalized the separation of Jews in air and gas defense through a 1937 Air Defense Law and an additional decree in the following year.[18] In a rather illustrative scene from the summer of 1942, one Jewish man reportedly asked a German official where he could obtain gas masks for his family. Making it clear that the Nazi air raid community was sincere in its exclusions, the official responded, "Gas masks are surely pointless for Jews."[19]

Pesticides and the Metaphors of Disinfection

As cultural histories of pesticides have made clear, the processes of "disinfection" and "extermination" maintained strong associations with poison gas in the years following World War I.[20] Since 1917, Fritz Haber and his pest control units had used hydrocyanic acid to de-louse soldiers and trenches. This wartime experimentation led to the adoption of Zyklon B as one of the preferred German pesticide in the postwar years. Throughout the 1920s and 1930s, the German pesticide industry's research always remained tied to the greater desire for new chemical weapons applications. If men such as Haber could find civilian uses for poison gas, then this could potentially justify the development of chemical weapons as a mere stepping stone in the long history of scientific and technical progress.

[17] Torrie, *"For Their Own Good,"* 1939–1945, 129.
[18] Süss, *Death from the Skies*, 38–39.
[19] Marion A. Kaplan, *Between Dignity and Despair: Jewish Life in Nazi Germany* (New York: Oxford University Press, 1998), 224.
[20] Edmund P. Russell, "'Speaking of Annihilation': Mobilizing for War against Human and Insect Enemies, 1914–1945," *Journal of American History* 82, no. 4 (1996): 1509.

As advertisements for synthetic household pesticides began to flood daily newspapers and magazines, they pulled on the metaphors of human violence to plainly express their stated purpose.[21] Because vermin often carried disease, attacked pets and children, and were notoriously difficult to kill, pesticide companies saw value in explicitly describing or alluding to the violent effects of their products. For instance, during a 1920 "Rat week" in Steglitz, authorities used a pesticide called "Pogrom" in an attempt to rid the city of vermin. Couching pest control as a matter of existential conflict, another German company named their product *Rattenkrieg* (rat war).[22]

The ability of vermin to destroy crops and spread life-threatening diseases was increasingly pitched as a biological struggle that pitted humans against "lesser" members of the animal kingdom. In a public announcement from 1920, German health officials averred this pseudo-Darwinian struggle, writing, "the rat plague is a major threat to the health of the people (*das Volk*)."[23] When viewed as a fight for life itself, few Germans decried this new application for previously vilified chemical weapons. Clearly, any weapon could be used on what seemed to be an inherently lesser form of life, especially when it supposedly threatened the collective health of the German people.

Scholars have long been interested in tracing what constituted "lesser forms of life" in interwar Germany in order to understand the importance of Nazi conceptual ties between vermin and Jews leading up to, and during, the Holocaust.[24] In a process of rhetorical dehumanization, the Nazis infamously relied on descriptive terms and metaphors for swine, rodents, and/or insects to describe the Jews and other non-Aryan peoples in Germany.[25] Of course, these nonhuman descriptors for Jews (and others) both predated the Nazis' rise to power and spread beyond the ranks of the party, ultimately serving to normalize the assertion that Jews and Germans were biologically different and forever in competition. For instance, in decrying the historical influence of the Jews in Germany, the nineteenth-century German orientalist Paul de Lagarde wrote, "One does not have dealings with pests and parasites: one does not rear them and cherish them; one destroys them as speedily as possible."[26] Similarly,

[21] Bakterielle Ratten und Mause bekämpfung, 1926 (BArch-B) R/154/.
[22] Ratten IV, 1920– (BArch-B) R/154/. [23] Ibid.
[24] Alex Bein, *The Jewish Question: Biography of a World Problem*, trans. Harry Zohn (Cranbury, NJ: Associated University Press, 1990), 378.
[25] Andreas Musolff, "What Role Do Metaphors Play in Racial Prejudice? The Function of Antisemitic Imagery in Hitler's Mein Kampf," *Patterns of Prejudice* 41, no. 1 (2007): 25.
[26] von Hippel, *The Chemical Age*, 232.

in a 1925 letter, the former Kaiser Wilhelm II wrote, "Jews and mosquitos, a nuisance that humankind must get rid of some way or other."[27]

Throughout the interwar period, similar vermin metaphors proliferated.[28] Anti-Semites saw parasites as ideal parallels for the Jews, who they claimed to be siphoning off the resources of Germany at an accelerating pace. After the Nazi rise to power, these metaphors and analogies swiftly and violently entered mainstream political discourse. Historian Johann Chapoutot has studied this process, arguing, "What had been a virulent and hate-filled – but still largely metaphorical – discourse [on Jews as vermin] in the nineteenth century became a literal truth in Nazi Germany."[29] And with the increased use of infestation and parasitism metaphors came fittingly violent rhetorical solutions. For instance, an SS man who was tracking down Jews in 1935 wrote, "One does not hunt rats with a revolver, but with poison gas."[30]

By the 1930s, the conceptual link between Jews, vermin, and poison gas was evoked even more during the amplification of anti-Semitic violence. For example, as Jews were evicted from their homes in the late 1930s, it was not uncommon for the Nazi authorities to fumigate the confiscated Jewish residences before they were given to Aryan families.[31] The use of terms such as "fumigation," "disinfection," "cleansing," and "extermination" posited such discriminatory actions as the necessary elimination of the supposedly weak and parasitic in the Nazis' proclaimed struggle for life, space, and power.[32] Heinrich Himmler later claimed that:

> We are the first to have resolved the issue of blood with concrete actions … The same thing goes for anti-Semitism as for de-lousing. Destroying lice is not an ideological question. It is a matter of cleanliness. Anti-Semitism, in the same way, has not been a worldview, but an issue of hygiene – indeed, one that will soon be resolved. We will soon be rid of our lice … Then it will be over and done with, in all of Germany.[33] (Figure 8.1)

[27] David Reynolds, *The Long Shadow: The Great War and the Twentieth Century* (London: Simon & Schuster, 2013), 292.

[28] Boria Sax, *Animals in the Third Reich: Pets, Scapegoats, and the Holocaust* (New York: Continuum, 2000), 160.

[29] Johann Chapoutot, *The Law of Blood: Thinking and Acting as a Nazi*, trans. Miranda Richmond Mouillot (Cambridge, MA: Belknap Press, 2018), 400.

[30] Koonz, *The Nazi Conscience*, 246.

[31] Wildt, *Hitler's Volksgemeinschaft and the Dynamics of Racial Exclusion*, 110, 203.

[32] Raul Hilberg, *The Destruction of European Jews* (New Haven: Yale University Press, 2003), 1097.

[33] Chapoutot, *The Law of Blood*, 400.

Figure 8.1 A political cartoon from the December 1917 edition of the Nazi publication *Der Stürmer*. The title reads: "When the Vermin Are Dead, the German Oak Will Flourish Once More." The gassed rats are labeled as "stock exchange," "the press," and "trusts" while the branches of the German oak tree are labeled "industry, agriculture, commerce, the arts, business, the sciences, social welfare, civil service, and workers."

Source. Der Stürmer, December 1917. Courtesy of Randall Bytwerk, German Propaganda Archive, Calvin University

The Turn to Gassing: The Ritual Cleansing of the National Body

In October of 1939, the head of the Nazi Criminal Police, Arthur Nebe, asked his department's chemist, Albert Widmann, if they maintained the necessary amounts of poison gas "to kill animals in human form: that means the mentally ill, whom one can no longer describe as human and for whom no recovery is in sight."[34] This request was made in the initial stages of the T4 program, which, directed by the Criminal Technical Institute, had begun to euthanize mentally and physically disabled Germans in late 1938. Widmann's response was to send the mid-level SS functionary, August Becker, to the BASF factories in Ludwigshafen in order to procure cylinders of carbon monoxide. In early trials, the gas was channeled through metal pipes into rooms where the victims would suffocate.[35] The apparent success of these early tests led to the subsequent employment of more than one hundred technicians to operate six gassing installations located at Bernburg, Brandenburg, Grafeneck, Hadamar, Hartheim, and Sonnenstein.[36]

In 1941, the T4 Program was stopped due to public protest.[37] Friends and family members of the physically and mentally disabled demanded an end to the program, thus securing a circumscribed place in the *Volksgemeinschaft* for those deemed by Nazis to be unfit. This would suggest that the Nazi employment of poison gas was never entirely decided by biological categories of national belonging. And it appears that Nazi officials, much like Wehrmacht officers, recognized that poison gas could never be fully controlled and dispensed. Reflecting on the weapon's inherent caprice after visiting the scene of a 1936 underground railway fire in Berlin, Joseph Goebbels wrote, "A huge heap of ruins. One gets an impression from it of what a future fire, air, and gas war will look like. A gruesome vision!"[38] With such fears in mind, chemical weapons

[34] Michael Burleigh, *Death and Deliverance: "Euthanasia" in Germany, c.1900 to 1945* (Cambridge: Cambridge University Press, 1994), 119.

[35] Ulrike Winkler and Gerrit Hohendorf, "The Murder of Psychiatric Patients by the SS and the Wehrmacht in Poland and the Soviet Union, Especially in Mogilev, 1929–1945," in *Mass Violence in Nazi-Occupied Europe*, eds. Alex J. Kay and Stahel David (Bloomington: Indiana University Press, 2018), 157.

[36] These installations and the chemists/physicians who ran them were responsible for the deaths of 70,273 people. Dagmar Herzog, *Unlearning Eugenics: Sexuality, Reproduction, and Disability in Post-Nazi Europe* (Madison: University of Wisconsin Press, 2018), 44.

[37] Götz Aly and Susanne Heim, *Architects of Annihilation: Auschwitz and the Logic of Destruction* (Princeton: Princeton University Press, 2015), 164.

[38] Toby Thacker, *Joseph Goebbels: Life and Death* (New York: Palgrave Macmillan, 2009), 196.

could not simply be levied against either the external or internal enemies of the Reich without a certain level of technological protection and assumed risk. Thus, for many Nazis, poison gas served not as a true wonder weapon but rather as a form of ritual cleansing that could determine the survivors and masters of a chemically impregnated world.

In this formulation, reminiscent of the post-World War I works of Ernst Jünger, all Germans would inevitably have to pass through poison gas in order to prove their merit as supposedly technologically superior people. This vision fit with the RLB's insistence that Germans must avoid evacuation in any future war, instead remaining in their hometowns and fortifying themselves for aero-chemical attack. It would further explain the Nazis' willingness to apply gas to particular internal populations, namely those who had no hope of real chemical protection. Historian Paul Weindling previously suggested this highly symbolic interpretation, writing that "Immersion in a lethal gas was a baptism of destruction and cleansing – a macabre rite of purification in the name of the extinction of parasites and the conservation of resources."[39]

The full extent of this ideological construction was perhaps best revealed in the popular 1933 Nazi film, *Hitler Youth Quex*. The film's plot followed a family in which the communist father habitually beat the mother. In an effort to escape this seemingly hopeless situation, the mother attempted to commit suicide with her son by opening the stove gas line. The mother dies, but the son survives and ultimately goes on to join the Hitler Youth and to sacrifice himself in the battles for Nazi political control. Sacrifice is the major recurrent theme in the film, and it is only through the boy's survival of gas that he can become a martyr for the Nazi cause.[40] Thus, to the Nazi mind, poison gas could serve as a technological test in which truly righteous and strong Germans could forge a path to national rebirth. If constructed correctly, the weak and corrupt would surely perish.

On December 8, 1941, the SS first used three hermetically sealed vans to gas local Jews at Chełmno.[41] As the war in the East became increasingly predicated on mobility, the SS expanded what they now saw as an efficient method for murdering hundreds of thousands of Jews,

[39] Weindling, *Epidemics and Genocide in Eastern Europe*, 91.

[40] Jay W. Baird, *To Die for Germany: Heroes in the Nazi Pantheon* (Bloomington: Indiana University Press, 1990), 123.

[41] The possibility of using gas on Jews in the East was first discussed by Heinrich Himmler and Odilo Globočnik on October 13, 1941. During that same autumn, gas trucks were tested on Soviet POWs at the Sachsenhausen concentration camp. Beth A. Griech-Polelle, *Anti-Semitism and the Holocaust: Language, Rhetoric and the Traditions of Hatred* (London: Bloomsbury, 2017), 184.

Roma, and mentally ill people on the Russian front.[42] In 1942, the SS erected more permanent versions of these carbon monoxide chambers at the Belzec, Sobibor, Maidanek, and Treblinka extermination camps. Many of the same SS technicians employed in the T4 Program were now sent to Poland in order to help run these top-secret installations.[43] There, prisoners were told to enter large shower rooms so that they could be "disinfected." Once the victims were inside, camp guards would seal the room and use large diesel engines to pump in carbon monoxide gas. These early gas chambers could kill and cremate up to 150 people at a time.[44]

Relying still more on the technical knowledge of pest control, the SS designed their early gas installations based on Degesch delousing chambers.[45] However, these carbon monoxide gassing operations were not always as seamless and industrialized as the SS had hoped. Unreliable engines struggled to produce the gas and the technicians often failed to sufficiently seal the chambers. According to postwar testimony, these faulty installations could sometimes take up to half an hour to kill their victims. As early as October 1941, SS technicians were instructed to find a more efficient method for the mass gassing of Soviet prisoners and Jews. Accordingly, they began testing Zyklon B at the Sachsenhausen and Birkenau concentration camps, relying on the expertise of Bruno Tesch and Gerhard Peters, two chemists involved in the production and sale of the pesticide.[46] In early experiments, Zyklon B demonstrated the ability to kill large numbers of humans at a faster pace and lower cost.[47]

In 1942, SS engineers at Auschwitz began constructing 10 cubic meter gas chambers with built-in ventilation systems and chutes for the introduction of Zyklon B pellets.[48] While these gassing chambers remained fairly crude in design, the use of hydrocyanic acid for genocidal killing has remained the central image of the Nazis' attempt to standardize,

[42] Saul Friedlander, *The Years of Extermination: Nazi Germany and the Jews, 1939–1945* (New York: HarperCollins, 2007), 16.

[43] Nicholas Stargardt, *The German War: A Nation under Arms, 1939–45* (New York: Basic Books, 2015), 252.

[44] Henry Friedlander, *The Origins of Nazi Genocide: From Euthanasia to the Final Solution* (Chapel Hill: University of North Carolina Press, 1995), 96–97.

[45] Michael Thad Allen, "The Devil in the Details: The Gas Chambers of Birkenau, October 1941," *Holocaust and Genocide Studies* 16, no. 2 (2002): 193.

[46] Now under 42 percent IG Farben control, the Degesch subsidiary received increased military contracts between 1939 and 1945, as Zyklon B was used to delouse military quarters, uniforms, and supplies. Between 1943 and 1944, Degesch sold 3.6 million Reichsmark worth of fumigants, mostly to the SS. Hayes, "The Chemistry of Business-State Relations in the Third Reich," 73.

[47] Weindling, "The Uses and Abuses of Biological Technologies," 296.

[48] Allen, "The Devil in the Details," 194.

mechanize, and even institutionalize murder. Once victims were ushered inside these so-called disinfection chambers and the gas was released, suffocation would usually occur within twenty minutes.[49] Scholars estimate that, once the gas chambers were fully operational, several thousand people could be murdered in a single day.[50]

Initial Interpretations and Final Consequences

Since the end of World War II, the Nazi use of chemical weapons in the Holocaust has encouraged scholars to evaluate the role of technocratic thinking in the engineering of genocide.[51] While studies of the intimate killing in Eastern Europe pose important questions about the specific motives for violence in a comparatively short historical period, analyses of Nazi industrialized murder can raise corollary questions about the possible connections between violence and science/technology throughout the modern era. Indeed, some scholars investigating this second line of inquiry have suggested that an overwhelming excitement for cutting-edge scientific research and/or a belief that all science eventually leads to human progress ultimately stifled the Nazi technocrat/technician's ability to empathize with human suffering and pain.[52]

Since Fritz Haber's early chlorine experiments, German gas scientists repeatedly justified poison gas research and production with utopian visions of peace and agnostic approaches to scientific ethics. As late as 1941, the German chemist Curt Wachtel defended nationalized poison gas research to an American audience in a book entitled *Chemical Warfare*. In it, he wrote, "A science guilty of creating the means of immeasurable destruction and human suffering may be considered more than fully exculpated through the great contributions to public hygiene and welfare which resulted from the same research on war gases."[53] Wachtel continued, "science itself is neutral; it equally serves the aims of the good and the evil. The scientist is devoted to science. The people and government determine whether science should be used for destruction, defense, or construction."[54] Plenty more interwar German chemists

[49] N/2503/ (BArch-B).
[50] Naomi Baumslag, *Murderous Medicine: Nazi Doctors, Human Experimentation, and Typhus* (Westport, CT: Praeger, 2005), 62.
[51] At the same time, it is important to acknowledge the fact that most genocidal violence in the Holocaust was not conducted in the gas chambers. Instead, it was intimate violence enacted at close quarters. Donald Bloxham, *The Final Solution: A Genocide* (Oxford: Oxford University Press, 2009), 29.
[52] Szöllösi-Janze, "The Scientist as Expert," 19.
[53] Wachtel, *Chemical Warfare*, 12.
[54] Ibid., 44.

and engineers would echo Wachtel's claims, arguing that the creation of poison gas was not inherently immoral and that the only ethical question was how Germany would ultimately employ their new weapons. As historian Peter Kenez has revealed, this widespread scientific position proved similarly attractive to Nazi technocrats and technicians. Kenez writes:

> The use of gas for killing people was not a crossing of yet another moral boundary, it was simply a change of method, and from the Nazi point of view a more efficient way of carrying out their assigned task. The Nazis were experimenting: how to kill as many people as possible and as quickly as possible.[55]

Undeniably, poison gas could actually appear rather humane to the SS Einsatzkommando who had witnessed numerous shootings and mass starvation in the East.[56] Even the Auschwitz commandant Rudolf Höss found poison gas to be psychologically relieving, later writing: "I was always appalled by shootings, particularly when I thought of the women and children ... Now I was relieved that we were going to be spared these bloodbaths."[57]

In Höss' writings we can clearly see the danger of justifying the development of technologies such as poison gas based on metrics of scientific efficiency. This line of reasoning ignores the fact that poison gas was intentionally developed, from its conception, to elicit human pain and suffering. In this case, the authorial intentions behind the technological creation and use of poison gas were not particularly indeterminate or contingent. The voices of the soldiers in the trenches of World War I made the physical and psychological destruction of gas appallingly clear. Furthermore, claims about the implicit neutrality of such objects of science and technology repeatedly relied on a narrow conception of the term "use." Technologies such as poison gas were (and are) not merely tools, equal in ethical standing with a hammer or chisel. Rather, in its development, poison gas was fully integrated into the moral considerations surrounding warfare and violence.[58] Thus, Germany's development and use of poison gas during World War I fashioned a political, social, and cultural climate in which the atmospheric extermination of human life became an

[55] Peter Kenez, *The Coming of the Holocaust: From Antisemitism to Genocide* (New York: Cambridge University Press, 2013), 264.

[56] Thomas Kühne, *Belonging and Genocide: Hitler's Community, 1918–1945* (New Haven: Yale University Press, 2010), 84.

[57] Michael Thad Allen, "Introduction: A Bureaucratic Holocaust – Toward a New Consensus," in *Networks of Nazi Persecution: Bureaucracy, Business, and the Organization of the Holocaust*, eds. Gerald D. Feldman and Wolfgang Seibel (New York: Berghahn, 2005), 264.

[58] Winner, *The Whale and the Reactor*, 12.

ever-present possibility. Roughly twenty years later, in the hands of men driven toward technical precision in the implementation of genocide, these technologies again expressed their fully destructive capabilities and assisted in a political and ethical system that championed quick and sterile death above all else.[59]

While men such as the gas specialists certainly did not work to engineer the Holocaust, they did promote the companies such as Degesch, Dräger, and Auer that provided the technical means necessary to create the gas chambers.[60] Furthermore, they reified the envisioned future gas war and supported a chemical arms race that naturally fed into Nazi Germany's national, ethnic, and racial antagonisms.[61] Without drawing direct lines of historical causation, this should encourage us to evaluate the overlap in values between Nazi idealogues, ancillary scientists and engineers such as the gas specialists, and German civilians at large. Historian Michael Thad Allen has already shown the ways in which the technocratic desire to "modernize" the Third Reich was very much tied to more troubling Nazi ideological convictions.[62] For many German scientists and engineers, including many of the gas specialists, Germanic science and technology was a product of an essentialized national or racial identity. Poison gas, these men then claimed, was an inherently German weapon that relied on the powers of rational Germanic thinking and thus, operated in the favor of the German nation.[63]

Such scientific and technological chauvinism was certainly part of the larger Nazi attempt to recast human history along the lines of national, ethnic, and racial struggle. As the Nazis attempted to carve out greater privileges for the Germans in a modern world mediated by increasingly deadly technologies, they recognized the political need to assert German superiority and inevitable victory. This was the prevailing logic behind the attempted creation of a "Nazi chemical subject" through RLB propaganda. However, poison gas, a weapons technology that redefined the concept of danger in the modern world, was inherently slippery.[64] World

[59] Katz, "On the Neutrality of Technology," 420.

[60] Jean Claude Pressac, "Engineering Mass Murder at Auschwitz," in *Death by Design: Science, Technology, and Engineering in Nazi Germany*, eds. Eric Katz (New York: Pearson, 2006), 46.

[61] Geyer, "How the Germans Learned to Wage War," 48.

[62] Michael Thad Allen, *The Business of Genocide: The SS, Slave Labor, and the Concentration Camps* (Chapel Hill: University of North Carolina Press, 2002), 32.

[63] Russell, "'Speaking of Annihilation'," 1519.

[64] A sense of control and advantage was clearly central to the employment of chemical weapons during and after World War I, particularly in colonial contexts. Michelle Bentley, *Syria and the Chemical Weapons Taboo: Exploiting the Forbidden* (Manchester: Manchester University Press, 2016), 3.

War I had already revealed the difficulties involved in safely harnessing the destructive powers of poison gas. Once released into the air, gas had no master, and it could move in unpredictable ways. Thus, it was immensely difficult for the Nazis to convince the German people, let alone themselves, that complete national control was possible.

For this reason, it is interesting that most Nazi leaders, including Adolf Hitler, rarely spoke of poison gas in their public addresses. Hermann Göring's March 3, 1933 speech at a Nazi rally in Frankfurt was one of the few to be soaked in the language of gas.[65] Charging the communists with poisoning the German nation, Göring encouraged his supporters to:

> Save the Germans in this bitter struggle with the eternally destructive poisons of Marxism, to make the German a valuable member of his nation again, to plant again the will to resist the tiring and lethargic suffocation of Marxism.[66]

For technophiles such as Göring and the other leaders of the RLB, the procurement of protective technologies such as the gas mask would allow the Germans to dominate this chemically precarious world. Playing on the widespread fears of aero-chemical attack, the RLB's search for an effective civilian gas mask was not a humanitarian response to the evils of chemical warfare but rather a technical quest for survival and mastery in the poisoned future that they imagined. Visions of terror and protection were continually intertwined, feeding into each other and creating a political arena in which ideological claims of chemical superiority were the most compelling forms of assurance.

While Göring was clearly willing to assert German ascendency over poisonous chemicals for political gain, other Nazi leaders were not so openly confident. Their silence on the matter can be read as a particular attention to the possibility that chemical weapons might slip from German control. For this reason, the nonpublicized use of gas in the Holocaust can be read as part of the Nazis' deference to this seemingly arcane technology.[67] By offering selected victims to the caprices of poison gas and industrialized violence, the Nazis used the gas chambers as a form of controlled reconciliation with a dangerous and unknown chemical future. As such, the gas chambers were the shadow side of the gas mask, ensuring chemical death for those who maintained no technological protection.

[65] See also Hitler's address to the German people on September 1, 1939.

[66] Hermann Göring, *Hermann Göring: Reden und Aufsätze* (München: Zentralverlag der NSDAP, 1942), 21–22.

[67] Kaiser, "Wie die Kultur einbrach," 214.

Conclusion

While 1945 clearly demarcates an important curtailment of Germany's unique ties to chemical weapons, it certainly did not signify the nation's final confrontation with harmful chemicals or even chemical weapons. The Germans remained within a broader "Chemical Modernity" into which the world had been increasingly plunged since the first industrial production of chemical gases. Indeed, in this longue durée formulation of a Chemical Modernity, it is important to note the extent to which Europeans struggled with environmental concerns such as industrial smoke abatement and arsenic dye poisoning in the nineteenth century.[1] But it was the chemical weapons of twentieth-century atmospheric terrorism that made such concerns far more apparent and pressing. The eminent historian Eric Hobsbawm spoke to this historical shift in consciousness when he remembered that "The image of fleets of airplanes dropping bombs on cities and of nightmare figures in gas masks, tapping their way like blind people through the fog of poison gas, haunted my generation."[2]

However, this distinct vision of aero-chemical gassing never materialized in World War II. At the same time, the incendiary bombs that the Americans dropped on both Germany and Japan had a closely related chemical history, tied again to the German war effort in World War I. Even when the remnants of this bombing were finally cleared, Germans then had to politically grapple with both the atomic bomb and napalm, still newer weapons that often conceptually resembled poison gas' method of aerial atmo-terrorism. As the postwar world continued to unfold, Germans on both sides of the Iron Curtain uniquely felt the

[1] Jean-Baptiste Fressoz, "Beck Back in the 19th Century: Towards a Genealogy of Risk Society," *History and Technology* 23, no. 4 (2007), 343.

[2] Eric Hobsbawm, *The Age of Extremes: A History of the World, 1914–1991* (New York: Pantheon Books, 1994), 35.

immediacy of the nuclear threat.[3] Situated in a potential ground-zero for an atomic war between the United States and the Soviet Union, Germans now frequently referred to their historical experience with chemical and incendiary weapons to decry the new geopolitical situation. In open-ended American opinion polls, for instance, West Germans claimed that if nuclear weapons were used, "Germany would become a desert. Everything would be lost," and that "any future war will be even more cruel."[4] Several respondents further noted their fears of radioactive dust, citing their "everlasting memories of the last war."[5]

In these moments of renewed German fear over atmospheric weapons, several distinct remnants of the interwar world resurfaced. The journal *Gasschutz und Luftschutz* rebranded itself as *Ziviler Luftschutz*, remaining under the direction of Rudolf Hanslian and Heinrich Paetsch until 1952.[6] Focusing now on atomic warfare, the journal did not appear substantially different in tone or content and it continued to encourage West Germans to create communities of defense that sounded quite similar to the ideal collectives of the *Reichsluftschutzbund*.[7] Atomic warfare drills such as "duck and cover" revealed that both atomic and chemical warfare required a routinized process of mass psychological comfort through discipline and training. At the least, nuclear protection experts argued that such civil defense measures provided people with a sense of preparation and security. Similar rhetoric, reminiscent of the earlier German gas specialists, led to the subsequent translation of the interwar works of Erich Hampe, who became an international figure in civil defense during the 1950s.

The American use of napalm similarly concerned the many West Germans who protested the Vietnam War in tandem with the nuclear arms race. Utilizing their own national memory of World War II fire bombings, protestors pointed to the unique cruelty of incendiary chemicals. Over the next four decades, napalm, nuclear weapons, and nuclear power served as some of the central political foci for lingering German concerns over atmospheric destruction. However, in 1990, a more tangible connection to the history of chemical weapons was unearthed when a geophysicist

[3] By May 1955, the Americans had placed their first of many atomic weapons in West Germany, soon to be the most densely nuclearized country in the world. Jeffrey Boutwell, *The German Nuclear Dilemma* (Ithaca: Cornell University Press, 1990), 14.

[4] Hans Speier, *German Rearmament and the Atomic War: The Views of German Military and Political Leaders* (Evanston, IL: Row, Peterson and Company, 1957), 249.

[5] Report no. 216 "German Reactions to Atomic Weapons" (UIUC).

[6] *Ziviler Luftschutz*, 16 (1), 1952.

[7] Edward M. Geist, *Armageddon Insurance: Civil Defense in the United States and Soviet Union, 1945–1991* (Chapel Hill: University of North Carolina Press, 2019), 154.

surveying the Elbe River found large magnetic anomalies near Hamburg. This led to the discovery of two large barges that had been sunk by Hugo Stoltzenberg during the interwar "removal" of German chemical weapons. These barges contained compromised barrels of phosgene and mustard gas, which had been slowly seeping into the river over the past sixty years.[8] The discovery led to a careful survey of all previous Stoltzenberg chemical sites, the chemical remediation of the soil, and the restriction of fishing in certain areas.[9] In sum, in the wake of World War II, many Germans became increasingly aware and skeptical of toxic chemicals and airborne inhalants, or what the sociologist Soraya Boudia has usefully termed the "residues" of modern industrial chemistry.[10]

Certainly, relatively few fully question or reject the chemistry behind industries such as pharmaceuticals, but the days of a broad societal consensus on chemistry's naturally assumed progress appear numbered. Over the first half of the twentieth century, chemical weapons had proved to many Germans that their potential destruction of both human bodies and the environment outweighed any potential benefits. This stimulated a new awareness of the chemical construction of the environment and its many potential threats in Germany and around the world. For the most perceptive witnesses, the experience of total war and its environmentally disastrous aftermath began to reveal humanity's limits in chemically manipulating its environment.[11]

These brief examples of postwar German chemical encounters are meant to remind us that the story of German chemical weapons between 1915 and 1945 sits within a much longer historical context that continues up to the present day. Since World War II, chemical weapons have been used in numerous contexts, particularly in Middle Eastern military conflicts, and tear gas remains one of the predominant methods of crowd control throughout the world.[12] The example of tear gas points to the ways in which the definition of chemical weapons remains fuzzy and

[8] Sinking drums in the North Sea, Baltic Sea, or Atlantic Ocean had long been the accepted means of chemical weapons disposal for Germany, Great Britain, France, and the United States. Simone Müller, "Cut Holes and Sink 'em': Chemical Weapons Disposal and Cold War History as a History of Risk," *Historical Social Research* 41, no. 1 (2016): 270.

[9] A German engineering firm has identified a total of 200 areas in Germany that may still be affected by chemical weapons production. Garret, "Hugo Stoltzenberg and Chemical Weapons Proliferation," 23.

[10] Soraya Boudia et al., "Residues: Rethinking Chemical Environments," *Engaging Science, Technology, and Society* 4 (2018): 165.

[11] Simo Laakkonen, Richard Tucker, and Timo Vuorisalo, eds., *The Long Shadows: A Global Environmental History of the Second World War* (Corvallis: Oregon State University Press, 2017), 324.

[12] Anna Feigenbaum, *Tear Gas: From the Battlefields of World War I to the Streets of Today* (London: Verso, 2017), 1–2.

malleable. When intentionally directed, even environmental pollutants and industrial poisons often appear quite similar to traditionally defined chemical weapons.[13] This reveals that since the Chemical Revolution, scientific forays into pharmaceuticals, fertilizers, plastics, and other chemically synthesized products have all been intricately tied to the history of chemical weapons development. Thus, the atmospheric threats of a much broader and enduring "Chemical Modernity" remain pressing concerns for humanity today. From terrorist plots such as the 1995 Tokyo subway Sarin attack to carbon dioxide emissions and even cigarette smoke, the composition of the air we breathe continually, and perhaps increasingly, occupies our minds.[14] Don DeLillo's acclaimed 1985 novel *White Noise* attempts to narrativize this often enigmatic yet perennial modern concern. When the small college town of Blacksmith is affected by what is termed the "airborne toxic event," the townspeople in DeLillo's novel, much like the citizens of Hamburg in 1928, are forced to come to terms with the modern synthetic chemicals that had previously escaped their notice. DeLillo's narrator states that:

> A sharp and bitter stink filled the air ... Whatever caused the odor, I sensed that it made people feel betrayed ... The odor drove us away but beneath it and far worse was the sense that death came two ways, sometimes at once, and how death entered your mouth and nose, how death smelled, could somehow make a difference to your soul.[15]

In projecting toward the future, the perceptive difference between industrially produced chemicals and the ostensibly natural environment seems increasingly important. In a contemporary moment of greater concern for large-scale environmental/biological processes such as climate change and global pandemics, the invisible toxins of the modern world appear progressively threatening.[16] With imminent chemical danger looming in the minds of many contemporaries, this distinctly German story invites us to re-write the history of the twentieth century with an eye

[13] Throughout the Cold War, Great Britain and the United States attempted to create weapons that intentionally blurred the distinction between environmental pollutants and chemical/biological weapons. For instance, in Britain's offensives against communist Malaysian insurgents during the 1950s, the British military tested chemicals that could destroy crops and starve the enemy. Jacob Darwin Hamblin, *Arming Mother Nature: The Birth of Catastrophic Environmentalism* (Oxford: Oxford University Press, 2013), 67.

[14] Joseph Masco, *The Theater of Operations: National Security Affect from the Cold War to the War on Terror* (Durham: Duke University Press, 2014), 21.

[15] Don DeLillo, *White Noise* (New York: Viking, 1985), 240.

[16] Scott Christianson, *Fatal Airs: The Deadly History and Apocalyptic Future of Lethal Gases That Threaten Our World* (Santa Barbara: Praeger, 2010), 157.

toward the ways in which the atmosphere has, at least conceptually, become the battleground in a newly imagined "struggle for existence."[17]

Yet, as this study has shown, the often-ephemeral poison gases were not the only technologies that helped to create a German imaginary in which both atmo-terrorism and industrialized genocide could be both conceived and realized. If we examine the larger technological web surrounding chemical weapons development, then we realize the innumerable connections between weapons of human annihilation and the technologies, such as the gas mask, that were deployed to combat them. Brought into the Nazis' value-laden technological order in 1930s, the gas mask demanded the acknowledgment and implicit acceptance of a modernity filled with poison gas. Simultaneously, gas masks encouraged the development of more deadly and insidious chemical weapons and a vast array of other protective devices. Gas suits, gas blankets, and gas shelters further attempted to encapsulate and seal off human beings, protecting them from the world that they had all, to varying degrees, helped to envision and create.

Interestingly, the uncanniness of the gas mask's physical shape often seems to reveal the device's obscured historical connections. Even decades after World War II, American military doctrine claimed that "even in theory the gas mask is a dreadful thing. It stands for one's first flash of insight into man's measureless malignity against man."[18] But regardless of such potentially visceral repugnance, humanity always runs the risk of treating technologies such as gas masks as magical protective talismans in moments of seemingly imminent toxic exposure.[19] Such technological honing of the individual mind is certainly disconcerting, but self-inflicted discipline pales in comparison to collective imposition. Large-scale civilian regulation was certainly at hand in Nazi Germany, where gas mask discipline and training were legally enforced.

The communal rituals that featured the gas mask in the Third Reich were highly indicative of the broader Nazi relationship to the technological. Reflecting on other examples of Nazi technoscience,

[17] Dorothee Brantz, "Environments of Death: Trench Warfare on the Western Front, 1914–1918," in *War and the Environment: Military Destruction in the Modern Age*, ed. Charles E. Closmann (College Station: Texas A&M Press, 2009), 81.

[18] Frank L. Smith, *American Biodefense: How Dangerous Ideas about Biological Weapons Shape National Security* (Ithaca: Cornell University Press, 2014), 79.

[19] In doing so, we also run the risk of forgetting that technological armoring frequently exacerbates social and economic inequalities in moments of severe environmental degradation. The most at-risk people, according to commonly employed socio-economic metrics, are often the most exposed to deadly contaminants. See, for example, Dorceta E. Taylor, *Toxic Communities: Environmental Racism, Industrial Pollution, and Residential Mobility* (New York: New York University Press, 2014).

historian Tiago Saraiva has argued that emblematic fascist political designs were envisioned in, and enacted through, the technologically assisted cultivation of agricultural products.[20] As this present study has shown, the Nazis did likewise at the chemical level, attempting to tame and cultivate volatile toxic gases to fit within the larger fascist collective. Like this study, Saraiva has further revealed the importance of a perceived national insecurity for such coercive fascist endeavors. In the philosopher Michel Foucault's vocabulary, such actions were deeply biopolitical in that they literally attempted to "take control of life, to manage it."[21]

Foucault saw the Nazis as the ultimate historical example of a modern regime underpinned by an attempt at total state biopower. By dramatizing the omnipresence of death and destruction, the Nazis were able to begin the regulation and governance of both life and death. Foucault writes:

> Risking one's life, being exposed to total destruction, was one of the principles inscribed in the basic duties of the obedient Nazi, and it was one of the essential objectives of Nazism's policies. It had to reach the point at which the entire population was exposed to death ... [this] was the only way it could truly constitute itself as a superior race and bring about its definitive regeneration once other races had been either exterminated or enslaved forever.[22]

In coordination with the power of widespread fear, this ideological worldview encouraged many Germans to hand over decisions on life and death to the regime.[23] In this same vein, the state's distribution of survival technologies such as the gas mask, which still always allowed for sacrificial German death, was intimately tied to the controlled gassing of supposedly inferior races.[24] Through the institutionalized and ritualized allocation of gas death, the Nazis thus attempted to redefine terms such as "humanity" and "life." Under their partially realized schemas of state biopower, in their most extreme and violent form, human life under the gas mask would no longer be spoken about in qualitative terms. Indeed, the dangers of genocide and total war encouraged a view of life (and death) as purely quantitative – the bare states of living and dead were all that truly mattered. In turn, this simplification of life could in theory turn human

[20] Saravia, *Fascist Pigs*, 14.
[21] Michel Foucault, *"Society Must Be Defended" Lectures at the Collège de France, 1975–1976*, trans. David Macey (New York: Picador, 1997), 261–262.
[22] Ibid., 259–260. [23] Fritzsche, *Life and Death in the Third Reich*, 5.
[24] Michel Foucault, *The History of Sexuality*, vol. 1, trans. Robert Hurley (New York: Vintage Books, 1990), 149–150.

beings into objects of the technological state, ready to be reserved, reallocated, reconstituted, or destroyed.[25]

Perhaps even more troubling, Saraiva and Foucault have both made it clear that less radical regimes of technoscientific rationality are regularly implemented in ostensibly democratic political systems. The increasing desire to control risk in a seemingly ever-more dangerous (chemically and otherwise) modern world, the continued existence of which is predicated on complex protective technologies that always maintain the possibility of failure, seemingly demands progressively centralized control of life and death. Recognizing the biopower at work in various national Covid-19 measures, the philosopher Giorgio Agamben has publicly expressed his fear of the growing tendency to use the state of exception as the paradigm for centralized governance, thus normalizing restrictions and suspensions of daily life.[26]

While Agamben treads into the dangerous waters of science denial to refute the dangers of Covid-19, this study is less concerned with a scientific consensus, which may or may not be reached. The greater concern here is the potential foreclosure of other possible political responses to a chemical or biological atmospheric threat and the wholesale adherence to the often naïve solutions of technoscientific protection.[27] For, in the technoscientific solution, there is always the danger of reaching what historian Thomas P. Hughes called "technological momentum," or the tendency of a given technology to become integral to the society that created or incorporated it.[28] At this point, the technology, and often its relational web of technologies, becomes exceedingly difficult to manage and/or control. In fact, it is normalized in the structures of daily life, thus becoming constitutive of human subjectivity itself.

As this particularly German historical story has attempted to demonstrate, in the search for mental and physical security, humanity runs the constant danger of embracing the chemically and biologically dangerous world that it wishes to technologically pacify or decontaminate. Atmospheric anxiety and bio-chemical fear can distract us from the deep and interlocking connections between modern protective technologies

[25] Todd Presner, *Mobile Modernity: Germans, Jews, Trains* (New York: Columbia University Press, 2007), 225.

[26] Giorgio Agamben, "The Invention of Epidemic," February 26, 2020, www.journal-psychoanalysis.eu/coronavirus-and-philosophers/.

[27] Eva Horn, *The Future as Catastrophe: Imagining Disaster in the Modern Age* (New York: Columbia University Press, 2018), 236.

[28] Thomas Hughes, "Technological Momentum," in *Does Technology Drive History? The Dilemma of Technological Determinism*, eds. Merritt Roe Smith and Leo Marx (Cambridge, MA: MIT Press, 1994), 101–113.

and the man-made dangers that continually threaten the contemporary environment. This would suggest that the greatest threats to our atmospheric health are not just the airborne toxins but also the ubiquitous form of technoscientific protection that has helped to manufacture the biologically and chemically precarious world in which we now live.

Bibliography

Published Secondary Material

Aftalion, Fred, and Otto Theodor Benfey. *A History of the International Chemical Industry*. Philadelphia: University of Pennsylvania Press, 1991.

Agamben, Giorgio. "The Invention of Epidemic," February 26, 2020. www.quodlibet.it/giorgio-agamben-l-invenzione-di-un-epidemia.

Alexander, Jennifer Karns. *The Mantra of Efficiency: From Waterwheel to Social Control*. Baltimore: Johns Hopkins University Press, 2008.

Allen, Michael Thad. *The Business of Genocide: The SS, Slave Labor, and the Concentration Camps*. Chapel Hill: University of North Carolina Press, 2002.

"The Devil in the Details: The Gas Chambers of Birkenau, October 1941." *Holocaust and Genocide Studies* 16, no. 2 (2002): 189–216.

Aly, Götz, and Susanne Heim. *Architects of Annihilation: Auschwitz and the Logic of Destruction*. Princeton: Princeton University Press, 2015.

Amend, Niko. "Die Sanitätsdienstliche Versorgung der Gasversehrten Deutschen Soldaten des Ersten Weltkrieges." Dissertation, Universität Ulm, 2014.

Apel, Dora. "Cultural Battlegrounds: Weimar Photographic Narratives of War." *New German Critique* 76, Special Issue on Weimar Visual Culture (Winter 1999): 49–84.

Armitage, John. "On Ernst Jünger's 'Total Mobilization': A Re-evaluation in the Era of the War on Terrorism." *Body and Society* 9, no. 4 (2003): 191–213.

Babington, Anthony. *Shell-Shock: A History of the Changing Attitudes to War Neurosis*. London: Leo Cooper, 1997.

Baird, Jay W. *To Die for Germany: Heroes in the Nazi Pantheon*. Bloomington: Indiana University Press, 1990.

Balfour, Sebastian. *Deadly Embrace: Morocco and the Road to the Spanish Civil War*. Oxford: Oxford University Press, 2002.

Bartov, Omer. *Murder in Our Midst: The Holocaust, Industrial Killing, and Representation*. New York: Oxford University Press, 1996.

Bauer, Richard. *Fliegeralarm: Luftangriffe auf München 1940–1945*. Munich: Heinrich Hugendubel Verlag, 1997.

Bauman, Zygmunt. *Modernity and the Holocaust*. Ithaca: Cornell University Press, 1989.

Baumslag, Naomi. *Murderous Medicine: Nazi Doctors, Human Experimentation, and Typhus*. Westport, CT: Praeger, 2005.

Beck, Matthias, and Beth Kewell. *Risk: A Study of Its Origins, History and Politics*. Hackensack, NJ: World Scientific Publishing, 2014.
Beck, Ulrich. *Risk Society: Towards a New Modernity*. Translated by Mark Ritter. London: Sage, 1992.
Beck, Ulrich, ed. *World at Risk*. Translated by Ciaran Cronin. Malden, MA: Polity Press, 2009.
Beer, John J. "Coal Tar Dye Manufacture and the Origins of the Modern Industrial Research Laboratory." *Isis* 49, no. 2 (1958): 123–131.
Bein, Alex. *The Jewish Question: Biography of a World Problem*. Translated by Harry Zohn. Cranbury, NJ: Associated University Press, 1990.
Bennett, Tony, and Patrick Joyce, eds. *Material Powers: Cultural Studies, History and the Material Turn*. New York: Routledge, 2010.
Bennette, Rebecca Ayako. *Diagnosing Dissent: Hysterics, Deserters, and Conscientious Objectors in Germany during World War One*. Ithaca: Cornell University Press, 2020.
Bentley, Michelle. *Syria and the Chemical Weapons Taboo: Exploiting the Forbidden*. Manchester: Manchester University Press, 2016.
Bevan, John. *The Infernal Diver: The Lives of John and Charles Deane, Their Invention of the Diving Helmet, and Its First Application to Salvage, Treasure Hunting, Civil Engineering and Military Uses*. London: Submex, 1996.
Birdsall, Carolyn. *Nazi Soundscapes: Sound, Technology and Urban Space in Germany, 1933–1945*. Amsterdam: Amsterdam University Press, 2012.
Birn, Donald S. "Open Diplomacy at the Washington Conference of 1921–2: The British and French Experience." *Comparative Studies in Society and History* 12, no. 3 (1970): 297–319.
Bloxham, Donald. *The Final Solution: A Genocide*. Oxford: Oxford University Press, 2009.
Bogousslavsky, Julien, and Laurent Tatu. "French Neuropsychiatry in the Great War: Between Moral Support and Electricity." *Journal of the History of the Neurosciences* 22 (2013): 144–154.
Böttcher, Welf, and Martin Thoemmes. *Heinrich Dräger: Eine Biographie*. Neumünster: Wachholtz Verlag, 2011.
Boudia, Soraya, Angela N. H. Creager, Scott Frickel, et al. "Residues: Rethinking Chemical Environments." *Engaging Science, Technology, and Society* 4 (2018): 165–178.
Bourke, Joanna. "The Body in Modern Warfare: Myth and Meaning 1914–1945." In *What History Tells: George L. Mosse and the Culture of Modern Europe*, edited by Stanley Payne, David J. Sorkin, and John Tortorice, 202–223. Madison: University of Wisconsin Press, 2004.
Fear: A Cultural History. London: Virago Press, 2005.
"Pain, Sympathy and the Medical Encounter between the Mid Eighteenth and the Mid Twentieth Centuries." *Historical Research* 85, no. 229 (2012): 430–454.
"Bodily Pain, Combat, and the Politics of Memoirs: Between the American Civil War and the War in Vietnam." *Histoire Sociale* 46, no. 91 (2013): 43–61.

Bousquet, Antoine. "Ernst Jünger and the Problem of Nihilism in the Age of Total War." *Thesis Eleven* 132, no. 1 (2016): 17–38.

Boutwell, Jeffrey. *The German Nuclear Dilemma*. Ithaca: Cornell University Press, 1990.

Braker, Regina. "Helene Stocker's Pacifism in the Weimar Republic: Between Ideal and Reality." *Journal of Women's History* 13, no. 3 (2001): 70–97.

Brauch, Hans Günter, and Rolf-Dieter Müller, eds. *Chemische Kriegführung – Chemische Abrüstung: Dokumente Und Kommentare*. Berlin: Berlin Verlag, 1985.

Brinkhaus, Jörn. *Luftschutz Und Versorgungspolitik: Regionen Und Gemeinden in NS-Staat, 1042–1944/45*. Bielefeld: Verlag für Regionalgeschichte, 2010.

Brose, Eric Dorn. *The Kaiser's Army: The Politics of Military Technology in Germany during the Machine Age, 1870–1918*. Oxford: Oxford University Press, 2001.

Buck-Morss, Susan. "Aesthetics and Anaesthetics: Walter Benjamin's Artwork Essay Reconsidered." *October* 62 (Autumn 1992): 3–41.

Buggeln, Marc. *Slave Labor in Nazi Concentration Camps*. Translated by Paul Cohen. Oxford: Oxford University Press, 2014.

Burleigh, Michael. *Death and Deliverance: "Euthanasia" in Germany, c.1900 to 1945*. Cambridge: Cambridge University Press, 1994.

Cabanes, Bruno. *The Great War and the Origins of Humanitarianism, 1918–1924*. Cambridge: Cambridge University Press, 2014.

Canning, Kathleen, Kerstin Barndt, and Kristin McGuire, eds. *Weimar Publics/Weimar Subjects: Rethinking the Political Culture of Germany in the 1920s*. New York: Berghahn Books, 2010.

Carrington, Tyler. *Love at Last Sight: Dating, Intimacy, and Risk in Turn-of-the-Century Berlin*. Oxford: Oxford University Press, 2019.

Cassidy, David C. *Uncertainty: The Life and Science of Werner Heisenberg*. New York: W.H. Freeman, 1992.

Channell, David F. *A History of Technoscience: Erasing the Boundaries between Science and Technology*. London: Routledge, 2017.

Chapoutot, Johann. *The Law of Blood: Thinking and Acting as a Nazi*. Translated by Miranda Richmond Mouillot. Cambridge, MA: Belknap Press, 2018.

Charles, Daniel. *Master Mind: The Rise and Fall of Fritz Haber, the Nobel Laureate Who Launched the Age of Chemical Warfare*. New York: HarperCollins, 2005.

Chickering, Roger, and Stig Förster, eds. *Great War, Total War: Combat and Mobilization on the Western Front, 1914–1918*. Cambridge: Cambridge University Press, 2000.

Christianson, Scott. *Fatal Airs: The Deadly History and Apocalyptic Future of Lethal Gases That Threaten Our World*. Santa Barbara: Praeger, 2010.

The Last Gasp: The Rise and Fall of the American Gas Chamber. Berkeley: University of California Press, 2010.

Citino, Robert M. *The Path to Blitzkrieg: Doctrine and Training in the German Army, 1920–39*. Boulder, CO: Lynne Rienner, 1999.

Closmann, Charles E., ed. *War and the Environment: Military Destruction in the Modern Age*. College Station: Texas A&M Press, 2009.

Cohen, Deborah. *The War Come Home: Disabled Veterans in Britain and Germany, 1914–1939*. Los Angeles: University of California Press, 2001.

Cook, Robert E. "The Mist that Rolled into the Trenches: Chemical Escalation in World War I." *Bulletin of the Atomic Scientists* 27, no. 1 (January 1971): 34–38.

Cook, Tim. "Creating the Faith: The Canadian Gas Services in the First World War." *Journal of Military History* 62, no. 4 (October 1998): 755–786.

Cornwell, John. *Hitler's Scientists: Science, War, and the Devil's Pact*. New York: Penguin Press, 2003.

Corum, James S. *The Roots of the Blitzkrieg: Hans von Seeckt and German Military Reform*. Lawrence: University of Kansas Press, 1992.

Crew, David F. *Bodies and Ruins: Imagining the Bombing of Germany, 1945 to the Present*. Ann Arbor: University of Michigan Press, 2017.

Croddy, Eric. *Chemical and Biological Warfare: A Comprehensive Survey for the Concerned Citizen*. New York: Springer-Verlag, 2002.

Crouthamel, Jason. "Nervous Nazis: War Neurosis, National Socialism and the Memory of the First World War." *War & Society* 21, no. 2 (2013): 55–75.

"War Neurosis versus Savings Psychosis: Working-Class Politics and Psychological Trauma in Weimar Germany." *Journal of Contemporary History* 37, no. 2 (2002): 163–182.

Das, Santanu. "'An Ecstasy of Fumbling': Gas Warfare, 1914–18 and the Uses of Affect." In *The Edinburgh Companion to Twentieth-Century British and American War Literature*, edited by Adam Piette and Mark Rawlinson, 396–405. Edinburgh: Edinburgh University Press, 2012.

Touch and Intimacy in First World War Literature. Cambridge: Cambridge University Press, 2005.

Davis, Belinda. *Home Fires Burning: Food, Politics, and Everyday Life in World War I Berlin*. Chapel Hill: University of North Carolina Press, 2000.

Deist, Wilhelm, ed. *The German Military in the Age of Total War*. Warwickshire: Berg, 1985.

Dove, Richard, and Stephen Lamb, eds. *German Writers and Politics 1918–39*. London: Macmillan Press, 1992.

Dräger, Lisa, ed. *Von Biermaschine zum Rettungswesen: Die Aufbaujahre des Drägerwerks*. Lübeck: DrägerDruck, 2007.

Eckart, Wolfgang. "Aesculap in the Trenches: Aspects of German Medicine in the First World War." In *War, Violence and the Modern Condition*, edited by Bernd Hüppauf, 177–193. New York: Walter de Gruyter, 1997.

Medizin und Krieg: Deutschland 1914–1924. Paderborn: Ferdinand Schöningh, 2014.

Ede, Andrew. "Science Born of Poison, Fire and Smoke: Chemical Warfare and the Origins of Big Science." In *The Romance of Science: Essays in Honour of Trevor H. Levere*, edited by Jed Buchwald and Larry Stewart. Cham, Switzerland: Springer International, 2017.

Ede, Andrew. "The Natural Defense of a Scientific People: The Public Debate over Chemical Warfare in Post-WWI America." *Bulletin of the History of Chemistry* 27, no. 2 (2002): 128–135.

Eksteins, Modris. "All Quiet on the Western Front and the Fate of a War." *Journal of Contemporary History* 15, no. 2 (April 1980): 345–366.

Rites of Spring: The Great War and the Birth of the Modern Age. Boston: Houghton Mifflin, 1989.

Ellul, Jacques. *The Technological Society*. New York: Vintage Books, 1964.

Emery, Theo. *Hellfire Boys: The Birth of the U.S. Chemical Warfare Service and the Race for the World's Deadliest Weapons*. New York: Little, Brown, 2017.

Encke, Julia, *Augenblicke der Gefahr: Der Krieg und die Sinne. 1914–1934*. München: Wilhelm Fink Verlag, 2006.

Enghigian, Greg, and Matthew Paul Berg, eds. *Sacrifice and National Belonging in Twentieth-Century Germany*. College Station: Texas A&M Press, 2002.

Evans, Richard J. *The Coming of the Third Reich*. New York: Penguin, 2003.

The Third Reich in Power: 1933–1939. New York: Penguin, 2005.

Evers, Kai. "Gassing Europe's Capitals: Planning, Envisioning, and Rethinking Modern Warfare in European Discourses of the 1920s and 1930s." In *Visions of Europe: Interdisciplinary Contributions to Contemporary Cultural Debates*, edited by Gail K. Hart and Anke S. Biendarra, 65–83. Frankfurt: Peter Lang, 2014.

"Risking Gas Warfare: Imperceptible Death and the Future of War in Weimar Culture and Literature." *Germanic Review* 89, no. 3 (2014): 269–284.

Violent Modernists: The Aesthetics of Destruction in Twentieth-Century German Literature. Evanston, IL: Northwestern University Press, 2013.

Faber, David. *Munich, 1938: Appeasement and World War II*. New York: Simon & Schuster, 2008.

Faith, Thomas Ian. *Behind the Gas Mask: The U.S. Chemical Warfare Service in War and Peace*. Urbana: University of Illinois Press, 2014.

Faith, Thomas. "'As Is Proper in Republican Form of Government': Selling Chemical Warfare to Americans in the 1920s." *Federal History* 2 (2010): 28–41.

Feigenbaum, Anna. *Tear Gas: From the Battlefields of World War I to the Streets of Today*. London: Verso, 2017.

Feldman, Gerald D., and Wolfgang Seibel, eds. *Networks of Nazi Persecution: Bureaucracy, Business, and the Organization of the Holocaust*. New York: Berghahn Books, 2005.

Fischer, William B. *The Empire Strikes Out: Kurd Lasswitz, Hans Dominik, and the Development of German Science Fiction*. Bowling Green: Bowling Green State University Popular Press, 1984.

Fitzgerald, Gerard J. "Chemical Warfare and Medical Response during World War I." *American Journal of Public Health* 98, no. 4 (April 2008): 611–625.

Förster, Stig, ed. *An Der Schwelle Zum Totalen Krieg: Die Militärische Debatte Über Den Krieg Der Zukunft 1919–1939*. Paderborn: Ferdinand Schöningh, 2002.

Foucault, Michel. *"Society Must Be Defended" Lectures at the Collège de France, 1975–1976*. Translated by David Macey. New York: Picador, 1997.

The History of Sexuality. Translated by Robert Hurley. Vol. 1. New York: Vintage Books, 1990.

Franklin, Jane, ed. *The Politics of Risk Society*. Cambridge: Polity Press, 1998.

Freemantle, Michael. *Gas! Gas! Quick, Boys! How Chemistry Changed the First World War*. Brimscombe Port, Stroud: The History Press, 2012.

The Chemists' War 1914–1918. Cambridge: The Royal Society of Chemistry, 2015.

Fressoz, Jean-Baptiste. "Beck Back in the 19th Century: Towards a Genealogy of Risk Society." *History and Technology* 23, no. 4 (2007): 333–350.

Friedlander, Henry. *The Origins of Nazi Genocide: From Euthanasia to the Final Solution*. Chapel Hill: University of North Carolina Press, 1995.

Friedlander, Saul. *The Years of Extermination: Nazi Germany and the Jews, 1939–1945*. New York: HarperCollins, 2007.

Friedrich, Bretislav. "Fritz Haber Und Der 'Krieg Der Chemiker.'" *Physik in Unserer Zeit* 46, no. 3 (May 2015): 118–125.

Friedrich, Bretislav, Dieter Hoffmann, Jürgen Renn, Florian Schmaltz, and Martin Wolf, eds. *One Hundred Years of Chemical Warfare: Research, Deployment, Consequences*. New York: Springer, 2017.

Fritzsche, Peter. *An Iron Wind: Europe under Hitler*. New York: Basic Books, 2016.

A Nation of Fliers: German Aviation and the Popular Imagination. Cambridge, MA: Harvard University Press, 1992.

"Did Weimar Fail?" *The Journal of Modern History* 68, no. 3 (1996): 629–656.

Life and Death in the Third Reich. Cambridge, MA: Belknap Press, 2008.

"Machine Dreams: Airmindedness and the Reinvention of Germany." *American Historical Review* 98, no. 3 (1993): 685–709.

"Nazi Modern." *Modernism/Modernity* 3, no. 1 (1996): 1–22.

The Turbulent World of Franz Göll: An Ordinary Berliner Writes the Twentieth Century. Cambridge, MA: Harvard University Press, 2011.

Garon, Sheldon. "Defending Civilians against Aerial Bombardment: A Comparative/Transnational History of Japanese, German, and British Home Fronts, 1918–1945." *The Asia-Pacific Journal* 14, no. 23 (2016). https://apjjf.org/2016/23/Garon.html.

"On the Transnational Destruction of Cities: What Japan and the United States Learned from the Bombing of Britain and Germany in the Second World War." *Past & Present* 247, no. 1 (2020): 235–271.

Garrett, Benjamin. "Hugo Stoltzenberg and Chemical Weapons Proliferation." *The Monitor* 1, no. 2 (Spring 1995): 11, 23.

Geist, Edward M. *Armageddon Insurance: Civil Defense in the United States and Soviet Union, 1945–1991*. Chapel Hill: University of North Carolina Press, 2019.

Geroulanos, Stefanos, and Todd Meyers. *The Human Body in the Age of Catastrophe: Brittleness, Integration, Science, and the Great War*. Chicago: University of Chicago Press, 2018.

Gerstenberger, Katharina, and Tanja Nusser, eds. *Catastrophe and Catharsis: Perspectives on Disaster and Redemption in German Culture and Beyond*. Rochester, NY: Camden House, 2015.

Geyer, Michael. "How the Germans Learned to Wage War: On the Question of Killing in the First and Second World Wars." In *Between Mass Death and Individual Loss: The Place of the Dead in Twentieth-Century Germany*, edited by Alan Confino, Paul Betts, and Dirk Schuman, 25–50. New York: Berghahn Books, 2008.

Giddens, Anthony. *The Consequences of Modernity*. Oxford: Polity Press, 1990.

Gil, Isabel Capeloa. "The Visuality of Catastrophe in Ernst Jünger's Der Gefährliche Augenblick and Die Veränderte Welt." *KulturPoetik* 10, no. 1 (2010): 62–84.

Girard, Marion. *A Strange and Formidable Weapon: British Responses to World War I Poison Gas*. Lincoln: University of Nebraska Press, 2008.

Goeschel, Christian. *Suicide in Nazi Germany*. Oxford: Oxford University Press, 2009.

Goodbody, Axel. *Nature, Technology and Cultural Change in Twentieth-Century German Literature: The Challenge of Ecocriticism*. New York: Palgrave Macmillan, 2007.

Grady, Tim. *A Deadly Legacy: German Jews and the Great War*. New Haven: Yale University Press, 2017.

Grayzel, Susan R. *At Home and Under Fire: Air Raids and Culture in Britain from the Great War to the Blitz*. Cambridge: Cambridge University Press, 2012.

"'Macabre and Hilarious', The Emotional Life of the Civilian Gas Mask in France during and after the First World War." In *Total War: and Emotional History*, edited by Claire Langhammer, Lucy Noakes, and Claudia Siebrecht, 40–59. Oxford: Oxford University Press, 2020.

The Age of the Gas Mask: How British Civilians Faced the Terrors of Total War. Cambridge: Cambridge University Press, 2022.

"The Baby in the Gas Mask: Motherhood, Wartime Technology, and the Gendered Division between the Fronts during and after the First World War." In *Gender and the First World War*, edited by Christa Hämmerle, Oswald Überegger, and Birgitta Bader Zaar, 127–144. New York: Palgrave Macmillan, 2014.

Griech-Polelle, Beth A. *Anti-Semitism and the Holocaust: Language, Rhetoric and the Traditions of Hatred*. London: Bloomsbury, 2017.

Guillemin, Jeanne. *Biological Weapons: From the Invention of State-Sponsored Programs to Contemporary Bioterrorism*. New York: Columbia University Press, 2006.

Haapamaki, Michele. *The Coming of the Aerial War: Culture and the Fear of Airborne Attack in Inter-War Britain*. New York: I.B. Taurus, 2014.

Habeck, Mary R. "Technology in the First World War: The View from Below." In *The Great War and the Twentieth Century*, edited by Jay Winter, Geoffrey Parker, and Mary R. Habeck, 99–131. New Haven: Yale University Press, 2000.

Haber, L. F. *The Chemical Industry 1900–1930: International Growth and Technological Change*. Oxford: Clarendon Press, 1971.

Haber, L.F. *The Chemical Industry during the Nineteenth Century*. New York: Oxford University Press, 1958.

The Poisonous Cloud: Chemical Warfare in the First World War. Oxford: Clarendon Press, 1986.

Hachtmann, Rüdiger. *Wissenschaftsmanagement Im "Dritten Reich": Geschichte Der Generalverwaltung Der Kaiser-Wilhelm-Gesellschaft*. Göttingen: Wallstein, 2007.

Hager, Thomas. *The Alchemy of Air: A Jewish Genius, a Doomed Tycoon, and the Scientific Discovery That Fed the World but Fueled the Rise of Hitler*. New York: Harmony, 2008.

Hallett, Christine E. *Containing Trauma: Nursing Work in the First World War*. Manchester: Manchester University Press, 2009.

Hamblin, Jacob Darwin. *Arming Mother Nature: The Birth of Catastrophic Environmentalism*. Oxford: Oxford University Press, 2013.

Hammond, James W. *Poison Gas: The Myths versus Reality*. Westport, CT: Greenwood Press, 1999.

Harris, Henry. "To Serve Mankind in Peace and the Fatherland in War: The Case of Fritz Haber." *German History* 10, no. 1 (1992): 24–38.

Harrison, Mark. *The Medical War: British Military Medicine in the First World War*. Oxford: Oxford University Press, 2010.

Hasegawa, Guy R. *Villainous Compounds: Chemical Weapons and the American Civil War*. Carbondale: Southern Illinois University Press, 2015.

Hayes, Peter. "Carl Bosch and Carl Krauch: Chemistry and the Political Economy of Germany, 1925–1945." *The Journal of Economic History* 47, no. 2 (1987): 353–363.

"The Chemistry of Business-State Relations in the Third Reich." In *Business and Industry in Nazi Germany*, edited by Francis R. Nicosia and Jonathan Huener, 66–80. New York: Berghahn Books, 2004.

From Cooperation to Complicity: Degussa in the Third Reich. Cambridge: Cambridge University Press, 2004.

Industry and Ideology: IG Farben in the Nazi Era. Cambridge: Cambridge University Press, 2001.

Heim, Susanne, Carola Sachse, and Mark Walker, eds. *The Kaiser Wilhelm Society under National Socialism*. Cambridge: Cambridge University Press, 2009.

Herf, Jeffrey. *Reactionary Modernists: Technology, Culture, and Politics in Weimar and the Third Reich*. Cambridge: Cambridge University Press, 1984.

Herzog, Dagmar. *Unlearning Eugenics: Sexuality, Reproduction, and Disability in Post-Nazi Europe*. Madison: University of Wisconsin Press, 2018.

Hett, Benjamin Carter. *Burning the Reichstag: An Investigating into the Third Reich's Enduring Mystery*. Oxford: Oxford University Press, 2014.

Heynen, Robert. *Degeneration and Revolution: Radical Cultural Politics and the Body in West Germany*. Boston: Brill, 2015.

Hilberg, Raul. *The Destruction of European Jews*. New Haven: Yale University Press, 2003.

Hippel, Frank A. von. *The Chemical Age: How Chemists Fought Famine and Disease, and Changed Our Relationship with the Earth*. Chicago: University of Chicago Press, 2020.

Hirschfeld, Gerhard and Gerd Krumeich. *Deutschland im Ersten Weltkrieg*. Frankfurt: S. Fischer Verlag, 2013.

Hobsbawn, Eric. *The Age of Extremes: A History of the World, 1914–1991*. New York: Pantheon Books, 1994.

Hoffmann, Klaus. *Otto Hahn: Achievement and Responsibility*. New York: Springer, 2001.

Hogg, Ian. "Bolimov and the First Gas Attack." In *Tanks and Weapons of World War I*, edited by Bernard Fitzsimons, 17–19. New York: Beekman House, 1973.

Holman, Brett. *The Next War in the Air: Britain's Fear of the Bomber, 1908–1941*. London: Routledge, 2014.

Horn, Eva. *The Future as Catastrophe: Imagining Disaster in the Modern Age*. Translated by Valentine Pakis. New York: Columbia University Press, 2018.

Hughes, Thomas. "Technological Momentum." In *Does Technology Drive History? The Dilemma of Technological Determinism*, edited by Merritt Roe Smith and Leo Marx, 101–113. Cambridge, MA: MIT Press, 1994.

Hugill, Peter J. *Global Communications since 1844: Geopolitics and Technology*. Baltimore: Johns Hopkins University Press, 1999.

Hull, Isabel V. *A Scrap of Paper: Breaking and Making International Law during the Great War*. Ithaca: Cornell University Press, 2014.

Absolute Destruction: Military Culture and the Practices of War in Imperial Germany. Ithaca: Cornell University Press, 2005.

Huyssen, Andres. "Fortifying the Heart – Totally: Ernst Jünger's Armored Texts." *New German Critique* 59 (1993): 3–23.

Hynes, Samuel. *The Soldiers' Tale: Bearing Witness to Modern War*. New York: Penguin Press, 1997.

James, Jeremiah, Thomas Steinhauser, Dieter Hoffmann, and Bretislav Friedrich. *One Hundred Years at the Intersection of Chemistry and Physics: The Fritz Haber Institute of the Max Planck Society 1911–2011*. Boston: De Gruyter, 2011.

Jansen, Sarah. "Histoire d'un Transfert de Technologie: De l'étude Des Insectes à La Mise Au Point Du Zyklon B." *La Recherche* 340 (2001): 55–59.

"Schädlinge": Geschichte Eines Wissenschaftlichen Und Politischen Konstrukts, 1840–1920. Frankfurt: Campus Verlag, 2003.

Jarausch, Konrad, and Michael Geyer. *Shattered Past: Reconstructing German Histories*. Princeton: Princeton University Press, 2003.

Jeffreys, Diarmuid. *Hell's Cartel: IG Farben and the Making of Hitler's War Machine*. New York: Metropolitan Books, 2008.

Jelavich, Peter. "German Culture in the Great War." In *European Culture in the Great War: The Arts, Entertainment, and Propaganda, 1914–1918*, edited by Aviel Roshwald and Richard Stites, 32–58. Cambridge: Cambridge University Press, 1999.

Joerges, Bernward. "Do Politics Have Artefacts?" *Social Studies of Science* 29, no. 3 (1999): 411–431.

Johler, Reinhard, Christian Marchetti, and Monique Scheer, eds. *Doing Anthropology in Wartime and War Zones. World War I and the Cultural Sciences in Europe*. Bielefeld: Transcript Verlag, 2010.

Johnson, Ann. *Hitting the Brakes: Engineering Design and the Production of Knowledge*. Durham: Duke University Press, 2010.

Johnson, Ian Ona. *Faustian Bargain: The Soviet-German Partnership and the Origins of the Second World War*. Oxford: Oxford University Press, 2021.

Johnson, Jeffrey A. "Academic, Proletarian … Professional? Shaping Professionalization for German Industrial Chemists, 1887–1920" in *German Professions, 1800–1950*, edited by Konrad Jarausch and Geoffrey Cocks, 123–142. New York: Oxford University Press, 1990.

Johnson, Jeffrey A. "The Scientist behind Poison Gas: The Tragedy of the Habers." *Humanities* 17, no. 5 (1996): 25–29.

The Kaiser's Chemists: Science and Modernization in Imperial Germany. Chapel Hill: University of North Carolina Press, 1990.

Johnson, Ryan Mark. "A Suffocating Nature: Environment, Culture, and German Chemical Warfare on the Western Front." Dissertation, Temple University, 2013.

Jones, Edgar. "Terror Weapons: The British Experience of Gas and Its Treatment in the First World War." *War in History* 21, no. 3 (July 2014): 355–375.

Kaes, Anton. *Shell Shock Cinema: Weimar Culture and the Wounds of War*. Princeton: Princeton University Press, 2009.

Kaiser, Gerhard. "Wie Die Kultur Einbrach Giftgas Und Wissenschaftsethos Im Ersten Weltkrieg." *Merkur* 56 (2002): 210–220.

Kalthoff, Jürgen, and Martin Werner. *Die Händler des Zyklon B Tesch & Stabenow Eine Firmengeschichte zwischen Hamburg und Auschwitz*. Hamburg: VSA-Verlag, 1998.

Kamp, Michael. *Bernhard Dräger: Erfinder, Unternehmer, Bürger 1970 bis 1928*. Kiel/Hamburg: Wachholtz Verlag, 2017.

Kaplan, Marion A. *Between Dignity and Despair: Jewish Life in Nazi Germany*. New York: Oxford University Press, 1998.

Kater, Michael H. *Hitler Youth*. Cambridge, MA: Harvard University Press, 2004.

Katz, Eric, ed. *Death by Design: Science, Technology, and Engineering in Nazi Germany*. New York: Pearson, 2006.

Katz, Eric. "On the Neutrality of Technology: The Holocaust Death Camps as a Counter-Example." *Journal of Genocide Research* 7, no. 3 (September 2005): 409–421.

"The Nazi Engineers: Reflections on Technological Ethics in Hell." *Science and Engineering Ethics* 17, no. 3 (2011): 571–582.

Kaufmann, Doris. "'Gas, Gas, Gaas!' The Poison Gas War in the Literature and Visual Arts of Interwar Europe." In *One Hundred Years of Chemical Warfare: Research, Deployment, Consequences*, 169–187. Springer, n.d.

Kay, Alex J., and Stahel David, eds. *Mass Violence in Nazi-Occupied Europe*. Bloomington: Indiana University Press, 2018.

Keegan, John. *The Face of Battle: A Study of Agincourt, Waterloo, and the Somme*. London: Pimlico, 2004.

Kenez, Peter. *The Coming of the Holocaust: From Antisemitism to Genocide*. New York: Cambridge University Press, 2013.

Kershaw, Ian. *Hitler, 1889–1936: Hubris*. London: Allen Lane, 1998.

The Nazi Dictatorship: Problems and Perspectives of Interpretation. London: Edward Arnold, 1985.

Kidd, Ian James. "Oswald Spengler, Technology, and Human Nature." *The European Legacy* 17, no. 1 (2012): 19–31.

Kitchen, Martin. "Militarism and the Development of Fascist Ideology: The Political Ideas of Colonel Max Bauer, 1916–18." *Central European History* 8, no. 3 (1975): 199–220.

Kittler, Wolf. "From Gestalt to Ge-Stell: Martin Heidegger Reads Ernst Jünger." *Cultural Critique* 69 (2008): 79–97.

Klemperer, Klemens von. *German Incertitudes 1914–1945: The Stones and the Cathedral*. Westport, CT: Praeger, 2001.

Kluge, Alexander. *The Air Raid on Halberstadt on 8 April 1945*. Translated by Martin Chalmers. London: Seagull, 2014.

Knell, Hermann. *To Destroy a City: Strategic Bombing and Its Human Consequences in World War II*. Cambridge, MA: De Capo Press, 2003.

Knoll, Michael. *Die Drägerwerke Im Dritten Reich*. Norderstedt: GRIN Verlag, 2004.

Knust, Herbert. "George Grosz: Literature and Caricature." *Comparative Literature Studies* 12, no. 3 (1975): 218–247.

Koonz, Claudia. *The Nazi Conscience*. Cambridge, MA: Belknap Press, 2003.

Kramer, Alan. *Dynamic of Destruction: Culture and Mass Killing in the First World War*. Oxford: Oxford University Press, 2007.

Kramer, Nicole. *Volksgenossinnen an Der Heimatfront. Mobilisierung, Verhalten, Erinnerung*. Göttingen: Vandenhoeck & Ruprecht, 2011.

Krause, Jonathan. "The Origins of Chemical Warfare in the French Army." *War in History* 20, no. 4 (2013): 545–556.

Kreike, Emmanuel. *Scorched Earth: Environmental Warfare as a Crime against Humanity and Nature*. Princeton: Princeton University Press, 2021.

Kühne, Thomas. *Belonging and Genocide: Hitler's Community, 1918–1945*. New Haven: Yale University Press, 2010.

Kunz, Rudibert, and Rolf-Dieter Müller. *Giftgas gegen Abd El Krim: Deutschland, Spanien und der Gaskrieg in Spanisch-Marokko, 1922–1927*. Freiburg: Verlag Rombach Freiburg, 1990.

Laakkonen, Simo, Richard Tucker, and Timo Vuorisalo, eds. *The Long Shadows: A Global Environmental History of the Second World War*. Corvallis: Oregon State University Press, 2017.

Leed, Eric J. *No Man's Land: Combat and Identity in World War I*. Cambridge: Cambridge University Press, 1979.

Legro, Jeffrey W. *Cooperation under Fire: Anglo-German Restraint during World War II*. Ithaca: Cornell University Press, 1995.

Leitner, Gerit von. *Wollen Wir Unsere Hände in Unschuld Waschen? Gertrud Woker (1878–1968) Chemikerin & Internationale Frauenliga 1915–1968*. Berlin: Weidler Buchverlag, 1998.

Lemke, Bernd, ed. *Luft- Und Zivilschutz in Deutschland Im 20. Jahrhundert*. Potsdam: Militärgeschichtliches Forschungsamt, 2007.

Lemke, Bernd. *Luftschutz in Grossbritannien Und Deutschland 1923 Bis 1939*. München: Oldenbourg Verlag, 2005.

Lemmerich, Jost, and Ann Hentschel. *Science and Conscience: The Life of James Franck*. Stanford: Stanford University Press, 2011.

Leonhard, Jörn. *Pandora's Box: A History of the First World War*. Translated by Patrick Camiller. Cambridge, MA: Harvard University Press, 2018.

Lepick, Olivier. *La Grande Guerre Chimique, 1914–1918*. Paris: Presses Universitaires de France, 1998.

Lerner, Paul. *Hysterical Men: War, Psychiatry, and the Politics of Trauma in Germany, 1890–1930*. Ithaca: Cornell University Press, 2003.

Lethen, Helmut. *Cool Conduct: The Culture of Distance in Weimar Germany*. Translated by Don Reneau. Berkeley: University of California Press, 2002.

Lewin Sime, Ruth. *Lise Meitner: A Life in Physics*. Berkeley: University of California Press, 1996.

Lindner, Stephan H. *Inside IG Farben: Hoechst during the Third Reich*. Translated by Helen Schoop. Cambridge: Cambridge University Press, 2008.

Linhardt, Andreas. "Die Fachzeitschrift 'Gasschutz Und Luftschutz' Unter Dem Einfluss Des Nationalsozialismus," n.d. www.bbk.bund.de/SharedDocs/Downloads/BBK/DE/FIS/DownloadsDigitalisierteMedien/FachzeitschriftNotfallvorsorge/GasschutzundLuftschutz/Ausarbeitung_Dr_Linhardt_PDF.pdf?__blob=publicationFile.

Linton, Derek. "The Obscure Object of Knowledge: German Military Medicine Confronts Gas Gangrene during World War I." *Bulletin of the History of Medicine* 74, no. 2 (2000): 291–316.

Lloyd, Nick. *The Western Front: A History of the Great War, 1914–1918*. New York: W.W. Norton, 2021.

Lorentz, Bernhard. *Industrieelite und Wirtschaftspolitik 1928–1950: Heinrich Dräger und das Drägerwerk*. Paderborn: Ferdinand Schöningh, 2001.

Loughran, Tracey. "Shell-Shock and Psychological Medicine in First World War Britain." *Social History of Medicine* 22, no. 1 (2009): 79–95.

Lüdtke, Alf, ed. *Everyday Life in Mass Dictatorship: Collusion and Evasion*. New York: Palgrave Macmillan, 2016.

Macleod, Roy, and Jeffrey Allan Johnson, eds. *Frontline and Factory: Comparative Perspectives on the Chemical Industry at War, 1914–1924*. Dordrecht: Springer, 2006.

MacLeod, Roy M. "Chemistry for King and Kaiser: Revisiting Chemical Enterprise and the European War." In *Determinants in the Evolution of the European Chemical Industry, 1900–1939*, edited by Anthony S. Travis, Harm G. Schröter, Ernst Homburg, and Peter J. T. Morris, 23–49. Dordrecht: Kluwer Academic, 1998.

Macrakis, Kristie. *Surviving the Swastika: Scientific Research in Nazi Germany*. New York: Oxford University Press, 1993.

Maier, Helmut. *Chemiker Im "Dritten Reich" Die Deutsche Chemische Gesellschaft Und Der Verein Deutscher Chemiker Im NS-Herrschaftsapparat*. Weinheim: Wiley-VCH, 2015.

Rüstungsforschung in Der Kaiser-Wilhelm-Gesellschaft Und Das Kaiser-Wilhelm-Institut Für Metallforschung, 1900–1945/48. Göttingen: Wallstein, 2007.

Main, Steven J. "Gas on the Eastern Front during the First World War (1915–1917)." *Slavic Military Studies* 28, no. 1 (2015): 99–132.

Malfoy, Jordan. "Britain Can Take It: Chemical Warfare and the Origins of Civil Defense in Great Britain, 1915–1945." Dissertation, Florida International, 2018.

Martinetz, Dieter. *Der Gaskrieg 1914–1918: Entwicklung, Herstellung, und Einsatz chemischer Kampfstoffe*. Bonn: Bernard & Graefe Verlag, 1996.

Masco, Joseph. *The Theater of Operations: National Security Affect from the Cold War to the War on Terror*. Durham: Duke University Press, 2014.

Meise, Klaus. "Der Giftgas-Unfall am 20. Mai 1928: Phosgen-Explosion in Der Chemischen Fabrik Stoltzenburg Am Hovenweg." *Die Insel: Zeitschrift Des Vereins Museum Elbinsel Wilhelmsburg* 5314 (2012): 63–65.

Mercelis, Joris. *Beyond Bakelite: Leo Baekeland and the Business of Science and Invention*. Cambridge, MA: MIT Press, 2020.

Mitcham, Carl. *Steps toward a Philosophy of Engineering: Historico-Philosophical and Critical Essays*. London: Rowman & Littlefield International, 2020.

Möbius, Torben. "World War II Aerial Bombings of Germany: Fear as Subject of National Socialist Governmental Practices." *Storicamente* 11, no. 21 (2015): 1–21.

Moore, William. *Gas Attack! Chemical Warfare 1915–19 and Afterwards*. London: Leo Cooper, 1987.

Moshenka, Gabriel. "Gas Masks: Material Culture, Memory, and the Senses." *The Journal of the Royal Anthropological Institute* 16, no. 3 (2010): 609–628.

Mosse, George. *Fallen Soldiers: Reshaping the Memory of the World Wars*. New York: Oxford University Press, 1990.

Müller, Rolf-Dieter, "Die deutschen Gaskriegsvorbereitungen 1919–45: Mit Giftgas zur Weltmacht?" *Militärgeschichtliche Zeitschrift* 27, no. 1 (1980): 25–54.

Das Tor zur Weltmacht: Die Bedeutung der Sowjetunion für die Deutsche Wirschafts- und Rüstungspolitik zwischen den Weltkriegen. Boppard am Rhein: Harald Boldt Verlag, 1984.

"Total War as a Result of New Weapons? The Use of Chemical Agents in World War I." In *Great War, Total War: Combat and Mobilization on the Western Front, 1914–1918*, edited by Roger Chickering and Stig Förster. Cambridge: Cambridge University Press, 2000.

Müller, Simone. "'Cut Holes and Sink 'em': Chemical Weapons Disposal and Cold War History as a History of Risk." *Historical Social Research* 41, no. 1 (2016): 263–284.

Murdoch, Brian. *German Literature and the First World War: The Anti-War Tradition*. Farnham: Ashgate, 2015.

Murmann, Johann Peter. "Knowledge and Competitive Advantage in the Synthetic Dye Industry, 1850–1914: The Coevolution of Firms, Technology, and National Institutions in Great Britain, Germany, and the United States." *Enterprise and Society* 1, no. 4 (2000): 699–704.

Murphy, David Thomas. *The Heroic Earth: The Flowering of Geopolitical Thought in Weimar Germany, 1924–1933*. Kent: Kent State University Press, 1990.

Musolff, Andreas. "What Role Do Metaphors Play in Racial Prejudice? The Function of Antisemitic Imagery in Hitler's Mein Kampf." *Patterns of Prejudice* 41, no. 1 (2007): 21–43.

Neaman, Elliot Y. *A Dubious Past: Ernst Jünger and the Politics of Literature after Nazism*. Berkeley: University of California Press, 1999.

Neuner, Stephanie. *Politik Und Psychiatrie: Die Staatliche Versorgung Psychisch Kriegsbeschädigter in Deutschland 1920–1939*. Göttingen: Vandenhoeck & Ruprecht, 2011.

Nevis, Thomas. *Ernst Jünger and Germany: Into the Abyss, 1914–1945*. Durham: Duke University Press, 1996.

Overy, R. J. *The Bombers and the Bombed: Allied Air War over Europe 1940–1945*. New York: Viking, 2013.

Overy, Richard. *The Bombing War: Europe 1939–1945*. London: Penguin, 2013.

Palmer, Scott W. *Dictatorship of the Air: Aviation Culture and the Fate of Modern Russia*. Cambridge: Cambridge University Press, 2006.
Panchasi, Roxanne. *Future Tense: The Culture of Anticipation in France between the Wars*. Ithaca: Cornell University Press, 2009.
Paxman, Jeremy, and Robert Harris. *A Higher Form of Killing: The Secret Story of Chemical and Biological Warfare*. New York: Hill and Wang, 1982.
Pehnke, Andreas. *Der Hamburger Schulreformer Wilhelm Lamszus (1881–1965) Und Seine Antikriegsschrift "Giftgas Über Uns": Erstveröffentlichen Des Verschollen Geglaubten Manuskripts von 1932*. Beucha: Sax Verlag, 2006.
Perrow, Charles. *Normal Accidents: Living with High-Risk Technologies*. New York: Basic Books, 1984.
Perry, Heather R. *Recycling the Disabled: Army, Medicine, and Modernity in WWI Germany*. Manchester: Manchester University Press, 2014.
Peukert, Detlev. *Inside Nazi Germany: Conformity, Opposition, and Racism in Everyday Life*. Translated by Richard Deveson. New Haven: Yale University Press, 1987.
Pfeiler, W. M. K. *War and the German Mind: The Testimony of Men of Fiction Who Fought at the Front*. New York: Columbia University Press, 1941.
Philpott, William. *War of Attrition: Fighting the First World War*. New York: Overlook Press, 2014.
Plumpe, Werner. *Carl Duisberg 1861–1935: Anatomie Eines Industriellen*. München: C.H. Beck, 2016.
German Economic and Business History in the 19th and 20th Centuries. London: Palgrave Macmillan, 2016.
Poore, Carol. *Disability in Twentieth-Century German Culture*. Ann Arbor: University of Michigan Press, 2007.
Presner, Todd. *Mobile Modernity: Germans, Jews, Trains*. New York: Columbia University Press, 2007.
Preston, Diana. *A Higher Form of Killing: Six Weeks in World War I That Forever Changed the Nature of Warfare*. New York: Bloomsbury, 2015.
Before the Fallout: From Marie Curie to Hiroshima. New York: Walker & Company, 2005.
Price, Richard. "A Genealogy of the Chemical Weapons Taboo." *International Organization* 49, no. 1 (Winter 1995): 73–103.
Price, Richard M. *The Chemical Weapons Taboo*. Ithaca: Cornell University Press, 1997.
Pritchard, Sara B., and Carl A. Zimring. *Technology and the Environment in History*. Baltimore: Johns Hopkins University Press, 2020.
Pruemm, Karl. *Die Literatur des Soldatischen Nationalismus der 20er Jahre*. Kronberg: Taunus, 1974.
Reid, Fiona. *Broken Men: Shell Shock, Treatment and Recovery in Britain 1914–30*. New York: Continuum, 2010.
Medicine in First World War Europe: Soldiers, Medics, Pacifists. London: Bloomsbury, 2017.
Reynolds, David. *The Long Shadow: The Great War and the Twentieth Century*. London: Simon & Schuster, 2013.

Rieger, Bernhard. *Technology and the Culture of Modernity in Britain and Germany 1890–1945*. Cambridge: Cambridge University Press, 2005.

Rose, Shelley E. "The Penumbra of Weimar Political Culture: Pacifism, Feminism, and Social Democracy." *Peace & Change* 36, no. 3 (2011): 313–343.

Rosenwald, Lawrence. "On Modern Western Antiwar Literature." *Raritan* 34, no. 1 (2014): 155–173.

Ross, Corey. "Mass Culture and Divided Audiences: Cinema and Social Change in Inter-War Germany." *Past & Present* 193 (November 2006): 157–195.

Rotter, Andrew J. *Hiroshima: The World's Bomb*. New York: Oxford University Press, 2008.

Russell, Edmund P. "'Speaking of Annihilation': Mobilizing for War against Human and Insect Enemies, 1914–1945." *Journal of American History* 82, no. 4 (1996): 1505–1529.

Saint-Amour, Paul K. "Air War Prophecy and Interwar Modernism." *Comparative Literature Studies* 42, no. 2 (2005): 130–161.

Tense Future: Modernism, Total War, Encyclopedic Form. Oxford: Oxford University Press, 2015.

Sammartino, Annemarie H. *The Impossible Border: Germany and the East 1914–1922*. Ithaca: Cornell University Press, 2010.

Sannwald, Wolfgang. *Schiefertafel, Gasmaske Und Petticoat: Erlebte Dinge Und Erinnerungen Aus Dem Landkreis Tübingen*. Gomaringen und Tübigen: Gomaringer Verlag und Schwäbisches Tagblatt, 1994.

Saraiva, Tiago. *Fascist Pigs: Technoscientific Organisms and the History of Fascism*. Cambridge, MA: MIT Press, 2016.

Sax, Boria. *Animals in the Third Reich: Pets, Scapegoats, and the Holocaust*. New York: Continuum, 2000.

Scarry, Elaine. *The Body in Pain: The Making and Unmaking of the World*. New York: Oxford University Press, 1985.

Schivelbusch, Wolfgang, *The Railway Journey: The Industrialization of Time and Space in the Nineteenth Century*. Berkeley: University of California Press, 2014.

Schmaltz, Florian. "Neurosciences and Research on Chemical Weapons of Mass Destruction in Nazi Germany." Translated by Stefani Ross. *Journal of the History of the Neurosciences* 15, no. 3 (2006): 186–209.

Kampfstoff-Forschung Im Nationalsozialismus: Zur Kooperation von Kaiser-Weilhelm Instituten, Militär Und Industrie. Göttingen: Wallstein Verlag, 2005.

Schmidt, Ulf. *Secret Science: A Century of Poison Warfare and Human Experiments*. Oxford: Oxford University Press, 2015.

Schneider, Thomas F. "Narrating the War in Pictures: German Photo Books on World War I and the Construction of Pictorial War Narrations." *Journal of War and Culture Studies* 4, no. 1 (2011): 31–49.

Schütz, Erhard. "Wahn-Europa. Mediale Gas-Luftkrieg Szenarien Der Zwischenkriegszeit." In *Krieg in Den Medien*, edited by Heinz-Peter Preußler, 127–149. Amsterdam: Rodopi, 2005.

Schweer, Henning. *Die Geschichte der chemischen Fabrik Stoltzenburg bis zum Ende des zweiten Weltkrieges*. Diepholz: Verlag für Geschichte der Naturwissenschaften und der Technik, 2008.

Shephard, Ben. *A War of Nerves: Soldiers and Psychiatrists in the Twentieth Century*. Cambridge, MA: Harvard University Press, 2000.

Shephard, Ben H. *War in the Wild East: The German Army and Soviet Partisans*. Cambridge, MA: Harvard University Press, 2004.

Sloterdijk, Peter. *Terror from the Air*. Translated by Amy Patton and Steven Corcoran. Los Angeles: Semiotext, 2009.

Slotten, Hugh R. "Humane Chemistry or Scientific Barbarism? American Responses to World War I Poison Gas, 1915–1930." *Journal of American History* 77, no. 2 (September 1990): 476–498.

Smith, Frank L. *American Biodefense: How Dangerous Ideas about Biological Weapons Shape National Security*. Ithaca: Cornell University Press, 2014.

Smith, Jessica M. *Extracting Accountability: Engineers and Corporate Social Responsibility*. Cambridge, MA: MIT Press, 2021.

Smith, Susan L. *Toxic Exposures: Mustard Gas and the Health Consequences of World War II in the United States*. New Brunswick, NJ: Rutgers University Press, 2017.

Speier, Hans. *German Rearmament and the Atomic War: The Views of German Military and Political Leaders*. Evanston, IL: Row, Peterson and Company, 1957.

Stanley, Matthew. *Einstein's War: How Relativity Triumphed Amid the Vicious Nationalism of World War I*. New York: Dutton, 2020.

Stargardt, Nicholas. *The German War: A Nation under Arms, 1939–45*. New York: Basic Books, 2015.

Witnesses of War: Children's Lives under the Nazis. New York: Alfred A. Knopf, 2005.

Steen, Kathryn. *The American Synthetic Organic Chemicals Industry: War and Politics, 1910–1930*. Chapel Hill: University of North Carolina Press, 2014.

Stephenson, Charles. *The Admiral's Secret Weapon: Lord Dundonald and the Origins of Chemical Warfare*. Rochester, NY: Boydell Press, 2006.

Stephenson, Jill. *The Nazi Organization of Women*. Totowa, NJ: Croom Helm, 1981.

Stern, Fritz. *Dreams and Delusions: The Drama of German History*. London: Weidenfeld & Nicolson, 1988.

Einstein's German World. Princeton: Princeton University Press, 1999.

"Freunde im Widerspruch. Haber und Einstein." In *Forschung im Spannungsfeld von Politik und Gesellschaft, Geschichte, und Struktur der Kaiser*, edited by Rudolf Vierhaus and Bernhard von Brocke, 516–543. München: Deutsches Verlag-Anstalt, 1990.

Stoltzenberg, Dietrich. *Fritz Haber: Chemiker, Nobelpreisträger Deutscher, Jude*. Weinheim: VCH Verlagsgesellschaft, 1994.

Fritz Haber: Chemist, Nobel Laureate, German, Jew. Philadelphia: Chemical Heritage Press, 2004.

Strachan, Hew. *Financing the First World War*. Oxford: Oxford University Press, 2004.

Strathausen, Carsten. "The Return of the Gaze: Stereoscopic Vision in Jünger and Benjamin." *New German Critique* 80 (2000): 125–248.

Strohn, Matthias. *The German Army and the Defense of the Reich: Military Doctrine and the Conduct of the Defensive Battle 1918–1939*. Cambridge: Cambridge University Press, 2011.

Sumner, Judith. *Plants Go to War: A Botanical History of World War II*. Jefferson: McFarland & Company, 2019.

Süss, Dietmar. "Wartime Societies and Shelter Politics in National Socialist Germany and Britain." In *Bombing, States and Peoples in Western Europe, 1940–1945*, edited by Richard Overy, Claudia Baldoli, and Andrew Knapp. London: Continuum, 2011.

Death from the Skies: How the British and Germans Survived the Bombing in World War II. Translated by Lesley Sharpe and Jeremy Noakes. Oxford: Oxford University Press, 2011.

Süss, Dietmar, ed. *Deutschland Im Luftkrieg: Geschichte und Erinnerung*. München: Oldenbourg Verlag, 2007.

Szöllösi-Janze, Margit. "Losing the War, but Gaining Ground: The German Chemical Industry during World War I." In *The German Chemical Industry in the Twentieth Century*, edited by John Lesch, 91–123. Boston: Kluwer Academic, 2000.

"Pesticides and War: The Case of Fritz Haber." *European Review* 9, no. 1 (2001): 97–108.

Fritz Haber, 1868–1934: eine Biographie. München: C.H. Beck Verlag, 1998.

Tammen, Helmuth. *Die I.G. Farbenindustrie Aktiengesellschaft (1925–1933): Ein Chemiekonzern in der Weimarer Republik*. Berlin: H. Tammen, 1978.

Taylor, Dorceta E. *Toxic Communities: Environmental Racism, Industrial Pollution, and Residential Mobility*. New York: New York University Press, 2014.

Thacker, Toby. *Joseph Goebbels: Life and Death*. New York: Palgrave Macmillan, 2009.

Thompson, Peter. "From Gas Hysteria to Nuclear Fear: A Historical Synthesis of Chemical and Atomic Weapons." *Historical Studies in the Natural Sciences* 52, no. 2 (2022): 223–264.

"The Pale Death: Poison Gas and German Racial Exceptionalism, 1915–1945." *Central European History* 54 (2021): 273–296.

"Wardens of the Toxic World: German Women's Encounters with the Gas Mask, 1915–1945." *German Studies Review* 43, no. 2 (2020): 353–376.

Torrie, Julia S. *"For Their Own Good": Civilian Evacuations in Germany and France, 1939–1945*. New York: Berghahn Books, 2010.

Trauma, 1914–1945. Exeter: University of Exeter Press, 2009.

Travers, Martin. *Critics of Modernity: The Literature of the Conservative Revolution in Germany, 1890–1933*. New York: Peter Lang, 2001.

Trumpener, Ulrich. "The Road to Ypres: The Beginnings of Gas Warfare in World War I." *The Journal of Modern History* 47 (1975): 460–480.

Turnbull, Neil. "Heidegger and Jünger on the 'Significance of the Century': Technology as a Theme in Conservative Thought." *Writing Technologies* 2, no. 2 (2009): 9–34.

Tubach, Frederic C. *German Voices: Memories of Life during Hitler's Third Reich*. Berkeley: University of California Press, 2011.

Ulrich, Berndt, and Benjamin Ziemann, eds. *German Soldiers in the Great War: Letters and Eyewitness Accounts*. Translated by Christine Brocks. Barnsley: Pen and Sword, 2010.

Uziel, Daniel. *The Propaganda Warriors: The Wehrmacht and the Consolidation of the German Home Front*. Bern: Peter Lang AG, 2008.

van Bergen, Leo. "The Poison Gas Debate in the Inter-War Years." *Medicine, Conflict and Survival* 24, no. 3 (2008): 174–187.

Vellacott, Jo. "Feminism as If All People Mattered: Working to Remove the Causes of War, 1919–1929." *Contemporary European History* 10, no. 3 (2001): 375–394.

von Bruch, Rüdiger, and Brigitte Kaderas, eds. *Wissenschaften und Wissenschaftspolitik Bestandsaufnahmen zu Formationen, Brüchen und Kontinuitäten im Deutschland des 20. Jahrhunderts*. Stuttgart: Franz Steiner Verlag, 2002.

Voskuhl, Adelheid. "Engineering Philosophy: Theories of Technology, German Idealism, and Social Order in High-Industrial Germany." *Technology and Culture* 57, no. 4 (2016): 721–752.

Vourkoutiotis, Vasilis. *Making Common Cause: German-Soviet Secret Relations, 1919–22*. New York: Palgrave Macmillan, 2007.

Wagener, Hans. *Understanding Erich Maria Remarque*. Columbia: University of South Carolina Press, 1991.

Watson, Alexander. *Enduring the Great War: Combat, Morale and Collapse in the German and British Armies, 1914–1918*. Cambridge: Cambridge University Press, 2008.

Weber, Thomas. *Hitler's First War: Adolf Hitler, The Men of the List Regiment, and the First World War*. Oxford: Oxford University Press, 2010.

Weindling, Paul. *Epidemics and Genocide in Eastern Europe 1890–1945*. Oxford: Oxford University Press, 2000.

"The Uses and Abuses of Biological Technologies: Zyklon B and Gas Disinfestation between the First World War and the Holocaust." *History and Technology* 11 (1994): 291–298.

Weisenfeld, Gennifer. "Gas Mask Parade: Japan's Anxious Modernism." *Modernism/Modernity* 21, no. 1 (January 2014): 179–199.

Westermann, Edward B. *Flak: German Anti-Aircraft Defenses, 1914–1945*. Lawrence: University of Kansas Press, 2001.

Wette, Wolfram. *The Wehrmacht: History, Myth, Reality*. Cambridge, MA: Harvard University Press, 2006.

Wettstein, Adrian E. "Urban Warfare Doctrine on the Eastern Front." In *Nazi Policy on the Eastern Front, 1941: Total War, Genocide, and Racialization*, edited by Alex J. Kay, Jeff Rutherford, and David Stahel, 45–73. Rochester: University of Rochester Press, 2012.

Wexler, Joyce. "The New Heroism." *Literature & The Arts* 27 (2015): 1–12.

Wietzker, Wolfgang. *Giftgas im Ersten Weltkrieg: Was konnte die deutsche Öffentlichkeit wissen?* Saarbrücken: Verlag Dr. Müller, 2007.

Wildt, Michael. *Hitler's Volksgemeinschaft and the Dynamics of Racial Exclusion: Violence Against Jews in Provincial Germany, 1919–1939*. Translated by Bernhard Heise. New York: Berghahn Books, 2012.

Winner, Langdon. *Autonomous Technology: Technics-out-of-Control as a Theme in Political Thought*. Cambridge, MA: MIT Press, 1977.

Winter, Denis. *Death's Men: Soldiers of the Great War*. London: Allen Lane, 1978.
Wright, Jonathan. *Germany and the Origins of the Second World War*. Basingstoke: Palgrave Macmillan, 2007.
Ziemann, Benjamin. *Contested Commemorations: Republican War Veterans and Weimar Political Culture*. Cambridge: Cambridge University Press, 2013.

Published Primary Material

Ahrends, Otto. *Mit dem Regiment "Hamburg" in Frankreich 1914–1916*. München: Verlag von Ernst Reinhardt, 1929.
Akademisches Wissenschaftliches Arbeitsamt. *A.W.A. Gasschutz-Lehrgang*. Berlin: Akademisches Wissenschaftliches Arbeitsamt e.V., 1932.
Albrecht Dorn, and Staszewsky, eds. *Kriegstagebuch Der 6. Kompagnie 18. Rgl. Sächs.Inf.Regt. Nr. 192*. Leipzig: Buchhandlung Sebr. Fändrich, 1930.
Alexander, Axel. *Die Schlacht über Berlin*. Berlin: Verlag "Offene Worte," 1933.
Alexander, Hans. *Der Völkermord Im Kommenden Giftgas-Kriege*. Wiesbaden: Westverlag D. Reisdorf & Co, 1926.
Aschoff, Ludwig. *Über anatomische und histologische Befunde bei "Gas"-Vergiftungen*. Berlin: Reichsdruckerei, 1916.
Auer-Gesellschaft. *Schulungs- Und Lehrmittel Für Den Gasschutz*. Berlin: Auergesellschaft Aktiengesellschaft, 1941.
Bab, Julius. "Giftgas und Zensur." *Die Hilfe* 35 (1929): 169–171.
Bauer, Max. *Der Grosse Krieg in Feld Und Heimat*. Tübingen: Osiander'sche Buchhandlung, 1922.
Becher, Johannes, *(CHCl=CH)3As (Levisite) oder Der einzig gerechte Krieg*. Berlin: Aufbau-Verlag, 1968.
Benary, Albert. *Luftschutz*. Leipzig: Verlag von Philipp Reclam, 1933.
Benjamin, Dora Sophie. "Die Waffen von Morgen." *Vossische Zeitung* no. 303 (1925): 1–2.
Benjamin, Walter. "Theories of German Fascism: On the Collection of Essays War and Warrior, Edited by Ernst Jünger." *New German Critique* 17, Special Walter Benjamin Issue (Spring 1979): 120–128.
The Work of Art in the Age of Its Technological Reproducibility and Other Writings on Media. Edited by Michael Jennings, Brigid Doherty, and Thomas Levin. Cambridge, MA: Harvard University Press, 2008.
Besse, Carl. *Gaskampf Und Gasschutz*. Berlin: Offene Worte, 1932.
Biha, Otto. "Der Proletarische Massenroman. Eine Neue Eine-Mark-Serie des 'Internationalen Arbeiterverlages.' " *Die Rote Fahne* 178 (August 2, 1930).
Binding, Rudolf. *A Fatalist at War*. Translated by Ian F.D. Morrow. London: George Allen & Unwin, 1929.
Blücher, Evelyn. *An English Wife in Berlin*. London: Constable and Company, 1920.
Boyd, William. *With a Field Ambulance at Ypres*. New York: George H. Doran Co, 1916.
Broch, Hermann. *The Sleepwalkers*. New York: Vintage, 1996.
Bruchmüller, Georg. *Die Artillerie beim Angriff im Stellungskrieg*. Berlin: Offene Worte, 1926.

Bucholtz, Ferdinand, ed. *Der Gefährliche Augenblick*. Berlin: Junker und Dünnhaupt Verlag, 1931.

Burr, Hans. *Das Württemburgische Infatrie-Regiment Nr. 475 im Weltkrieg*. Stuttgart: C. Belsersche Verlagsbuchhandlung, 1921.

Büscher, Hermann. *Giftgas und wir?: Die Welt der Giftgase Wesen und Wirkung, Hilfe und Heilung*. Leipzig: Johann Ambrosius Barth, 1937.

Büscher, Hermann. *Grün- und Gelbkreuz: Spezielle Pathologie und Therapie der Körperschädigungen durch die chemischen Kampfstoffe der Grünkreuz- (Phosgen und Perchlorameisensäuremethylester [Perstoff]) und der Gelbkreuz-Gruppe (Dichloraethylsulfid und β-Chlorvinylarsindichlorid [Lewisit])*. Leipzig: Johann Ambrosius Barth Verlag, 1932.

Chemische Fabrik Stoltzenberg. "Instruktionskasten Fur Chemische Kampfstoffe Und Ihren Sanitatskoffer Fur Kampfgaserkrankungen," n.d. Leibniz-Informationszentrum Wirtschaft. http://zbw.eu/beta/p20/company/42006/00002/about.en.html.

Convention (II) with Respect to the Laws and Customs of War on Land (Hague, II) (29 Jul 1899), Organisation for the Prohibition of Chemical Weapons. www.opcw.org/cheimcal-weapons-convention/related-international-agreements/chemical-warfare-and-chemical-weapons/hague-convention-of-1899/.

Degea A.G. *Farbspritzen Und Gasschutz*. Berlin: Degea A.G., 1936.

Dehmel, Richard. *Zwischen Volk und Menschheit: Kriegstagebuch von Richard Dehmel*. Berlin: S. Fischer Verlag, 1919.

DeLillo, Don. *White Noise*. New York: Viking, 1985.

Denckler, Heinz, ed. *Gasschutz Und Gasmaske*. Berlin: Heinz Denckler Verlag, 1935.

Döblin, Alfred. *Berlin Alexanderplatz*. Translated by Eugene Jolas. New York: Frederick Ungar, 1961.

Doebel, Peter, ed. *"So Etwas Wie Weltuntergang": Günter Goebel Kriegstagebücher 1939–1945*. Mainz: C.P. Verlag, 2005.

Doering, Walther. *Luftschutz Und Gasschutz: Markbuch Für Den Wehrfähigen*, vol. 3. Breslau: Völkischer Verlag Walter Uttikal, 1934.

Dollen, Ingrid von der, ed. *Briefe an Die Eltern von Hermann Boeddinghaus 1898–1941*. Bad Honnef: KAT-Verlag, 2009.

Dominik, Hans. *Atomgewicht 500*. Berlin: Verlag Scherl, 1935.

Der Gas-Tod Der Grossstadt. Dortmund: Synergen Verlag, 2015.

Douhet, Giulio. *The Command of the Air*. Translated by Dino Ferrarri. Washington, DC: Air Force History and Museums Program, 1998.

Dräger Gasschutz Im Luftschutz: Individual-Gasschutz Kollektiv-Gasschutz; Charakter Des Chemischen Krieges Chemische Kampfstoffe; Organisation Des Luftschutzes Städtebau Und Luftschutz, vol. 2. Lübeck: Kommissionsverlag H.G. Rahtgens, 1933.

Dräger, Heinrich. *Heinrich Dräger Lebenserinnerungen*. Hamburg: Alfred Janssen, 1917.

Duisberg, Carl. *Meine Lebenserinnerungen*. Edited by Jesco v. Puttkamer. Leipzig: Philipp Reclam, 1933.

Ebeling, Fritz. *Geschichte Des Infanterie Regiments Herzog Friedrich Wilhelm von Braunschweig Nr 78 Im Weltkriege*. Berlin: Gerhard Stalling Verlag, 1924.

Ehrenburg, Ilya. *Men, Years – Life, III*. London: MacGibbon & Kee, 1964.
Einstein, Albert. *Einstein on Politics: His Private Thoughts and Public Stands on Nationalism, Zionism, War, Peace, and the Bomb*. Edited by David E. Rowe and Robert Schulmann. Princeton: Princeton University Press, 2007.
Elsner, Richard, ed. *Das Deutsche Drama: Ein Jahrbuch*. Berlin-Pankow: Verlag der Deutschen Nationalbühne, 1929.
Endres, Franz Carl. *Giftgaskrieg: Die Grosse Gefahr*. Zürich: Rascher & Cie, 1928.
Fendrich, Anton. *Mit Dem Auto an Der Front*. Stuttgart: Franckh'sche Verlagshandlung, 1918.
Fessler, Gebele, and Prandtl. *Gaskampstoffe Und Gasvergiftungen: Wie Schützen Wir Uns?* München: Verlag der Ärtzlichen Rundschau, 1931.
Finzen, Dieter. "Kriegstagebuch: Das Tagebuch von Dieter Finzen Aus Beiden Weltkriegen." *The War Diaries of Dieter Finzen* (blog), June 26, 2017. https://dieter-finzen.blogspot.com/search?q=gas.
Flury, Ferdinand, and Franz Zernik. *Schädliche Gase: Dämpfe, Nebel, Rauch- Und Staubarte*. Berlin: Springer-Verlag, 1931.
Foerster, Wolfgang, ed. *Wir Kämpfer im Weltkrieg: Feldzugsbriefe und Kriegstagebücher von Franktkämpfern aus dem Material des Reichsarchivs*. Berlin: Neufeld & Henius Verlag, 1929.
Frauenholz, Eugen von, ed. *Kronprinz Rupprecht von Bayern: Mein Kriegstagebuch*. München: Deutscher National Verlag, 1929.
Fredewess-Wenstrup, Stephanie. *Mutters Kriegstagebuch: Die Aufzeichnungen der Antonia Helming 1914–1922*. Münster: Waxmann, 2005.
Fried, Alfred Hermann. *Mein Kriegstagebuch*. Bremen: Donat Verlag, 2005.
Friedrich, Ernst. *Krieg dem Kriege!* Berlin: Freie Jugend, 1924.
Frisch, Otto. *What Little I Remember*. Cambridge: Cambridge University Press, 1979.
Frommhold, Hans. *Luftschutzraum-Bauweisen*. Berlin: Verlag Gasschutz und Luftschutz, 1939.
Gemeinhardt, K. "Gasschutz und Luftschutz der Zivilbevölkerung." *Archiv Der Pharmazie* 270, no. 4 (1932): 232–246.
Geyer, Hermann. *Der Weltkampf um Ehre und Recht: Der Seekrieg, der Krieg um die Kolonien, die Kampfhandlungen in der Türkei, der Gaskrieg, der Luftkrieg*. Band 4. Leipzig: Alleinvertrieb durch Ernst Finking, 1922.
Gobsch, Hans. *Wahn-Europa 1934*. Hamburg: Fackelreiter-Verlag, 1931.
Göring, Hermann. *Hermann Göring: Reden Und Aufsätze*. Edited by Erich Gritzbach. München: Zentralverlag der NSDAP, 1942.
Graus aus Schmechten, Karl. *Elektrokardiographie Untersuchungen bei Belastun älterer Personen unter der Gasmaske*. Höxter: Huxaria Druckerei und Verlagsanstalt, 1939.
Griesbach, Georg Erich. *Lost Wird Luftschutzhauswart*. Leipzig: A. Anton & Co, 1934.
Grimme, Hugo. *Der Reichsluftschutzbund: Ziele, Leistungen Und Organisation*. Berlin: Junker und Dünnhaupt Verlag, 1936.
Grothe, Heinz. *Das Fronterlebnis*. Berlin: Joachim Goldstein Verlag, 1932.
Haase-Lampe, Wilhelm. *Handbuch für das Grubenrettungswesen Sauerstoffretttungswesen und Gasschutz, Gerätebau und Organisation seit 1924*. Lübeck: H.G. Rahtgens, 1929.

Sauerstoffrettungswesen und Gasschutz Gerätebau und Organisation in ihrer internationalen Entwicklung: Gerätebau. Lübeck: H.G. Rahtgens, 1924.
Haber, Charlotte. *Mein Leben mit Fritz Haber*. Düsseldorf: Econ Verlag, 1970.
Haber, Fritz. *Fünf Vorträge Aus Den Jahren 1920–1923*. Berlin: Verlag von Julius Springer, 1924.
Haeuber, E., and G. Gassert. *Der Kampf Um Den Luftschutz*. Berlin: Verlag Deutscher Luftschutz, 1929.
Hahn, Otto. *Mein Leben: Die Erringerung des grossen Atomforschers und Humanisten*. München: Piper, 1986.
Hampe, Erich. *Der Mensch Und Die Gase: Einführung in Die Gaskunde Und Anleitung Zum Gasschutz*. Berlin: Räder-Verlag, 1932.
Der Mensch Und Die Gasgefahr: Einführung in Die Gaskunde Und Anleitung Zum Gasschutz. Berlin: Räder-Verlag, 1940.
Der Zivile Luftschutz Im Zweiten Weltkrieg: Dokumentation Und Erfahrungsberichte Über Aufbau Und Einsatz. Frankfurt: Bernard & Graefe Verlag, 1963.
Hänsel, Günter. *Die Spätfolgen von Kampfgasvergiftungen an den Lungen und ihre versorgungsrechtliche Bewertung*. Lippstadt: Thiele Lippstadt/Westf., 1934.
Hanslian, Rudolf. *Der chemische Krieg*. Berlin: E.S. Mittler & Sohn, 1927.
Der deutsche Angriff bei Ypern am 22. April 1915. Berlin: Verlag Gasschutz und Luftschutz, 1934.
Die Internationale Gasschutzkonferenz in Brüssel. München: Verlag für das gesamte Schiess- und Sprengstoffwesen, 1928.
Vom Gaskrieg Zum Atomkrieg: Die Entwicklung Der Wissenschaftlichen Waffen. Stuttgart: Verlag Chemiker Zeitung, 1951.
Harder-Hamburg, Hans. "Der Gaskrieg Im Völkerrecht." *Die Friedens-Warte* 27, no. 5 (1927): 137–139.
Hartley, Harold. "Report on German Chemical Warfare, Organisation and Policy, 1914–1918." Public Record Office, Kew, WO 33 (A 2715) 1072.
Hartung, Wilhelm. *Großkampf Männer und Granaten!* Berlin: Verlag Tradition Wilhelm Kolk, 1930.
Heiber, Helmut, and David M. Glantz, eds. *Hitler and His Generals: Military Conferences 1942–1945*. New York: Enigma, 2003.
Helbig, Hans, and E. Sellien. *Der Luftschutz in Schulen Und Hochschulen: Eine Erläuterung Der LDv 755/2 Mit Einer Sammlung Der Einschlägigen Gesetze, Verordnungen, Erlasse*. Berlin: Otto Stollberg, 1942.
Hiller, Kurt. *Der Sprung Ins Helle*. Leipzig: Wolfgang Richard Lindner Verlag, 1932.
Hirsch, Werner. *Die Rote Fahne: Kritik, Theorie, Feuilleton 1918–1933*. Edited by Manfred Brauneck. München: Wilhelm Fink Verlag, 1973.
Hitler, Adolf. *Mein Kampf*. Translated by Marco Roberto. Independent, 2017.
Hoffmann, Rudolf, ed. *Der deutsche Soldat: Briefe aus dem Weltkrieg*. München: Albert Langen/Georg Müller, 1937.
Hoffmann, Wilhelm. *Die deutschen Ärzte im Weltkriege. Ihre Leistungen und Erfahrungen*. Berlin: E.S. Mittler & Sohn, 1920.
Hugo Stoltzenberg Fabrik. "Gibt Es Ein Gaskampf?," n.d. Leibniz-Informationszentrum Wirtschaft. http://zbw.eu/beta/p20/company/42006/00001/about.en.html.

Hulbert, Harold S. "Gas Neurosis Syndrome." *American Journal of Insanity* LXXVII, no. 2 (October 1920): 213–216.

Hummel, Karl, and Friderich Schweinsberg. *Erste Hilfe Im Täglichen Leben Und Im Luftschutz*. München: Dr. Friedrich Schwinsberg, 1936.

Hunke, Heinrich. *Luftgefahr Und Luftschutz: Mit Besonderer Berücksichtigung Des Deutschen Luftschutzes*. Berlin: Verlag von E.S. Mittler & Sohn, 1935.

Imme, Theodor, *Die deutsche Soldatensprache der Gegenwart und ihr Humor*. Dortmund: Fr. Wilhelm Ruhfus, 1917.

Jaspers, Karl. *General Psychopathology*. Translated by J. Hoenig and Marian W. Hamilton. Vol. II. Baltimore: Johns Hopkins University Press, 1997.

J.P. "Die Erste Gasmaske." *Protar* 11, no. 3 (1936): 200–201.

Jünger, Ernst. "Die Technik in Der Zukunftsschlacht." *Militär Wochenblatt* 14, October 1, 1921.

"On Danger." In *New German Critique*, 59. New York: Telos Press, 1993, 27–33.

"War and Photography." In *New German Critique,* 59. New York: Telos Press, 1993, 24–27.

Blätter und Steine. Hamburg: Hanseatische Verlagsanstalt, 1941

Copse 125: A Chronicle from the Trench Warfare of 1918. Translated by Basil Creighton. New York: Zimmerman & Zimmerman, 1985.

Der Kampf Als Inneres Erlebnis. Berlin: E.S. Mittler & Sohn, 1922.

Die Totale Mobilmachung. Berlin: Verlag für Zeitkritik, 1931.

Jünger, Ernst, ed. *Die Unvergessenen*. Berlin: Wilhelm Andermann Verlag, 1928.

Feuer und Blut. Magdeburg: Frundsberg-Verlag, 1926.

On Pain. Translated by David C. Durst. London: Telos, 2008.

Sämtliche Werke, Band 7: Essays I. Stuttgart: Klett-Cotta, 1980.

Storm of Steel. Translated by Michael Hofmann. New York: Penguin, 2003.

Sturm. Translated by Alexis P. Walker. Candor, NY: Telos Press, 2015.

The Worker: Dominion and Gestalt. Translated by Dirk Leach. Albany: SUNY Albany Press, 1990.

Jünger, Ernst, ed. *Krieg Und Krieger*. Berlin: Junker und Dünnhaupt Verlag, 1930.

Jünger, Ernst, ed. *Luftfahrt Ist Not!* Leipzig: Wilhelm Andermann Verlag, 1930.

Kaiser, Georg, *Gas I*. Potsdam: Gustav Kiepenheuer, 1925.

Gas II. New York: Ungar, 1963.

Kalshoven, Hedda, ed. *Between Two Homelands: Letters across the Borders of Nazi Germany*. Translated by Peter Fritzsche and Hester Velmans. Urbana: University of Illinois Press, 2014.

Kerwin, Jerome G. "The German Reichstag Elections of July 31, 1932." *The American Political Science Review* 26, no. 5 (1932): 921–926.

Kessel, Jürgen, ed. *"Jezt gehts in die Männer mordende Schlacht..." Das Kriegstagebuch von Theodor Zuhöne 1914–1918*. Damme: Heimatvereins, 2002.

Kessler, Harry. *Berlin in Lights: The Diaries of Count Harry Kessler (1918–1937)*. Translated by Charles Kessler. New York: Grove Press, 1999.

Keun, Irmgard. *After Midnight*. Translated by Anthea Bell. Brooklyn: Melville House, 2011.

Kiesel, Helmuth, ed. *Ernst Jünger Kriegstagebuch 1914–1918*. Stuttgart: Klett-Cotta, 2010.

Knack. "Kampfgasvergiftungen." *Deutsche Medizinische Wochenschrifft* 43 (1917): 1246.

Knipfer, Kurt, and Erich Hampe. *"Der Zivile Luftschutz" Ein Sammelwerk Über Alle Fragen Des Luftschutzes*. Berlin: Verlagsanstalt Otto Stollberg, 1937.

Knipfer, Kurt, and Werner Burkhardt. *Luftschutz in Bildern: Eine Gemeinverständliche Darstellung Des Gesamten Luftschutzes Für Jeden Volksgenossen*. Berlin: Landsmann, 1935.

Kock, Oliver, ed. *Das Tagebuch des Leutnants Nilius: Ostern 1916 bis 21.2.1918*. Bayreuth: Scherzers-Militaer Verlag, 2013.

Köppen, Edlef. *Heeresbericht*. München: Deutsche Verlags-Anstalt, 2004.

Kraus, Karl. "Das Technoromantische Abenteuer." *Die Fackel* 474–483 (May 23, 1918).

Krohne, Rudolf. *Luftgefahr Und Luftschutzmöglichkeiten in Deutschland*. Berlin: Verlag Deutscher Luftschutz, 1928.

Krüger, Friedrich. *Der Luftschutz Für Jugend Und Schule*. Leipzig: Verlag der Dürr'schen Buchhandlung, 1933.

Laffert, Karl August von. *Giftküche*. Berlin: August Scherl, 1929.

Leers, Johann von. *Bomben Auf Hamburg*. Leipzig: R. Voigtländers Verlag, 1932.

Lichtenberg, Wilhelm. Luft- Und Gasschutz-Fibel Als Instruktion Für Die Hitler-Jugend Und B.D.M. *Reichsjugendführung*. Berlin, 1934.

Ludendorff, Erich. *Der Totale Krieg*. München: Ludendorffs Verlag, 1935.

Maier, Reinhold, ed. *Feldpostbriefe Aus Dem Ersten Weltkrieg 1914–1918*. Stuttgart: W. Kohlhammer Verlag, 1966.

Mehl, Arnold, *Schatten der aufgehenden Sonne*. Leipzig: Wilhelm Goldmann Verlag, 1935.

Meyer, Andreas. *It's Not the Fatherland's Fault!? – Keep the Letters for Later: Dr. Otto Meyer, from Gunner to Lieutenant, a German-Jewish War Career*. Andreas Meyer, 2013.

Meyer, E, E. Sellien, and Pol-Major Borowietz. *Schule Und Luftschutz*. München: Verlag von R. Oldenbourg, 1934.

Meyer, Julius. *Der Gaskampf Und Die Chemischen Kampfstoffe*. Leipzig: Verlag von S. Hirzel, 1925.

Mühsam, Erich. "Phosgen." Fanal: Ararchistische Monatsschrift *Jahrgang* 2, Nummer 9 (June 1928): 208–210.

Müller-Kiel, Ulrich. *Die Chemische Waffe: Im Weltkrieg Und Jetzt*. Berlin: Verlag Chemie, GMBH, 1932.

Müller, Carl Werner. *Verzicht auf Revanche: Das Kriegstagebuch 1914/18 des Divisionspfarrers der Landauer Gernison Dr. Anton Foohs*. Speyer: Pfälizische Gesellschaft zur Förderung der Wissenschaften, 2010.

Muntsch, Otto. "Beiträge Zur Behandlung der Hautschädigungen Durch Dichloräthylsulfid." In *Med. Dissertationen Der Bayer Julius-Maximilians-Universität Zu Würzburg*, 1–41. Würzburg: C.J. Becker Universitäts-Druckerei, 1928.

Leitfaden Der Pathologie Und Therapie Der Kampfstofferkrankungen. Leipzig: Georg Thieme, 1939.

Nestler, Waldus. "Collective and Individual Protection." In *Chemical Warfare: An Abridged Report of Papers Read at an International Conference at Frankfurt am Main*. London: Williams & Norgate, 1930.

Giftgas Über Deutschland. Berlin: Nebelhorn-Verlag, 1932.

Ohliger, Ernst. *Bomben Auf Kohlenstadt*. Oldenburg: Gerhard Stalling Verlag, 1935.

Olmütz-Luttein. *Die Gasmaske : Auszüge Aus Den in Den Gas- u. Luftschutzkursen in Olmütz-Luttein Vorgetragenen Vorträgen*. Olmütz-Luttein: Olmütz-Luttein Verlag Chema, 1936.

Orts-Gruppe Heidenheim des Reichsluftschutzbundes, ed. *Die Luftschutzgrundschule: Ein Laitfaden Für Die Schulung Im Luftschutz*. Stuttgart: W. Kohlhammer Verlag, 1937.

Owen, Wilfred. "Dulce et Decorum Est." In *Poems*. New York: Viking Press, 1992

Pfuhl, Karl Helmut. *Gaskrieg Und Völkerrecht*. Würzburg: Handelsdruckerei Würzburg, 1930.

Pick, Friedel. "Über Erkrankungen Durch Kampfgase." *Zentralblatt Für Innere Medizin* 39, no. 20 (1918): 305–310.

Plessner, Helmuth. *The Limits of Community: A Critique of Social Radicalism*. Translated by Andrew Wallace. Amherst, NY: Humanity Books, 1999.

Reck, Friedrich. *Diary of a Man in Despair*. Translated by Paul Rubens. New York: New York Review of Books, 2013.

Reichsluftschutzbund. *Luftschutz Ist Selbstbehauptungswille: Aufgaben Und Erfahrungen Über Die Ausbildung Im Zivilen Luftschutz*. Berlin: Reichsluftschutzbund RLB, 1936.

Reimann, Günther. *Giftgas in Deutschland: Die Machtstellung der I.G. Farbenindustrie A.G.* Berlin: Vereinigung Internationaler Verlagsanstalten, 1927.

Remarque, Erich Maria. *All Quiet on the Western Front*. Greenwich: Fawcett, 1967.

The Road Back. New York: Grosset & Dunlap, 1931.

Remy, Heinrich. *Denkschrift, Aufgaben des Chemikers im Luftschutz*, 16 May 1933.

Renn, Ludwig. *Krieg*. Leipzig: Philipp Reclam, 1979.

Riedel, Freider, ed. *Cornelius Breuninger: Kriegstagebuch 1914–1918*. Leinfelden-Echterdingen: Numea Verlag, 2007.

Rosten, Curt. *Was Man Vom Luftschutz Wissen Muss*. Berlin: Verlag Deutsche Kultur-Wacht, 1935.

Rothe, Hans. "Die Liebe mit der Gasmaske." *Die Literarische Welt* 7, no. 7 (1931): 5–6.

Sauter, Robert, ed. *Ich setze mich nieder und schreibe. Gregor: Eine Lebensgeschichte zum Ersten Weltkrieg*. Plaidt: Cardamina Verlag, 2014.

Schauwecker, Franz. *Aufbruch Der Nation*. Berlin: Frundsberg Verlag, 1930.

Endkampf 1918. Frankfurt: Verlag Moritz Diesterweg, 1936.

Im Todesrachen. Halle: Heinrich Diekmann Verlag, 1921.

So war der Krieg. Berlin: Frundsberg Verlag, 1929.

Schmidt, Walter. Das 7. *Badische Infantrie-Regiment Nr. 142 im Weltkrieg 1914/18*. Freiburg im Breisgau: Herder Verlag, 1927.

Schneider, Karl. *Blutgase und Kreislauf bei Arbeit under der Gasmaske*. Würzburg, Universitätsdruckerei H. Stürtz A.G. Würzburg, 1939.

Scholtz, Gerhard. *Tagebuch einer Batterie*. Potsdam: Rütten und Loening Verlag, 1941.

Schröder, Richard, Hans von Fichte, Karl Pomplum-Breslau, Klaus Schaefer, and Kurt Oskar Bark, eds. *Luftschutz : Die Deutsche Schicksalsfrage*. Stuttgart: Verlag Heinrich Plesken, 1934.

Schultz, Edmund, ed. *Die Veränderte Welt*. Breslau: Wilhelm Gottl. Korn Verlag, 1933.

Schuster, Harald. *Die Aufnahmeleistung von Atemfiltern und ihre Bedingtheiten*. Lübeck: Heinr. & Bernh. Draeger, 1936.

Schütt, K. *Die Chemischen Und Physikalischen Grundlagen des Luftschutzes in der Schule*. Berlin: C.J.E Volckmann Nachf., 1935.

Schütter, Theo, ed. *Feldpostbriefe Kriegstagebuch 1914–1918*. Werne: Ventura Verlag Magnus See, 2009.

Schwarte, Max. *Die militärischen Lehren des Grossen Krieges*. Berlin: Ernst Siegfried Mittler und Sohn, 1920.

Schweisheimer, W. "Die Medizinischen Grundlagen Des Giftgaskrieges." *Militär Wochenblatt* 17, no. 110 (1925): 577–582.

Shirer, William L. *Berlin Diary: The Journal of a Foreign Correspondent, 1934–1941*. Baltimore: Johns Hopkins University Press, 2002.

Speer, Albert. *Inside the Third Reich*. Translated by Richard and Clara Winston. New York: Simon and Schuster, 1970.

Spengler, Oswald. *Man and Technics: A Contribution to the Philosophy of Life*. Translated by Charles Francis Atkinson. New York: Greenwood Press, 1976.

Steuernagel, Konrad. *Kriegstagebuch 1914/1918 des II. Batls. Ref.=Inf.=Regts. Nr. 221*. Worms am Rhein, 1937.

Stoltzenberg-Bergius, Margarete. *Was Jeder Vom Gaskampf Und Den Chemischen Kampfstoffen Wissen Sollte*. Hamburg: Chemische Fabrik Stoltzenberg, 1930.

Stoltzenberg, Hugo. *Experimente Und Demonstrationen Zum Luftschutz*. Hamburg: Hugo Stoltzenberg Fabrik, 1933.

Stoltzenberg. Hugo. *Anleitung Zur Herstellung von Ultra Giften*. Hamburg: Norwi-Druck, 1930.

Teetzmann, Otto A. *Der Luftschutz Leitfaden Für Alle*. Berlin: Verlag des Reichsluftschutzbundes, 1935.

Tempelhoff, Friedrich von. *Gaswaffe Und Gasabwehr: Einführung in Die Gastaktik*. Berlin: Verlag von E.S. Mittler & Sohn, 1937.

The Avalon Project, Laws of War: Declaration on the Use of Projectiles the Object of Which is the Diffusion of Asphyxiating or Deleterious Gases; July 29, 1899, Published 2008. http://avalon.law.yale.edu/19th_century/dec99–02.asp.

The Treaty of Versailles. www.loc.gov/law/help/us-treaties/bevans/m-ust000002-0043.pdf.

Thuillier, Henry F. *Gas Im Nächsten Krieg*. Berlin: Albert Nauck & Co, 1939.

Tochenhausen, Friedrich von, ed. *Wehrgedanken: Eine Sammlung Wehrpolitischer Aufsätze*. Hamburg: Hanseatische Verlagsanstalt, 1933.

Tucholsky, Kurt. "Waffe gegen den Krieg." *Die Weltbühne*, 23 February 1926.

Unruh, Fritz von. *Opfergang*. Frankfurt: Frankfurter Societäts Druckerei, 1966.

Viktorit-Antiplyn, eds. *Was Ein Jeder Über Die Gasmaske Wissen Soll Und Muss*. Brünn: Viktoria Gummiindustrie, 1933.

Vogel, Bruno. *Es Lebe der Krieg!* Leipzig: Verlag die Wölfe, 1925.
Wachtel, Curt. *Chemical Warfare*. London: Chapman & Hall, 1941.
Weber, Josef. *Das Neue Handbuch Für Den Luftschutz: Ein Buch Für Jedermann!* Berlin: J. Weber Buch-Verlag, 1935.
Westman, Stephen. *Surgeon with the Kaiser's Army*. London: William Kimber, 1968.
Willstätter, Wilhelm. *Aus meinem Leben*. Winheim: Verlag Chemie, 1949.
Woker, Gertrud. "Die Gaskampfinteressenten." *Die Friedens-Warte* 27, no. 2 (February 1927): 41–42.
"Die Wahrheit Über Den Gaskrieg." *Die Friedens-Warte* 27, no. 1 (1927): 5–8.
"Erwiderung." *Die Friedens-Warte* 25, no. 9 (September 1925): 266–268.
"Im Zeichen der Wissenschaft dem Abgrund Entgegen. (Betrachtungen Zum Chemischen Krieg)." *Die Friedens-Warte* 25, no. 1 (January 1925): 12–20.
Der Kommende Giftgaskrieg. Leipzig: Ernst Oldenburg Verlag, 1932.
Wissenschaft Und Wissenschaftlicher Krieg. Zürich: Zentralstelle für Friedensarbeit, 1925.
Woker, Getrud. "Ueber Giftgaskrieg." *Die Friedens-Warte* 23, no. 11/12 (November 1923): 393–395.
Zech, Paul. *Von der Maas bis die Marne*. Leipzig: Griefenverlag zu Rudolstadt, 1986.
Zöberlein, Hans. *Der Glaube an Deutschland*. München: Zentralverlag der NSDAP, 1937.
Zuckmayer, Carl. *Second Wind*. New York: Doubleday, Doran and Co, 1940.
Zweig, Arnold. *Education Before Verdun*. London: Martin Secker & Warburg, 1936.
Zweig, Stefan. *Stefan Zweig Briefe 1932–1942*. Edited by Knut Beck and Jeffrey B. Berlin. Frankfurt: S. Fischer Verlag, 2005.

Archival Material

BASF Archiv (BASF)

IG A 866/1.
Pertinenzbestand P.7./6
Pertinenzbestand BASF PB/A.8.7.3./9

Bayer Archiv (BA)

2 Weltkrieg Chemische Waffen. 204-006.
Abwehr von Stinkgeschossen etc, 1916–. 201-008-001.
Direktions Abteilung Luft- und Gasschutz, 1934–1940. 329-0350-02.
Herstellung Und Lieferung von Geschoss-Fullung, 1914. 201-005-002.
Herstellung Und Lieferung von Geschoss-Fullungen III, 1914. 201-005-003.
Kranken Arztenwesen Korperbeschadigte, 1915–. 231.012.
Victor Lefebure, *The Riddle of the Rhine*, 1921. 201-043.

Bayerisches Hauptsaatsarchiv: Abteilung Kriegsarchiv (BHK)

Armeeoberkommando 6, 1917–1918, 2427.
Armee Oberkommando der 6. Armee Gasschutz, 1915–1918.
Betäubungs und Gas-Schutzmittel im Kriege, 1914/15. M Kr. 13826.
Gasschutzgerät, 1919–1920. Reichswehrbrigade 23, 127.
Gasschutzwesen, Allg.u.Besonderes, 1915–1918, Generalkommando I. Armee-Korps (WK) 1276.
Gasubungen Sanitätsamt, 1918. Stv GenKdo I. AK SanA 136.
Infanterie-Divisionen (WK) 1245, 1915–1918.

Brandenburgisches Landeshauptarchiv (BL)

Auergesellschaft Oranienburg, 1929. Rep 43 Eberswalde, Nr 129.

Bundesarchive Berlin-Lichterfelde (BArch-B)

Auergesellschaft, 1939. NS/19/ Sig 3446 Ort 51 Haus 901/EG.
Ausnutzung der deutschen chemischen Industrie fur eine entscheidungsuchende Kriegfuhrung, 1938. R/3113/.
Bakterielle Ratten und Mause bekämpfung, 1926. R/154/.
Beschaffung der Tragetaschen zür Gasmaske 24, 1932.
Deutsche Luftschutz und Gasschutz Korrespondenz, 1933. R/36/.
Gasschutz fur die Zivilbewolkerung, 1925–1933. R/43/I.
NS/6/ 1943.
Ratten IV, 1920–. R/154/.
Reichsfinanz-ministerium, 1935. S.
Stahlhelmzeitung "Luftschutz tut not" Briefe, 1932. S.
Vergiftungen durch Gasse, 1928. R/86/.
Vergiftungen durch Gase, 1932. R/86/.
Vertrag über die Herstellen von Gasschutzfiltern.
Zahlungsschwierigkeiten der chemischen Frabrik Dr Hugo Stoltzenberg, 1926. R 43-I/420.

Bundesarchiv Militärarchiv (BA-MA)

4. Armee, 1915. PH3/569 Lesefilmnummer M4521F.
Allgemeine Erfahrungen im Gaskampf, 1919. PH 14/226 Lesefilmnummer 4875F.
Anhaltspunkte fur die Prufung des Standes der Ausbildung in der Gasabwehr bei Besichtigungen Ubungen usw, 1940. RH/11/IV.
Anhaltspunkte fur die Prufung des Standes der Ausbildung in der Gasabwehr bei Besichtigungen Ubungen usw. RH/11/IV.
Armeeoberkommando 6, 1917–1918. 2427.

Befehle zum Schutz gegen Gas, 1925. RH/12/4.
Bericht der Kaiser Wilhelm Institut fur Arbeitsphysiologie, 1934. RH/12/4.
Bericht uber Beobachtungen an Gaskranken an der Westfront, 1917. PH/2/.
Besprechung bei Herrn Generalmajor Thomas am September 5, 1939. RH 12-9/25.
Bestimmungen Ausstattung des Heeres mit Gasschutzgerat- Beschaffung, Versendung, Verwaltung und Unterbringung, 1920. RH/12/4.
Bestimmungen Ausstattung des Heeres mit Gasschutzgerat, 1920–1924. RH/12/4.
Brief an den Feldsantitätschef, August 3, 1916. PH/2/.
Chef des Generalstabes des Feldheeres, 1915, PH3/1012 Lesefilmnummer 4576F.
Chemische Abteilung II, 1927. RH/12/4.
Die feindlichen Gasangriffe in der Somme-Schlacht, 1916. PH 5-II/19 Lesefilmnummer 4590F.
Dienstvorschrift fur den Gaskampf und Gasschutz, 1917. PH/2/.
Einzel und Sammelschutz, 1931–1932. RH/12/4.
Einzel-u. Sammelschutz, 1933–1936. RH/12/4.
Erfahrungen im Gaskampf, 1920. PH 14/229 Lesefilmnummer 4875F.
Erfahrungs aus den Gasangriffen der Englander am 25 September 1915. PH 6-I/36 Lesefilmnummer 4644F.
Feldartillerie-Regiment Nr 18 Gasschiessen aus dem Jahre, 1918, PH/21.
Formationen des ingeneurskorps und Pionierkorps der Preussischen Armee, 1917. PH/14/.
Gas und Entseuchungsraume, 1931–1936. RH/12/4.
Gas und Nebel im Volkerrecht, 1933–1936. RH/12/4.
Gasabwehrubung, 1937 RH/12/4.
Gasabwehrubungen, 1938 RH/12/9.
Gasangriff an der amerikanischen Front in Frankreich, 1918. RH/8/.
Gaskampf. Grundlegende Erlasse und Arbeitspläne II, 1937. RH/12/9.
Gasschutz Vorschriften und andere Vorschriften, 1938. RH/12/9.
Gasschutz-Volk, 1932–1933. RH/12/4.
Gasschutz-Volk, 1934–1936. RH/12/4.
Generalstabes des Heeres on Gas, 1915, PH3/252 Lesefilmnummer 4503F.
Geratetechnisches, Konstruktion, Erfindungen, Versuche, Formveranderungen usw, 1933–1936. RH/12/4.
Gorz 18 Hauptmann Ruge Stab 213.J.D., 1918. PH 8-I/56 Lesefilmnummer 4676F.
Hans von Seeckt, Vortrag von Gaskrieg. RH 2/2207, File HL 27 1923 1A.
Infanterie-Divisionen. WK. 1245, 1915–1918.
Kampfstoff Forschung Beiheft: Sitzungberichte, 1925–1931. RH/12/4.
Konstruktion, Erfindungen, Erfahrungen uber Gasschutzgerat, 1927–1929. RH/12/4.
Kriegserfahrungen im Gasschutz, 1919. PH 14/230 Lesefilmnummer 4875F.
Kriegsmedizin, 1918. PH 2/653.
Lichtbild-Erkundung englischer Gaswerfer, 1917. PH 5-II/267 Lesefilmnummer 4599F.
Literatur, Kriegserfahrungen, Auslandsnachrichten, 1934–1936. RH/12/4.
Luftschutzubungen mit Nebel, 1931–1936. RH/12/4.

Nachlass Gerhard Tappen, Kriegstagebuch, November 6–9, 1914.
Nederland I, 1924–1929. RH/12/4.
Neue Gasmaske, 1933. RH/12/4.
PH 2 Preussisches Kriegsministerium, 1918. PH2 282–302 Lesefilmnummer 4445F.
Polen IV, 1929–1931. RH/12/4.
Preussische Heeresgasschule Offiziers Ausbildungskurs 1, Febraury 1917 PH/21/.
Preussisches Kriegsministerium 139–197, 1917, PH/2/.
RH/12/9 Sig 21, 1931–1933.
RH/12/9 Sig 22.
RH/12/9 Sig 26, 1926.
RH/12/9 Sig 32
RH/12/9 Sig 33
RH/12/9 Sig 4, 1938.
RH/12/9 Sig 66, 1927–1928.
RH/12/9 Sig 69, 1931–1932.
RH/12/9 Sig 70, 1934.
RH/12/9 Sig 71, 1937–1938.
RH/12/9 Sig 72, 1944.
RH/12/9 Sig 85, 1916.
RH/12/9 USA III, 1927.
RH/12/9/20 1930.
Russland I, 1926. RH/12/4.
Russland II, 1927. RH/12/4.
Russland III, 1927–1939. RH/12/4.
Russland IV, 1929–1932. RH/12/4.
Schutzengraben-Markblatt Beachte fur den Gaskampf, 1915. PH 5-II/61 Lesefilmnummer 4590F.
Schutzmassnahmen in Unterkunften, 1935–1936. RH/12/4.
Sondergerat fur Gasschutz, 1931–1936. RH/12/4.
Tomka, 1930–1931. RH/12/4.
Truppenversuche, Erfahrungsberichte, 1933. RH/12/4.
Versuche mit Nebelgerat Ausfuhr vom Nebelgerat, 1930–1931. RH/12/4.
Versuche mit Nebelgerat usw Ausfuhr von Nebelgerat usw, 1923–1928. RH/12/4.
Verwaltung, Lagerung, Nachschub, Unterbringung, Besichtigung des Gasschutzgerats bei der Truppe, 1933–1936. RH/12/4.
Vorschriften uber Gasschutz, 1935–1936. RH/12/4.
Vortäge der Gasschuztlehrgänge Teil II, 1930. RH/12/9.
Vortrag uber Gasschiessen in Verteidigung und Angriff, 1917.PH 5-II/379 Lesefilmnummer 4605F.

Deutsches Tagebucharchiv Emmendingen (DTA)

Eva Schwedheem, Kriegsjahre in Köln, 1943/44 1943–1944.
Fritz Rider, Mit 18 als Freiwilliger an die Front, 1916–1918.
Hermann Knaüer Tagebuch. 1221.1.

Johannes Martini, "Errinerung aus dem Weltkireg 1914–1915."
Karin Husmann, Es bleibt unsagbar... Zwei Kinder erleben das Inferno Dresdens 1945 und die Kriegs- und Nachkriegsjahre, 1999.
Karl Minning Kriegstagebuch von, 1916.
Konstantin Kraatz, Kriegstagebuch, 1916–1919.
Ludwig Steinmetz Kriegstagebuch, 1918.
Otto Mallebrein, Mein Kriegstagebuch, 1914–1918.
Reinhardt Lewald Kriegstagebuch, 1914–1918.
Wilhelm Wittekindt, "Meine Kriegserlebnisse 1916–1918."

Deutsches Technikmuseumarchiv (TMA)

Deutscher Jägerbund, 1934. I.4.040 08039.
Die deutsche Volksgasmaske. III.2 08754.
Dräger Geräte Gesellschaft, 1933. III.2 08746.
Luftschütz in der Familie. II.Kleine 1907.
Reichsluftschutzbund betr. Luftschutz, Vortragsmaterial, 1944. KE 1656.
Reichsluftschutzbund Unterrichts-Anweisung. II.Kleine 1906.
Schutzraum. II.Kleine 1656.
Unnamed Private, Tagebucher 1941, 1942, 1943. II.Kleine 2090.

Evonik Industries Archives (EIA)

Berichte, Korrespondenz Berufsberichte Dr. Bauer (Röhm Bestand), 1932. RHT.2./6.
Berichte Korrespondenz von Bauer an Letcher und Strauss (Röhm Bestand), 1928–1933. RHT.2./6.
Degea Aktiengesellschaft (Auergesellschaft) Berichte, Korrespondenz, Degussa IW 24.5./2-3.
Degea Aktiengesellschaft (Auergesellschaft) Berichte, Korrespondenz, 1933. Degussa IW 24.5./2-3.

Institut fur Zeitgeschichte München (IZM)

Gasmaske Plakatt. PS 152.
Luftschutz ist Selbstschutz Plakatt, 1939. 11/ZGa 035.005.

Institut für Zeitgeschichte Stuttgart (IZS)

1914=1918 Dokumente M Wolfen.
Auszug aus dem Kriegstagebuch von Otto Gerhard Lt. im Res Reg 201 und ab Marz 15 bei den Gastruppen, 1956. N 60.14/ 1–16.
Deutscher Meldereiter mit Gasmaske. 12674/17.
Die Kriegstagebücher des Sebastian Heinlein, 11 Juli 1916. N04.3-6.
Linienmasken Neuve Chapelle Nr 106. Neg. Nr 60/30 aus N: Leibfried.
Otto Borggräfe, "Mein Tagebuch," 1914–1915. N10.3-7.

Landesarchiv Baden-Württemberg Generallandesarchiv Karlsruhe (LBWK)

Analgen zum Kriegstagebucher, April 1916–Juni 1916. 456 F 50, 145.
Anlagen zum Kriegstagabuch, Juli 1916–Aug 1916. 456 F 26, 22.
Erkrankungen durch Kampfgas, 1916. 456 F 115, 67.
Gas als Kampfstoff, 12 August 1916–27 April 1917. M 33/2 Bü 402.
Gasschutz, April 1916–Januar 1917. 456 F 86, 219.
Gasschutzmassnahmen, 1931–1932. 233, 12570.
Gefechtsakten des Gasdienst-Offiziers, Januar 1918–September 1918. 456 F 16, 423.
Gesuch des Oberingenieurs Karl Leuprecht in Karlsruhe Luft und Gasschutz betreffend, 1932. 231, 6516.
Kriegstagebuch Band II- 11. August bis 22. Dezember 1915. M 660/032 Nr 22.
Kriegstagebuch Band III 23 Dezember 1915 bis 12 November 1916. M 660/032 Nr 23.
Luftschutzdienst der Gruppe Karupfalz, 1935–1937. 465 c Zugang 1991–1997, 83.
Organisation (Luftschutz), 1934. 465 c Zugang 1991–1997, 14.
Verwaltungsunterlagen, 1917–1918. 456 F 114, 111.
Zur Kenntnis und Behandlung der Gasvergiftungen, 1917. M 635/1 Nr 963.

Landesarchiv Baden-Württemburg Hauptstaatsarchiv Stuttgart (LBWH)

Anlagen - Englischer Gasangriff Juli 1 1917–September 30 1918. M 411 Bd. 1276.
Anweisung für die Handhabung der Gasmaske 30; mit Abbildungen, 1936. M 635/1 Nr 1461.
Ausbildung der Johanniterschwestern im Gasschutz, 1932. P 7/2 Bü 466.
Gas/21 Juni 1918–30 Sept 1918. M 33/2 Bü 407.
Schutzmassnahmen gegen feindliche Fliegerangriffe Juni 1916–Mai 1917. M 77/1 Bü 629.

Landeskriminalamt Baden-Württemberg Ludwigsburg (LBWL)

Vorermittlungsverfahren Tatbestand: NS-Verbrechen in Nebenlagern des Konzentrationslagers Sachsenhausen Tatort: Nebenlager Auerwerk-Oranienburg, 1944–1945 (LBWL). EL 48/2 I.

Leo Baeck Institute (LBI)

Wilhelm Lustig, *Kriegstagebuch 1915–1918*. ME 6 MM 2.

Max Planck Gesellschaft (MPG)

Die chemische Industrie und der Krieg, 1920. Va 5, 734.

Franz Richardt, report, Dept. Va, Rep 5, 1494.
Fritz Haber Briefe, 1915. *Va* 5, 856.
Fritz Haber Briefe, 1916, *Va* 5, 857.
Fritz Haber, memorandum, September 18, 1917. Dept Va, Rep 5, 1616.
Fritz Haber to Carl Duisberg, August, September, and October 1915. Dept Va, Rep 5, 856, 961, 962.
Gasschutz Abteilungen CUF, 1918. Va 5, 526.
Gasschutz Lummitzsch, Otto: Meine Erinngerung an Geheimrat Haber, 1955. Va 5, 534.
Gastruppe S. 18-51 Manuscript für Buch "Otto Hahn- Mein Leben" 1915.
Johannes Jaenicke, memorandum on a conversation with Otto Hahn in January 1955, Dept. Va, Rep. 5, 1453.
Ministry of War, Dept. W8, October 1916, KWG, Dept I, Rep. 1a, 36, General Administration, 1153.
Mit der Länge des Kriegs wachsen, 1918. Va 5, 522.
Otto Lummitsch, "Meine Erinnerungen an Geheimrat Prof. Dr. Haber." Dept. Va, Rep 5, 1480.
Rundschreiben an die Mitglieder der Kameradschaft der Offiziere der ehemaligan Gestruppen, 1940.
Therapie der Kampfgas-vergiftungen, 1918. Va 5, 524.

Preußisches Geheimes Staatsarchiv Kulturbesitz (GStA)

Justizministerium Untersuchungsverfahren, 1929 I. HA Rep 84 a Justizministerium Nr 52373.
Luftschütz, 1939–1945. I HA Rep 178 B Nr 795.

Staatsbibliothek Berlin (SBB)

Adolf Harnack papers, KWG, 3.-5. Jahresbericht, pp. 3–5, Sect. IV, Box 23.
Die Gasmaske, 1937–1938. 4 Ona 30/98-9/10.
Hast du eine Gasmaske Volksgenosse? 1939. Einbl. 1939/45, 4422. H-J.
Jedem Volksgenossen seine Volksgasmaske! 1937. Einbl. 1937, 008 kl.
Luftkrieg und Gaskrieg. Einbl. 1939/45, 4319.125;125Vor; 125.2.Ex.

Staatsarchiv Freiburg (SF)

Luftschutz hier: Versorgung der Bevolkerung mit Gasmasken, 1944–1945. B 702/1, 4852.
Militar und Kriegssachen: Luftschutzmassnahmen in Heil- und Pflegeanstalten, 1933–1945. G 1215/3, 711.

Staatsarchiv Hamburg (SH)

Giftgas Unglück auf der Veddel, 1928 (SH) 135-1 I-IV 4069.

Phosgen Ermittlung, 1928. 311–312 IV Vuo II C 5 a II A 34.
Phosgen Giftgas Katastrophe auf der Weddel im Mai, 1928. 352–353 II N 55.

Staatsarchiv Lübeck (SL)

Luft und Gasschutzeinrichtungen bei Fliegergefahr, 1932. 18 4.6-6 Tiefbauamt 779.

Tübingen Universitätsarchiv (TU)

Franz Emmel, 1918. 669/30128.
Gaupp Akten, 1918. 308/89.

Universitätsbibliothek Heidelberg (UH)

Arthur Thiele, "Die fünf Sinne der Gasmaske" 1918. G 5442-2 Folio RES.

University of Illinois, Urbana-Champaign Library

German reactions to atomic weapons. Report no. 216, Q. 303.48243 am216.

Newspapers and Periodicals

Archiv für Schiffs und Tropenhygiene 19, 1915.
Berliner Illustrierte Zeitung, 1927, 1939.
Berliner Tageblatt, April 26, 1915.
Der Stürmer, 1927.
Die Gasmaske, 1929–1940.
Die Gasmaske: Feldzeitung der Armee Abteilung C, no 20 (2), 1917.
Die Luftwacht, 1927–1928.
Die Sappe, 1915–1918.
Die Sirene: Illustrierte Zeitschrift mit den Mitteilung des Reichsuftschutzbundes, 1934–1939.
Dräger-Vorträge: Atemschutz Wiederbelebung, vol. 3. 1933.
Draeger Gasschutz im Luftschutz, 1932.
Gasschutz und Luftschutz, 1931–1945.
Hammburger Stimmen, 1928.
Internationales Luftfahrt Archiv, 1933.
Kölnische Zeitung, June 26, 1915.
Kriegszeitung der 1. Armee, no 148, April 8, 1918.
Luftschutz-Rundschau, 1931–1933.
Militär Wochenblatt, 1915–1916, 1925.
Neuer Vorwärts, 1934.
Nouvelles Litteraires, 1930.
The London Times, 1915, 1928.

Vierteljahrsschrift für gerichtliche Medicin und öffentliches Sanitätswesen, vol. 59, 1920.
Ziviler Luftschutz, 1952.

Artwork

Dix, Otto. *Die Schlafenden von Fort Vaux (Gas-Tote)*. 1924. Etching, aquatint, and drypoint from a portfolio of fifty etching, aquatint, and drypoints, 1924, MoMA.
Dix, Otto. *Gas Victims (Templeux-La-Fosse, August 1916)*. 1924. Etching, aquatint, and drypoint from a portfolio of fifty etching, aquatint, and drypoints, (19.6 x 29 cm), 1924, 159.1934.3, MoMA, www.moma.org/s/ge/collection_ge/object/object_objid-87723.html.
Dix, Otto. *Lichtsignale,* 1917. Oil on canvas.
Dix, Otto. *Shock Troops Advance under Gas*. 1924. Etching, aquatint, and drypoint from a portfolio of fifty etching, aquatint, and drypoints. 19.3 x 28.8 cm, 1924, 159.1934.12, MoMA, www.moma.org/collection/works/63260.
Gilles, Barthel. *Self Portrait with Gas Mask*. 1929. egg tempura on wood. 50 x 39.5cm, 1929. https://journals.ub.uni-heidelberg.de/index.php/akb/article/viewFile/35042/28704.
Grosz, George. *Shut up and Do Your Duty*. 1927. Rotogravure on laid paper, 15.24 x 18.1 cm, 1927. Los Angeles County Museum of Art. https://collections.lacma.org/node/181506.
Grosz, George. *Soldier on Horse with Gas Mask*. 1936. Lithograph. 1936. Private.
Schlichter, Rudolf. Dada Dachatelier. 1920. Watervolor and pen on paper, 45.8 cm x 63.8 cm, Sammlung Karsch, Berlinische Galerie.

Films

Dubson, Michael. *Giftgas*, 1929.

Index

Printed in the United States
by Baker & Taylor Publisher Services